A Short University Course in

ELECTRICITY SOUND AND LIGHT

MAXWELL PRESS

A Short University Course in

ELECTRICITY SOUND AND LIGHT

BY

Robert Andrews Millikan, Ph.D

Winner of Nobel Prize for Physics

Associate Professor of Physics in the University of Chicago

AND

John Mills, A.M.

Instructor in Physics in the Western Reserve University

MAXWELL PRESS

Chennai Trichy NewDelhi

MAXWELL PRESS

ISBN 978-93-90063-37-6

Printed and bound in India
MJP 884

Maxwell Press
No. 44, Nallathambi Street,
Triplicane, Chennai 600 005

Publisher : C. Janarthanan

Publisher's Note

The legacy of a country is in its varied cultural heritage, historical literature, developments in the field of economy and science. The top nations in the world are competing in the field of science, economy and literature. This vast legacy has to be conserved and documented so that it can be bestowed to the future generation. The knowledge of this legacy is slowly getting perished in the present generation due to lack of documentation.

Keeping this in mind, the concern with retrospective acquiring of rare books has been accented recently by the burgeoning reprint industry. Maxwell Press is gratified to retrieve the rare collections with a view to bring back those books that were landmarks in their time.

In this effort, a series of rare books would be republished under the banner, "Maxwell Press". The books in the reprint series have been carefully selected for their contemporary usefulness as well as their historical importance within the intellectual. We reconstruct the book with slight enhancements made for better presentation, without affecting the contents of the original edition.

Most of the works selected for republishing covers a huge range of subjects, from history to anthropology. We believe this reprint edition will be a service to the numerous researchers and practitioners active in this fascinating field. We allow readers to experience the wonder of peering into a scholarly work of the highest order and seminal significance.

Maxwell Press

PREFACE

This book represents primarily an attempt to secure a satisfactory articulation of the laboratory and class-room phases of instruction in physics. It is an outgrowth of the conviction that in courses of intermediate grade in colleges, universities, and engineering schools a real insight into the methods of physics, and a thorough grasp of its foundation principles are not readily gained unless theory is presented in immediate connection with such concrete laboratory problems as are calculated to give the student a sound basis for intelligent theoretical work.

Nevertheless the book is intended to be much more than a laboratory manual. It represents an attempt to present a complete logical development, from the standpoint of theory as well as experiment, of the subjects indicated in the title. It is designed to occupy a half year of daily work, two hours per day, in either the freshman, sophomore, or junior years of the college or technical-school course. In the University of Chicago about one half of this time is devoted to class discussions, lecture-table demonstrations, quizzes, and problems, and the remainder to laboratory work. The course is preferably preceded by a similar course in mechanics, molecular physics, and heat, the two courses together constituting a year's work in college physics.

The method of treatment is throughout analytical rather than descriptive, although no mathematics beyond trigonometry is presupposed. It is assumed that the student has already had a beginning course in descriptive physics in the high school or elsewhere.

Most of the apparatus required is of the stock sort found in all moderately well-equipped college laboratories. A few special pieces have been designed (Figs. 60, 99, etc.) where for one reason or another existing forms seemed ill adapted to the needs of the course. In the University of Chicago the apparatus is not, in general, used in duplicate. Nine or ten experiments are commonly kept going at once, two pupils working together. The classes are limited to twenty-five.

The authors' thanks are due to Leeds & Northrup for the cut of their post-office-box bridge (Fig. 51 *b*), to Queen & Co. for the original of their tangent galvanometer (Fig. 22), to the American Instrument Company for the original of their voltmeter (Fig. 40), and to William Gaertner & Co. for the originals of the magnetometer (Fig. 14), the voltameter (Fig. 27), the electric calorimeter (Fig. 39), the ballistic galvanometer (Fig. 60), the earth inductor (Fig. 107), the ideal dynamo (Fig. 99), and the spectrometer (Fig. 194).

R. A. M.
UNIVERSITY OF CHICAGO J. M.

CONTENTS

ELECTRICITY, SOUND, AND LIGHT

CHAPTER I

MAGNETIC AND ELECTRIC FIELDS OF FORCE

1. Quantity of magnetism. It is well known that under certain circumstances bars of iron and of some other metals, when suspended so as to be free to rotate about a vertical axis, turn so as to point north and south. A body which possesses the property of so doing is called a *magnet.* If the two north-seeking ends of two such magnets are brought near together, they are found to repel each other. The same is true of the two south-seeking ends. But a north-seeking and a south-seeking end are found to attract each other. On account of these opposite characteristics the north-seeking ends of magnets are called N poles, the south-seeking, S poles. The above facts may then be stated in the general law: *Magnetic poles of like kind repel one another, of unlike kind attract one another.*

Different magnets of the same length placed at a given distance from a suspended magnet are observed to exert different forces upon it. The *quantities of magnetism* in the poles, or the *pole strengths,* are then arbitrarily taken as proportional to the forces exerted. That is, the force which a given pole M exerts, at a given distance, upon some standard pole is taken as the measure of the number of units of magnetism in M. It may then be proved experimentally, by the method used in Chapter II, that the force f exerted between any two poles M and m is directly proportional to the product mM and inversely proportional to the square of the distance r between them. The algebraic statement of this experimental relation is

$$f = k\frac{mM}{r^2}, \tag{1}$$

in which k is a factor of proportionality depending upon the choice of the unit of magnetism and upon the medium through which the force acts. It has been decided to choose this unit so that k equals unity for air; that is, the equation $f = mM/r^2$, applied to air, contains the definition of unit magnetic pole. Thus if two magnets are chosen for which $m = M$, and if r is taken equal to 1 cm., then m is by definition unity if f is found to be equal to 1 dyne. In other words, *a unit magnetic pole is a pole of such strength that when placed at a distance of 1 cm. from an equal pole it repels or attracts it with a force of 1 dyne.*

2. Quantity of electricity. It is also equally well known that when a glass rod has been rubbed with silk it attracts a pith ball, but after contact with the pith ball, repels it. Similarly, when ebonite has been rubbed with cat's fur it attracts a second pith ball, but after contact with the pith ball, repels it. Furthermore, the pith ball which has touched the rubbed glass and is repelled by it is attracted by the ebonite, while the ball which is repelled by the ebonite is attracted by the glass. On account of this behavior the pith balls are said to have been *electrified*, or to have received *charges of electricity*; and on account of the opposite characteristics of these charges the one is called *positive* and the other *negative.* These charges of electricity can be produced in other ways, but in every case it has been decided to call a charge positive when it is repelled by a glass rod which has been rubbed with silk, negative when it is repelled by an ebonite rod which has been rubbed with cat's fur. The above facts may then be stated in the general law: *Electrical charges of like sign repel one another, of unlike sign attract one another.* It need scarcely be said that in adopting these conventions and in setting up this law no assumption whatever has been made regarding the nature of electricity. It has merely been agreed to call a body *electrified*, or *charged* *with electricity*, which behaves toward pith balls or other objects as does the rubbed glass or the rubbed ebonite.

Definitions precisely similar to those used in the quantitative study of magnetism are adopted also in the quantitative study of electricity. Thus if q and Q are two electric charges the magnitudes of which are measured by the forces which they exert upon a third charge at a given distance from it, then it can be proved experimentally that

$$f \propto \frac{qQ}{r^2}.$$

As in magnetism, so in electricity, the unit of quantity is chosen so that for action between charges separated by air

$$f = \frac{qQ}{r^2}. \qquad (2)$$

This equation contains, then, the definition of unit charge. In words, *unit quantity of electricity (unit charge) is defined as that quantity which placed in air at a distance of 1 cm. from an equal quantity acts upon it with a force of 1 dyne.*

3. Electrical conduction. If a charged ebonite rod is rubbed over one end of a long metal body which rests upon sealing wax or glass, a pith ball placed near the remote end of the metal body will at once be attracted to it. If the metal body is replaced by one of glass, or wood, or almost any nonmetallic solid, no effect whatever is produced upon the pith ball. In view of experiments of this sort, it is customary to divide substances into two classes, *conductors* and *insulators*, or nonconductors, according to their ability to transmit electrical charges. Thus metals and solutions of salts and acids in water are all conductors of electricity, while porcelain, rubber, mica, shellac, wood, silk, vaseline, turpentine, paraffin, and oils generally are insulators. No hard and fast line, however, can be drawn between conductors and insulators, since substances can be found of all degrees of conductivity between that of sulphur, amber, or quartz, the best insulators, and that of silver and copper, the best conductors.

4. Distinctions between electricity and magnetism. The fact of conduction constitutes one of the most essential distinctions between electricity and magnetism. Electrically charged bodies lose their charges, in part at least, as soon as they are touched by conductors, but such treatment has no influence whatever upon magnetic poles. The phenomena of magnetism therefore show nothing which is at all analogous to the phenomenon of conduction in electricity. Furthermore, all bodies, conducting or nonconducting, can be strongly electrified by friction if they are mounted upon insulating supports, but only iron, steel, nickel, and some newly discovered alloys of copper, magnesium, and aluminum, called Heussler alloys, can be appreciably magnetized. Magnetism and electricity are then to be regarded as distinct phenomena. A peculiar relationship, however, which has been found to exist between them will be discussed in Chapter III.

5. Electrostatic induction. If a positively charged body *A* (Fig. 1) is placed in the neighborhood of an uncharged body *B*, which is supplied with pith balls or strips of paper *a*, *b*, *c*, as shown in the figure, the divergence of *a* and *c* will show that the ends of *B* have received electrical charges, while the failure of *b* to diverge will show that the middle of *B* is uncharged. Further, a positively charged glass rod will be found to repel *c* and attract *a*. The experiment illustrates the fact that the mere *influence* which an electric charge exerts upon a conductor placed in its neighborhood is able to produce electrification in that conductor, the remote end receiving a charge of sign like that of the original charge, while the near end has a charge of opposite sign. If *A* is removed the charges at *a* and *b* entirely disappear, and the conductor *B* is found to be altogether uncharged, thus showing that the total amount of positive electricity which appeared at one end of *B* must have been exactly equal to the total amount of negative which appeared at the other end. *The phenomenon of the appearance of equal and opposite electrical charges in the opposite ends of a conductor placed near a charged body is known as electrostatic induction*, and a conductor in this condition is said to be *electrically polarized.*

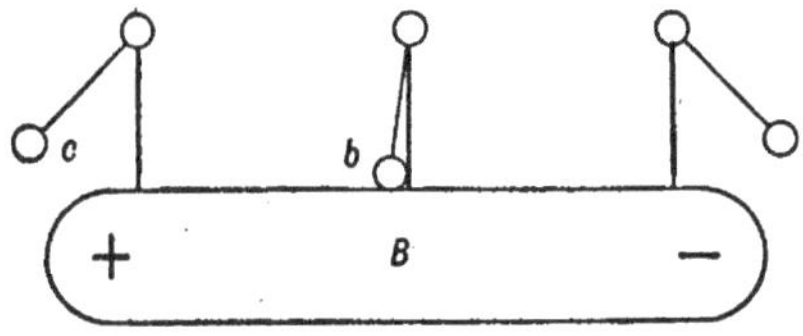

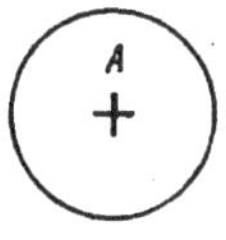

FIG. 1

6. Positive and negative electricities always appear in equal amount. That positive and negative electricities always appear in exactly equal amount, as well when the electrification is produced by friction as by induction, may be convincingly shown by attaching a piece of fur or flannel to the end of a strip of ebonite, rubbing with it the end of another similar strip, bringing the two together, without separating them, near a charged pith ball or other electroscope, then separating them and bringing each in succession near the electroscope. So long as they are together they will exhibit no electrification whatever, but when separated they will show charges of opposite sign. That these charges are exactly equal is shown by the fact that they exactly neutralized each other before the separation.

The test for the equality of the two charges may be made extremely delicate by inserting the two rubbed bodies together into a hollow metal vessel to which a gold-leaf electroscope is connected, as in Figure 2. So long as the two rubbed bodies are together the leaf will show no trace of divergence, but when one of them is removed the leaf will stand out in the position of the dotted line. When this rubbed body is replaced by the other the divergence will be of equal amount, but the charge will be of opposite sign, as can be shown by bringing a positively charged glass rod near the electroscope; for then, if the latter already had a positive charge, the divergence will be increased by the approach of the glass rod, but if it had a negative charge the divergence will be diminished.

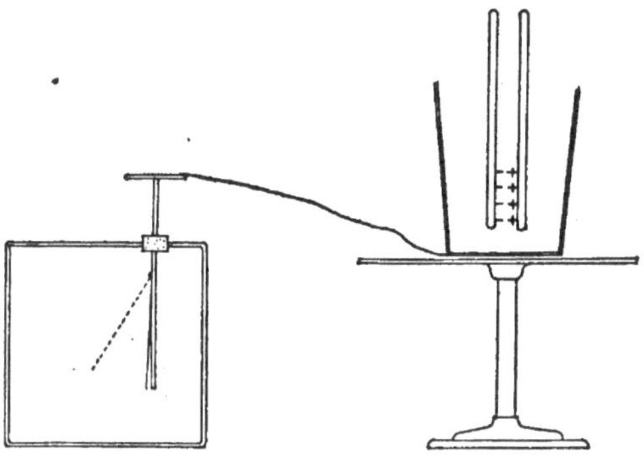

FIG. 2

7. Theories of electricity. During the nineteenth century the facts of electricity were most commonly described in terms of the so-called *two-fluid theory.* Although this theory is no longer regarded as corresponding closely to reality, and although it has perhaps never been generally accepted as anything more than a *convenient fiction*, it is so intimately related to the present nomenclature of the subject as to deserve careful attention. According to it all bodies contain equal amounts of two weightless electrical fluids, *positive electricity* and *negative electricity*. These fluids are self-repellent but mutually attractive, so that their effects completely neutralize one another in bodies in the normal condition. But if a conductor, i.e. a body in which the fluids are able to move about, is brought near a charged body, the fluid of like sign is driven to the remote end, while the fluid of unlike sign is drawn to the near end of the conductor. This furnishes the explanation of the phenomenon of induction.

In insulators the fluids are supposed not to be free to move from point to point, but friction between dissimilar substances causes electrical separation in the adjacent boundary layers of the dissimilar substances, the excess of positive which goes to one body being necessarily equal to the deficiency of positive, i.e. the excess of negative, which is left upon the other.

The modern modification of the two-fluid hypothesis is that which has been developed within the last decade by Drude and Riecke in Germany. It is identical with the older two-fluid theory, save that it replaces the weightless and continuous electrical fluids by equal numbers of positive and negative corpuscles, or *electrons*, which are assumed to be constituents of the atoms of all substances, but which, in the case of conductors, are continually becoming detached from the atoms; so that at a given instant there are always present free positive and free negative corpuscles which are able to move through the conductor in opposite dircotions under the influence of any outside electrical force. Conductors differ from insulators only in that the atoms of the latter do not lose their corpuscles to any appreciable extent.

The old-time rival of the two-fluid theory was the so-called *one-fluid theory*, originally due to Benjamin Franklin. It differed from the two-fluid theory only in regarding a positive charge as indicating an *excess*, a negative charge, a *deficiency* in a certain normal amount of one single, all-pervading electrical fluid, viz. positive electricity, which was self-repellent but strongly attracted by ordinary matter. In order to account for the mutual repulsions of negatively charged bodies, it was found necessary to assume that the particles of ordinary matter, when dissociated from electricity, repelled one another.

A modern modification of the one-fluid theory has recently come into prominence through the combined work of several physicists in high standing, notably Lord Kelvin and J. J. Thomson. According to this theory a certain amount of positive electricity is supposed to constitute the nucleus of the atom of every substance. About the center of this positive charge are grouped a number of very minute negatively charged corpuscles, or electrons, the mass of each of which is approximately $\frac{1}{2000}$ of that of the hydrogen atom. The sum of the negative charges of these electrons is supposed to be just equal to the positive charge of the atom, so that in its normal condition the whole atom is neutral or uncharged. But in the jostlings of the molecules of a conductor electrons are continually becoming detached from the atoms, moving about freely between the molecules, and then reëntering other atoms which have lost electrons. Therefore, at any given instant, there are always present in any conductor a large number of free negative electrons and an exactly equal number of atoms which have lost electrons, and which are therefore positively charged. Such a conductor would, as a whole, show no charge either of positive or of negative electricity. But the presence near it of a body charged, for example, negatively would cause the negatively charged electrons to stream away to the remote end, leaving behind them the positively charged atoms, which, in solids, are not supposed to be free to move appreciably from their positions. In the presence of a positively charged body, on the other hand, the electrons would be attracted to the near end, while the remote end would be left with the immovable positive atoms.

The only advantage of this theory over that which assumes the existence of two types of corpuscles is that, while there is much direct experimental evidence for the existence of negative corpuscles of about $\frac{1}{2000}$ the mass of the hydrogen atom, no direct evidence whatever for the existence of such positively charged corpuscles has as yet been brought to light. In general, wherever positively charged bodies appear they are found to be of atomic size. The negative corpuscles, on the other hand, are sometimes found as constituents of atoms, sometimes as independent detached bodies.

It will be seen that this last theory is like Franklin's in that it assumes but one *movable* kind of electrical matter, i.e. one *electrical fluid,* while it is unlike it in making this fluid negative instead of positive, and also in making it consist of discrete particles. It is like the two-fluid theory, however, in postulating the existence of two distinct entities called respectively positive and negative electricity. The positive electricity, however, plays quite the same rôle which in the old one-fluid theory was assigned to ordinary matter.

8. Fields of force. A *field of force* is simply a region in which force exists. It may be a magnetic, electrical, or gravitational field which is under consideration. The *strength of field* at any point in such a region is the number of units of force which unit quantity (be it mass, pole strength, or charge) experiences at the point considered. Thus the strength of field at a point 1 cm. distant from a unit pole (conceived as concentrated at a point) is unity. *Unit field is, then, a field in which unit quantity experiences 1 dyne of force.* For example, the strength of magnetic field at a given point in space is ten units if unit pole experiences 10 dynes of force when placed at this point.

The *direction* of a gravitational field at any point is defined as the direction in which a small quantity of matter would tend to move if placed in the field at the point considered. The direction of a magnetic field is defined as the direction in which an isolated *N* pole would move. The direction of an electric field is defined as the direction in which an isolated positive charge of electricity would move.

A *line of force* in any one of these fields is the direction in which a free mass, a free *N* pole, or a free *positive* charge would move if it had no inertia. It is convenient and customary, however, to conceive of as many *lines* drawn across any square centimeter taken at right angles to the direction of the force as the field possesses units of strength at the point considered. The line of force then becomes the unit of field intensity or strength; that is, in a gravitational field *a line* means a field strength such that a force of 1 dyne acts on every gram placed in it. Thus, since the earth's field strength at the surface is 980 dynes, it is customary to consider 980 lines of gravitational force as piercing each square centimeter of the earth's surface. In magnetic fields, in connection with which the convention is most commonly used, a line means a field strength such that a force of 1 dyne acts on each unit pole. It has received the special name of a *gauss.* In electrical fields the more usual term is *tube of force,* and the conception is that as many tubes of force cross any square centimeter at right angles to the direction of the field as there are units in the field strength at that point.

With these conventions it is evident that a uniform field is represented by a system of parallel lines of force, and conversely, that where the lines of force are parallel, the field has everywhere the same strength. A convergent system of lines represents a field of increasing strength; a divergent system of lines represents a field of decreasing strength. It need scarcely be said that a line of force has no objective reality. The representation of fields of force by lines is a matter of convenience only.

9. Gravitational potential. The term "potential" was first used in connection with gravitational forces. Potential is a characteristic of a point in space, not, in general, of a body. It may be looked upon merely as an abbreviation for the expression "the potential energy of unit mass at the point considered." Thus at a point *a* above the surface of the earth (Fig. 3) a gram of mass possesses a certain potential energy with reference to a point *b* on the surface of the earth. This potential energy is the amount of work that the gram of mass can do in falling to the earth. Since the term "potential" is merely an abbreviation for the potential energy of unit mass, it is evident that potential is measured in energy or work units. The potential of a point is unity when it requires one erg of work to bring unit mass from the point which is taken as the zero of potential up to the point considered. The potential energy of unit mass at the point *a* (see Fig. 3) is greater if the point *c*, below the

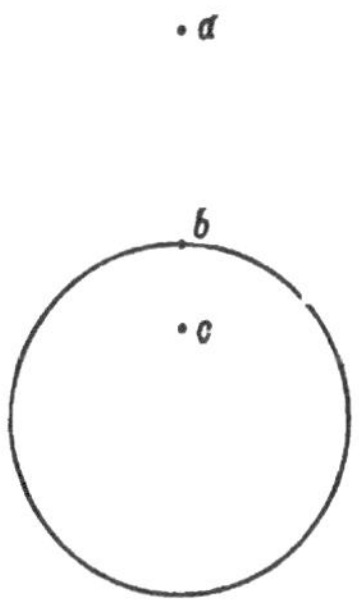

FIG. 3

surface of the earth, is taken as the point of reference instead of the point *b*. Thus it is evident that some point must be chosen arbitrarily from which to reckon the potential of other points.

A point which is infinitely distant from all attracting or repelling bodies is taken as the point from which *absolute potential* is reckoned, i.e. such a point is taken as the *absolute zero* of potential. The absolute potential, then, of any point in the universe is the number of ergs of work which must be done (by some outside agent) to bring unit mass from this absolute zero up to the point considered.

In general, we are more concerned with the difference in potential (usually written P.D.) between two points than with the absolute potentials of the points. The P.D. between two points is then defined as the amount of work which the external agent must do in order to carry the unit mass from the point of the lower to that of the higher potential; or, what amounts to the same thing, the P.D. is the amount of work which the acting force does in carrying unit mass from the point of higher to that of lower potential.

Now the force of gravitation is always an attractive, never a repellent, force. For all points which are at a finite distance from any astronomical body the work which the external agent must do in bringing unit mass from an infinite distance to any point is therefore less than nothing. That is, the work is done not by, but against, the action of the external agent. From the definition of absolute potential, as given above, it is evident that the gravitational potential of all points within a finite distance of any astronomical body must be negative.

10. Magnetic potential. The above definitions hold almost without change for magnetic forces, save that it is necessary to specify whether the unit quantity is an N pole or an S pole. *The magnetic potential of a point is defined as the amount of work which an external agent must do against the existing magnetic field in order to bring a unit N pole from infinity up to the point considered.* Thus the potential in the immediate neighborhood of an N pole is evidently positive, because the unit N pole is repelled, and hence the external agent must do work in order to bring up the pole from infinity. It is equally evident that the potential in the immediate neighborhood of an S pole is negative.

11. Electrical potential. Similarly, the electrical potential of a point is the amount of work required to bring unit positive charge from infinity up to the point considered. Points in the neighborhood of a positive charge have therefore a positive potential; those in the neighborhood of a negative charge have a negative potential. These definitions apply as completely in the study of so-called current electricity as in that of static electricity. *Under all circumstances the term P.D. means the amount of work required to carry unit positive charge between the two points considered.*

12. Equipotential surfaces. An equipotential surface is the locus of a system of points all of which have the same potential. Thus the gravitational equipotential surfaces about the earth are approximately spherical surfaces concentric with it, because the amount of work required to bring unit mass to within a certain distance of the center of the earth is the same, no matter from what side the earth is approached.

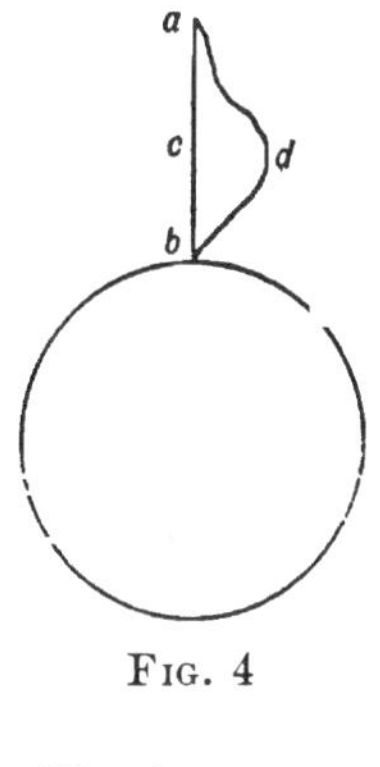

FIG. 4

Now it can be shown that *the direction of the field of force at any point is perpendicular to the equipotential surface passing through that point.* In order to prove this statement it is first necessary to show that the work done in carrying a body between any two points is independent of the path chosen. If a body is carried from *b* to *a* over the path *bda* (Fig. 4) a certain amount of work *w* is done upon it. Now suppose the work of the path *bda* is less than that of the path *bca*. Then when the body returns to *b* over the path *acb* it will give up a certain amount of energy *w′*, i.e. it will do a certain amount of work *w′*. As a net result of the operation we have expended an amount of work *w* and have received back a larger amount of work *w′*, and yet have brought everything back to the initial condition. We have therefore created an amount of energy $w' - w$. But according to the doctrine of the conservation of energy this is impossible. Hence *w′* cannot be greater than *w*. By reversing the operation it can be proved that *w* cannot be greater than *w′*. That is, the work of the path *bda* is equal to that of any other path *bca*.

Now let *nop* (Fig. 5) represent any equipotential surface. By the definition of such a surface the work required to move a unit body from any arbitrary zero, say *m*, over the distance *mn* is the same as that required to move it over the distance *mo*. But by the proposition just proved the work of the path *mn* is equal to the work of the path *mo* plus *on*. Hence the work corresponding to the path *on* must be zero. If it requires no work to move a body over an equipotential surface, the existing force can have no component along that surface, i.e. *the force must be everywhere normal to an equipotential surface.*

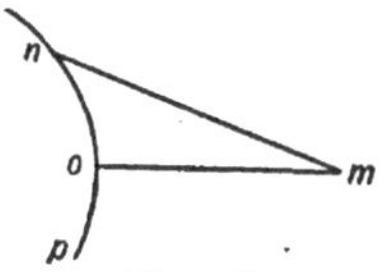

FIG. 5

13. Potential of a conductor in electrical equilibrium. That the electrical potentials of all points on or within a conductor in the static condition must be the same follows at once from the fact of conductivity. For as soon as a conductor *ab* (Fig. 6) is brought into the field of a positively charged body *c* the negative electricity within *ab* at once moves toward *b* and a positive charge appears at *a* until further movement of negative toward *b* is checked by the action of the negative accumulated at *b* and the positive accumulated at *a*. It is obvious, then, that *there can be electrical equilibrium within a conductor only when all electrical forces within the conductor have been reduced to zero.* No electrical force, then, can exist within a conductor in the static condition; hence no work can be required to move unit charge from one point to another within the conductor. It follows that *all points within or upon a conductor in the static condition must have the same potential.* The above reasoning holds as well for hollow as for solid conductors, provided no insulated charged bodies exist within the hollow portion.

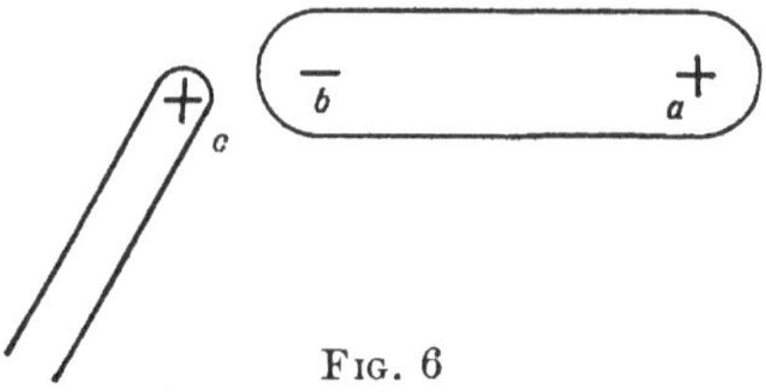

FIG. 6

The experimental verification of these conclusions was first made by Faraday, who covered a large box with tin foil and went inside with very delicate electroscopes. These remained wholly unaffected even when powerful electrical disturbances took place just outside the box. In laboratory practice it is now customary to screen delicate electrostatic instruments from external disturbances by surrounding them with sheet metal or wire gauze.

Since, then, all points on a conductor have the same potential, it has become customary to speak of the *potential of the conductor* rather than of the potential of points on the conductor, in spite of the fact that the term "potential" is one which in strictness characterizes points in space rather than bodies. Further, since the surface of a conductor in electrical equilibrium is always an equipotential surface, it follows that electrical lines of force always enter or leave a conductor normally to the surface.

14. Mapping equipotential surfaces. It is evident from section 12 that if it is possible to map the direction of the lines of force in any field, it must also be possible to map out the equipotential surfaces in that field. Thus, suppose it to be required to find the intersections with the plane of the paper of the equipotential surfaces about two isolated spheres whose centers are in the plane of

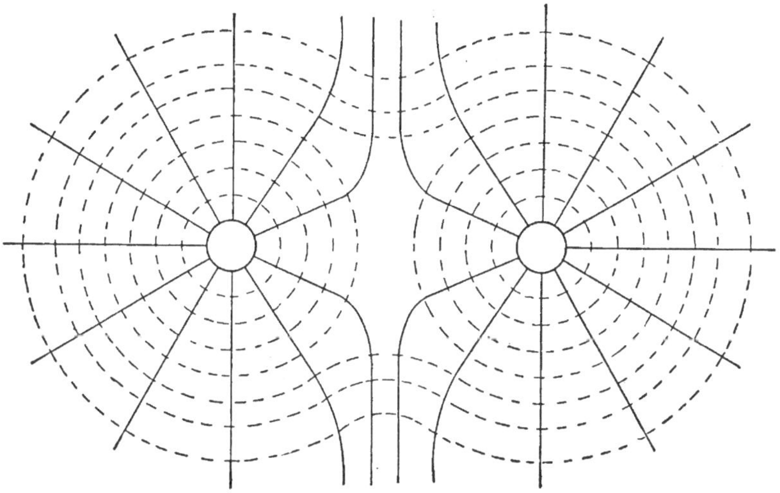

FIG. 7

the paper. A little consideration of the field of force about two such spheres will show that the general shape of the lines of force is that represented by the full lines of the sketch (Fig. 7). By drawing lines everywhere perpendicular to these force lines the equipotential lines represented by the dotted lines of the figure are obtained.

EXPERIMENT 1

(A) Object. To map completely a somewhat complicated magnetic field, and to draw the equipotential lines.

Directions. The field here plotted will be the combination of the field of the earth and the fields of two magnets. Two straight bar magnets (about one foot long) are to be placed vertically in a box the top of which (about three feet square) is shown in Figure 8. The magnets pass through the holes *a* and *b* and their upper ends are flush with the surface of the box. First see that the line connecting *a* and *b* is in the magnetic meridian; then, for the first exercise, place at *a* a north pole and at *b* a south pole. With brass pins attach to the box a sheet of paper large enough to cover the entire top.

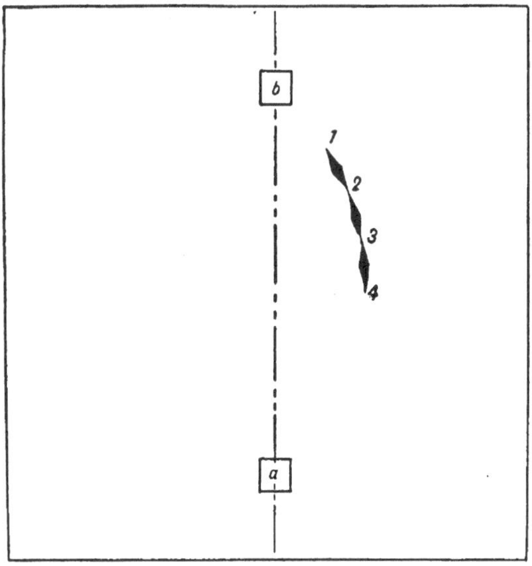

FIG. 8

The direction in which the north pole of a short compass needle points when the compass is placed in a magnetic field represents the direction of the magnetic force at the center of the compass needle; for the two poles are always under the action of forces which tend to move them in opposite directions along the line connecting them, and which would so move them if they could be detached from one another. Hence the following method of plotting the direction of the magnetic lines at once suggests itself. Set the needle at some point in the field and with a sharp pencil place two dots, e.g. *1* and *2* (Fig. 8) directly under the extremities of the needle. Move the needle along until the pole which stood over *1* now stands directly over *2*, and make a new dot *3* to indicate the position of the other pole. Proceed thus until the line either runs into a pole or leaves the paper. In this way plot enough lines to show a complete outline of the field. Plot with especial care the lines which pass near the singular points, i.e. points in which the needle will assume no definite position.

As a second exercise plot the field produced when both *a* and *b* are north poles. When the outline of these fields has been obtained estimate the directions of the intermediate lines wherever it is perfectly evident what their general course must be, and thus fill in the field thickly with force lines. Then upon each of the two fields draw, in red ink, a second system of lines which is everywhere roughly perpendicular to the first system, i.e. draw the equipotential lines.

(B) Object. To map the equipotential surfaces of an electrical field, and to draw the force lines.

Directions. In Figure 9 is shown a tray on the glass bottom of which is pasted a sheet of coördinate paper. The tray is filled to a depth of three or four millimeters with a solution of ammonium chloride or of any convenient salt. *A* and *B* are the two points which are to be maintained at different potentials and about which it is desired to find the equipotential lines which lie in the plane of the liquid. If both of the terminals of an

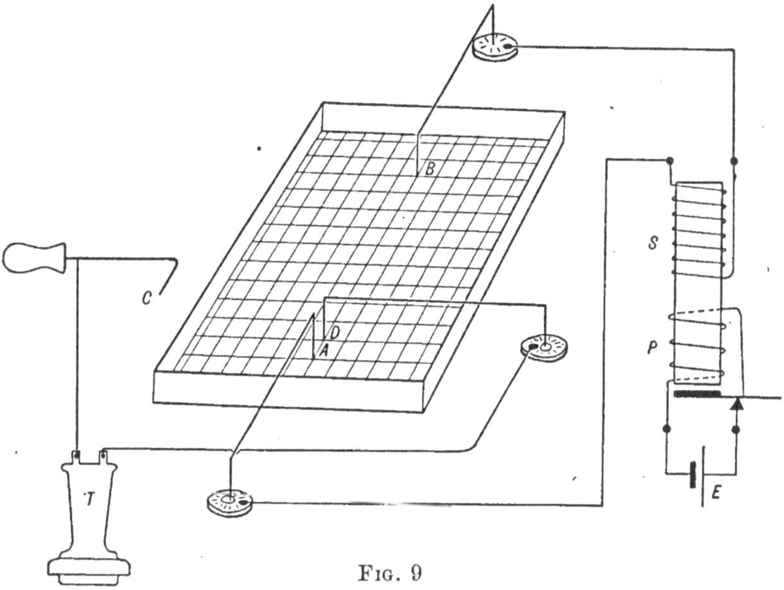

FIG. 9

instrument for detecting an electrical current lie on the same equipotential surface, no current passes through the instrument. If, however, the two terminals lie on surfaces of different potentials, a charge is urged through the instrument from the point of higher to that of lower potential, and the instrument indicates the fact. If one terminal be fixed and the other moved to various points of the field, the equipotential surface on which the first terminal lies will pass through all the positions of the second for which no current flows through the instrument. The telephone *T* is used for detecting a difference of potential between its terminals *C* and *D*.

Since the telephone receiver will respond by buzzing only as the result of an intermittent or alternating difference of potential, it is necessary that *A* and *B* be maintained at a constantly varying P.D. But whatever the P.D. between *A* and *B* the equipotential lines will preserve the same configuration, although, of course, assuming different absolute values. This varying P.D. is obtained by connecting *A* and *B* to the secondary *S* of a small induction coil, to the primary *P* of which is connected a storage battery or dry cell *E*.

Assume two lines on the coördinate paper as axes and locate points from these. Connect *A* and *B* as described and set them at two points about 20 cm. apart. Set the electrode *C* about 1 cm. from *A*, toward *B*. With the electrode *D* locate enough points for which there is no buzzing in the receiver to plot the equipotential line through the point at which *C* is placed. Move *C* to a position about one sixth of the remaining distance toward *B* and repeat. Continue, moving *C* another sixth, until the whole field has been explored. Plot on coördinate paper to one half scale the lines so found. The lines of force will form a system everywhere at right angles to these equipotential lines. Draw this system in red ink.

CHAPTER II

THE DETERMINATION OF THE STRENGTHS OF MAGNETIC FIELDS AND OF MAGNETIC POLES

15. Every magnet possesses equal amounts of *N* and *S* magnetism. The preceding chapter dealt only qualitatively with magnetic and electrical phenomena. The present chapter has to do with the exact measurement of magnetic quantities. One of the most important of these quantities is the so-called *magnetic moment* of a magnet. In order to gain a clear conception of the meaning of this term it is first desirable to consider why we believe that the *N* and *S* poles of any particular magnet are of exactly equal strength.

The earth behaves like a huge magnet, one pole of which is situated near the north geographical pole, the other near the south geographical pole. According to our convention the former of these is an *S* pole, the latter an *N* pole. Since both of the poles of the earth are very remote in comparison with the length of any ordinary magnet, it follows that if the north and south poles of a suspended magnet are of equal strength, then the attraction which either of the earth's poles exerts upon one of the ends of the suspended magnet cannot be sensibly different from the repulsion which this same pole exerts upon the other end of the magnet. Conversely, if the attraction which the northern pole of the earth exerts upon the *N* pole of a magnet is found to be equal to the repulsion which it exerts upon the *S* pole, then these two poles must be of equal strength. A similar method of reasoning may be applied to the action of the southern pole of the earth. Now experiment shows that a floating magnet experiences no motion of translation toward north or south, but only a motion of rotation about a vertical axis. Since this experiment may be performed with any magnet, it follows that the *N* and *S* poles of a magnet are always of equal strength.

16. Magnetic moment. The moment of force tending to produce rotation in a magnet placed at right angles to the lines of a uniform field, such as that due to the horizontal component of the earth's magnetic field, is the product of the strength of the field and the quantity known as the *magnetic moment* of the magnet. *Magnetic moment is defined as the product of the pole strength m and the distance l between the poles of the magnet.* The application of the term "magnetic moment" to this quantity is appropriate, because if the magnet were placed in a field of unit strength, in which case the force acting on each pole would be m dynes, and were placed at right angles to the direction of the field, as in Figure 10, the moment of force M acting to turn it into the direction of the lines of force would be given by the equation

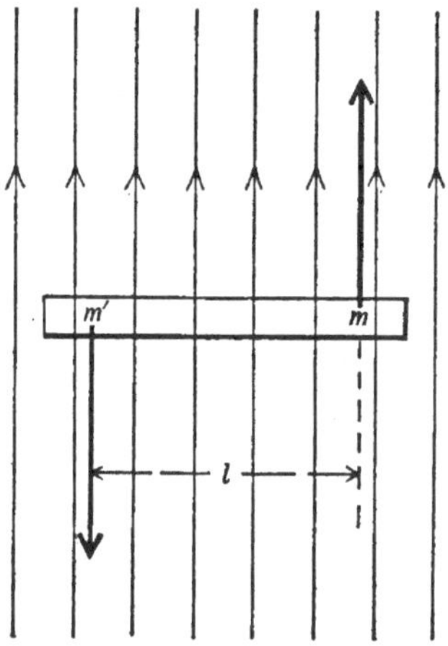

FIG. 10

$$M = ml. \tag{1}$$

17. Determination of the strength of a magnetic field. The particular field for which we shall now outline a method of determining the strength will be the horizontal component of the earth's magnetic field. The measurement of this field strength is intimately connected with the determination of the magnetic moment of a magnet. Neither the field strength H nor the magnetic moment M of the magnet used can be determined directly by a single experiment; but two experiments can be performed, each of which gives a relation between M and H. From these two relations both M and H can be found.

One experiment consists in suspending a large magnet mm' of unknown moment M in the field of which the strength H is sought, and observing the period of its oscillation. This gives an expression for MH, as will be shown in section 20.

The other experiment consists in placing the same magnet mm' of magnetic moment M in a position due east or west, as in

Figure 11 (p. 20), from a very small suspended magnet *bc* which hangs in a horizontal position in the earth's field. The suspended magnet *bc* is then acted upon simultaneously by two fields which are at right angles to each other: first, the earth's field H which exerts a moment, or couple, tending to swing the needle into a north-and-south position; and second, the magnetic field F due to the magnet mm' which exerts a couple tending to swing the needle into an east-and-west position. From the position of equilibrium assumed by the small magnet under the action of these two fields, an expression for M/H is obtained, as will be shown in section 19. But in order to obtain this expression, the strength of the field F must first be found in terms of the magnetic moment M of the large magnet.

18. Strength of field due to a magnet. The strength of the field produced at o (Fig. 11) by the magnet mm' is by definition the force which this field exerts at that point upon unit magnet pole. If, then, r is the distance from the middle of the magnet mm' to the suspended needle, the length of which is very minute in comparison with the distance r, then by the law of force given in section 1 the force which the pole m exerts upon unit pole at the given distance, namely $(r - l/2)$, is $\dfrac{m}{(r-l/2)^2}\cdot$ The other pole m' exerts an equal and opposite force of $-\dfrac{m'}{(r+l/2)^2}$ dynes. Since $m = m'$, the resultant force, that is the field strength F, is given by the equation

$$F = \frac{m}{\left(r-\frac{l}{2}\right)^2} - \frac{m}{\left(r+\frac{l}{2}\right)^2}$$

$$= \frac{m\left(r^2 + lr + \frac{l^2}{4} - r^2 + rl - \frac{l^2}{4}\right)}{\left(r+\frac{l}{2}\right)^2\left(r-\frac{l}{2}\right)^2} = \frac{2\,mlr}{\left(r^2 - \frac{l^2}{4}\right)^2},$$

or
$$F = \frac{2\,Mr}{\left(r^2-\frac{l^2}{4}\right)^2}.$$

It will be seen from this equation that if r is taken ten times as large as l, an error of one half of one per cent is introduced by neglecting $l^2/4$ in comparison with r^2. If r is twenty times as large as l, this error is only one eighth of one per cent. Thus it is evident that if r is taken sufficiently large, no observable error is introduced into F by writing simply

$$F = \frac{2\,Mr}{r^4} = \frac{2\,M}{r^3}. \tag{2}$$

This gives the field strength F due to the magnet mm' of moment M at the position at which the small magnet is suspended.

19. Determination of M/H. Returning now to the consideration of the deflection which is produced in the needle bc (Fig. 11) by the presence of the magnet mm', it is evident that when the needle takes up its position of rest it is in equilibrium under the

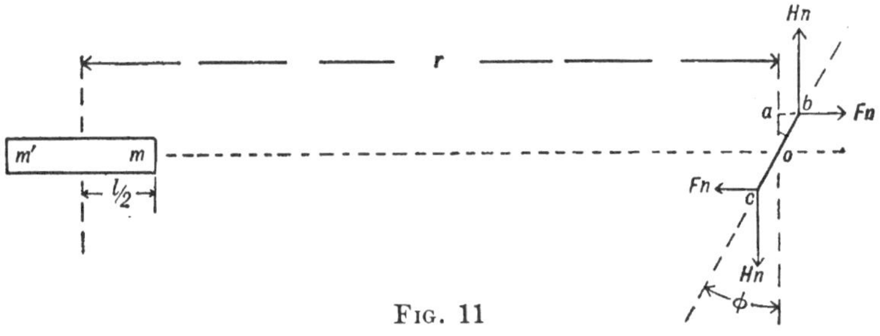

FIG. 11

action of two moments of force. If n is the pole strength of the suspended magnet bc, then, considering both poles, the moment of force due to H is $2\,Hn \times ab$. The moment of force due to F is $2\,Fn \times ao$. But $ab = bo \sin\phi$ and $ao = bo \cos\phi$. Hence

$$2\,Hn\,(bo)\sin\phi = 2\,Fn\,(bo)\cos\phi, \tag{3}$$

or

$$H\tan\phi = F; \tag{4}$$

or from (2)

$$\frac{M}{H} = \frac{r^3\tan\phi}{2}. \tag{5}$$

This gives the first relation between M and H in terms of the measurable quantities ϕ and r.

20. Determination of *MH*. The second relation between M and H is obtained from an observation of the period of vibration of the magnet mm' when it is suspended at the point for which H is to be determined, namely the position occupied above by the small magnet bc. In order to find this period it is first necessary to find what relation exists between the moment of force $\overline{Fh}$ which acts to restore the magnet to its north-and-south position at any instant at which, in the course of its vibration, it is displaced from this position by an angle θ. It is evident at once from Figure 12 that this restoring moment is given by

$$\overline{Fh} = 2\,Hm \times de = Hm \times df = Hml \sin\theta = MH \sin\theta. \qquad (6)$$

This equation shows that in general the restoring moment is not proportional to the angle of displacement θ, but rather to $\sin\theta$. If, however, we keep the angle of swing very small, $\sin\theta$ will not differ sensibly from θ, and we may write

$$\overline{Fh} = MH\theta, \quad \text{or} \quad \frac{\overline{Fh}}{\theta} = MH. \qquad (7)$$

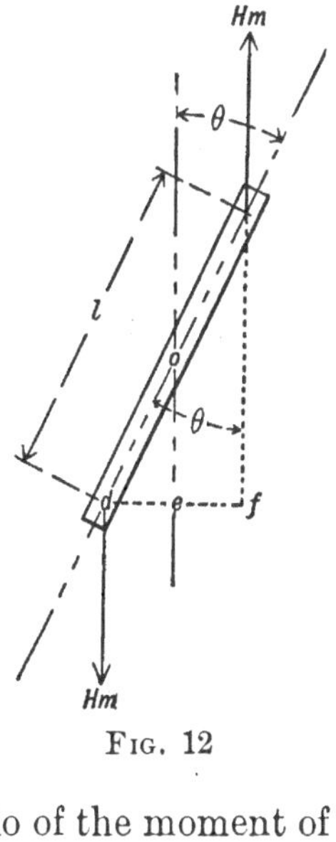

FIG. 12

Now *a simple harmonic vibration is defined as one in which the restoring moment of force acting upon the vibrating system is always proportional to the angle of displacement of the system from its position of rest.* We see, therefore, that so long as the amplitude of vibration of the magnet is small, its motion is a case of simple harmonic motion; and we may therefore apply to it the general formula for the period of any simple harmonic motion. If t represents the period of a half vibration, I the moment of inertia of the vibrating system, and $\overline{Fh}/\theta$ the force constant of the system (i.e. the constant ratio of the moment of force and the displacement), then the general equation for any simple harmonic motion of rotation is

$$t = \pi \sqrt{\frac{I}{\frac{\overline{Fh}}{\theta}}}.^{*} \qquad (8)$$

* See "Mechanics, Molecular Physics, and Heat," pp. 87–91; also pp. 74 and 75.

Since by equation (7) the force constant of the vibration under consideration is MH, we have at once as the expression for the period t_1 of a half swing of the magnet

$$t_1 = \pi\sqrt{\frac{I}{MH}}, \qquad \text{or} \qquad MH = \frac{\pi^2 I}{t_1^2}. \tag{9}$$

This equation gives us the second relation between M and H in terms of the measurable quantities t_1 and I.

21. Expression for *H*. From the two equations (5) and (9) either M or H can easily be obtained. Thus the division of (9) by (5) gives

$$H^2 = \frac{2\pi^2 I}{t_1^2 r^3 \tan\phi}. \tag{10}$$

22. Moment of inertia of the magnet. In order to determine experimentally I, the moment of inertia of the large magnet (mm' of Fig. 11), it is most convenient to lay upon it a brass ring of known mass W and known radius R (Fig. 14, p. 26); to adjust it until its center coincides with the axis of suspension; and then to take the period t_2 (for small vibrations) of the new system. The moment of inertia of this new system is now $I + I_0$, I_0 being the known moment of inertia of the ring. Hence the formula for the new period is

$$t_2 = \pi\sqrt{\frac{I + I_0}{MH}}. \tag{11}$$

The solution of (9) and (11) for I gives

$$I = \frac{I_0 t_1^2}{t_2^2 - t_1^2}. \tag{12}$$

The moment of inertia of the ring is simply WR^2, where W is the mass in grams, and R the mean radius in centimeters of the ring. For it is obvious that the general formula for moment of inertia, namely Σwr^2, in which w represents an element of mass and r its distance from the axis, reduces in the case of a ring, for which all the elements are at the same distance from the center, to WR^2.*

* The rigorous formula for the moment of inertia of a ring is $W\left(\frac{R_1^2 + R_2^2}{2}\right)$ in which R_1 and R_2 are the inner and outer diameters.

As a check upon the experimental determination of I it is well to take also the theoretical value, namely $\frac{W'(l^2+b^2)}{12}$,* where W' is the mass of the magnet, l its length, and b its width.†

23. Comparison of magnetic fields. From equation (9) it is evident that if t_1 and t_2 represent the half periods of the same

* See "Mechanics, Molecular Physics, and Heat," pp. 78–81.

† **Correction of the formulas for torsion of the suspending fibers.** Equation (3) was deduced upon the assumption that no forces other than magnetic ones were concerned in the equilibrium of the magnet when its deflection was represented by the angle ϕ (Fig. 11). But evidently the torsion of the suspending fiber, as well as the field H, opposes the deflection due to the field F. This torsional element cannot always be neglected. In order to take it into account we have only to consider that by Hooke's law there is a constant ratio between the restoring torsional moment $\overline{Fh'}$ and the angle of displacement. *It is customary to define this ratio as the moment of torsion of the suspension and to denote it by* T_0. Thus, by definition,

$$\frac{\overline{Fh'}}{\phi} = T_0, \qquad \text{or} \qquad \overline{Fh'} = T_0\phi. \tag{13}$$

Thus when the magnetic needle bc (Fig. 11) was deflected through an angle ϕ the restoring torsional couple was $T_0\phi$. Hence the rigorous equation of equilibrium was not (3), but rather

$$2\,Hn\,(bo)\sin\phi + T_0\phi = 2\,Fn\,(bo)\cos\phi. \tag{14}$$

Now T_0 can be obtained by a second experiment as follows: The pin to which the upper end of the supporting fiber is attached is always arranged in magnetometers so that it can be rotated through any desired angle about a vertical axis. If, then, this so-called *torsion head* (see t, Fig. 13), be rotated through, say, 180° (π radians), the needle, which was before in the magnetic meridian, will now be deflected by the torsion of the fiber alone through an angle of, say, α radians. Since the actual twist in the fiber is then $(\pi - \alpha)$ radians, the equation of equilibrium under these conditions is evidently

$$2\,Hn\,(bo)\sin\alpha = T_0(\pi - \alpha), \qquad \text{or} \qquad T_0 = \frac{2\,Hn\,(bo)\sin\alpha}{\pi - \alpha}. \tag{15}$$

Substituting this value for T_0 in (14) and dropping the common terms, we get

$$H\sin\phi + H\sin\alpha\left(\frac{\phi}{\pi - \alpha}\right) = F\cos\phi,$$

or

$$H\sin\phi\left(1 + \frac{\sin\alpha}{\sin\phi}\cdot\frac{\phi}{\pi - \alpha}\right) = F\cos\phi. \tag{16}$$

Now since the experiment should always be arranged so that both α and ϕ are small (if α is more than 5° the torsion head may be twisted through 90°

magnet in two different fields of horizontal intensities H_1 and H_2 respectively, then

$$\frac{H_2}{H_1} = \frac{t_1^2}{t_2^2}. \tag{23}$$

EXPERIMENT 2

(A) Object. To determine the ratio M/H.

Directions. The apparatus used, commonly known as a *magnetometer*, is shown in Figure 13.

I. Set the suspended magnetic needle nn' in the magnetic meridian. To do this set up a reading telescope and scale in front of the small mirror m and two or three meters distant from it. The mirror and

---

instead of 180°), no appreciable error is introduced by replacing $\sin\alpha/\sin\phi$ by α/ϕ. Equation (16) then reduces to

$$H \tan\phi \left(\frac{\pi}{\pi - \alpha}\right) = F. \tag{17}$$

Substitution in equation (5) gives

$$\frac{M}{H} = \frac{r^3 \tan\phi}{2} \cdot \frac{\pi}{\pi - \alpha}. \tag{18}$$

The final equation for H, namely (10), then becomes

$$H^2 = \frac{2\,\pi^2 I}{t_1^2 r^3 \tan\phi} \cdot \frac{\pi - \alpha}{\pi}. \tag{19}$$

Rigorously, t in equations (10) and (19) also needs correction for the torsion of the fiber which supports the magnet when its period is determined. But since the magnet mm' is large, if the fiber is properly chosen this correction is unnecessary. In case it should be needed the method by which it is obtained and applied may be seen from the following.

The correct equation for the half period is

$$t_1 = \pi\sqrt{\frac{I}{MH + T_0'}}, \quad \text{or} \quad t_1^2(MH + T_0') = \pi^2 I, \tag{20}$$

where T_0' is the moment of torsion of the thread which holds the large magnet. If twisting the torsion head of this magnet support through an angle of π radians causes the magnet to turn through β radians, we have, since β is very small, $T_0'(\pi - \beta) = MH\beta$. Substitution of this value of T_0' in the above equation (20) gives

$$MH\left(\frac{\pi}{\pi - \beta}\right) = \frac{\pi^2 I}{t_1^2}. \tag{21}$$

The combination of this with the equation (18) found above gives as the final corrected form for the value of H

$$H^2 = \frac{2\,I(\pi - \alpha)(\pi - \beta)}{t_1^2 r^3 \tan\phi}. \tag{22}$$

magnet are rigidly connected, the latter usually consisting of a bit of magnetized needle a few millimeters long attached by shellac directly to the back of the mirror. Focus* the telescope upon the reflected image of the scale. Then so adjust the torsion head t that equal deflections of the head to the right and to the left produce equal deflections upon the scale.

II. Place the magnet mm' upon the graduated crossbar of the magnetometer at such a distance east of the suspended magnet nn' that $r/l = 10$ (see p. 20); then read with the telescope and scale the deflection produced by turning mm' end for end. Repeat this operation several times, and thus determine the order of accuracy of the observation.

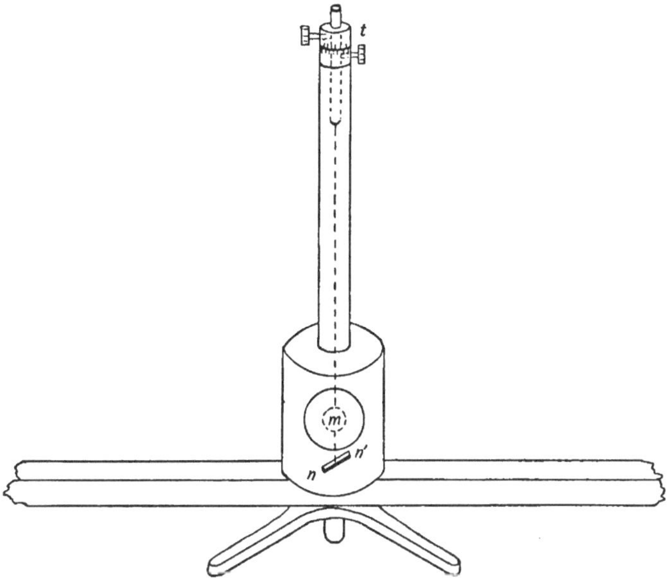

FIG. 13

III. Make the same observation when mm' is at an equal distance to the west of the suspended magnet. From the mean of these two deflections determine $\tan\phi$. This is found with sufficient accuracy by dividing the deflection, measured in centimeters, by four times the distance L from the scale to the mirror of the suspended magnet. The reason for this will be clear when it is remembered that the beam of light turns through twice as large an angle as the mirror, and that turning the magnet mm' end for end produces a deflection of 2θ in the mirror.

* See directions for focusing the telescope, as given on page 64.

IV. Remove mm' and determine α (if necessary) as outlined in the theory of footnote †, page 23. The angle α is found in radians by dividing the scale deflection by $2L$. Take the mean of the two results obtained by twisting the torsion head equal distances in opposite directions.

V. Obtain a value for $\tan\phi$ for another position of mm' such that $r/l = 14$. Calculate the value of M/H for each deflection.

(B) Object. To determine the value of MH with the same magnet that was used in **(A)**, and for the same locality, and by combination of the results with those of **(A)** to find H.

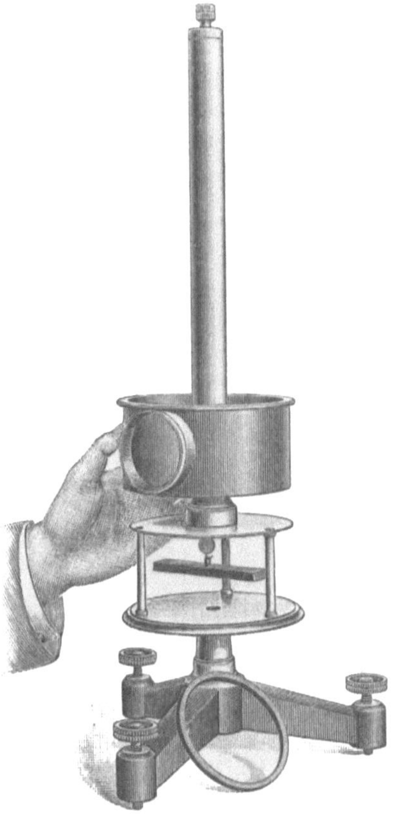

FIG. 14

Directions. I. Eliminate torsion from the suspending fiber of mm' by turning the torsion head (Fig. 14) until the stirrup comes to rest approximately in the meridian. Then suspend mm' from the fiber by placing it in this stirrup so that it rests accurately horizontal. Take the period of the magnet with the stop watch, using a telescope and scale. Make three observations of the time of thirty half oscillations. The amplitude should not exceed 3°, and the instants of passage through the mid-points, *not through the end points*, of the swing should be observed.

II. From the dimensions of the magnet mm', and from its mass as found by a rough balance, determine I. If time permit check this by the experimental method of section 22.

III. Calculate H from each of the mean values of $\tan\phi$ and the corresponding value of r.

(C) Object. To find the value of H in a different locality by comparison with the value found in **(B)**.

Directions. Set up the magnet and suspension of **(B)** at a point to be indicated by the instructor (the proposed location of the tangent galvanometer of Experiment 3). Make three observations of the time of thirty half oscillations. From equation (23) and the value of H_1 and t_1 obtained in **(B)** calculate H_2 for this position.

EXAMPLE

(A) The distance in room No. 9 of mm' west of the small magnet mounted on the mirror of the magnetometer was 100 cm. The distance of the telescope and scale from the mirror was 413 cm. The first reversal of the magnet changed the scale reading from 20.25 cm. to 42.55 cm., the second from 42.70 to 20.40, and the third from 20.40 to 42.50, thus making the mean deflection at 100 cm., 22.20 cm., and the mean value of $\tan\phi$ $22.20 \div (4 \times 413)$. Similarly the mean value of $\tan\phi$ at a distance of 140 cm. was $8.12 \div (4 \times 413)$. Twisting the torsion head through 180° produced a change in the scale reading of 3.1 cm., thus making the value of α $3.1 \div (2 \times 413)$ radians.

(B) Three observations on the time of thirty half swings of mm' when placed in the position occupied by nn' above gave 148.4 sec., 148.6 sec., and 148.4 sec. Hence $t_1 = 4.947$. The mass of the magnet was 61.29 g., its length 10 cm., and its breadth 1.3 cm. The value of I computed from these dimensions was 518.6. When a 40 g. ring of 4 cm. mean radius was added the half period was 7.40 sec. The value of I computed from these observations was 517.8. H computed from the observations at $r = 100$ was .1764; from the observations at $r = 140$, .1757, thus giving a mean value in room No. 9 of .1760. The value of M was 1185, and the number of units of magnetism in each pole was 150, approximately.

(C) The half period of the magnet in room No. 19 was 4.818 sec. Hence the mean value of H in room No. 19 was .1856 gausses.

CHAPTER III

MEASUREMENT OF ELECTRIC CURRENTS

24. Definition of the electric current. The charges (or quantities) of electricity described and defined on page 2, section 2, by means of certain attractive and repellent properties which they possess, are found to exhibit new properties as soon as they are set into motion. *An electric charge in motion is called an electric current.*

There are, however, two ways in which the charge may move. Suppose, for instance, that the given charge is contained upon the small round body A (Fig. 15), and that the body with its charge is rapidly carried to B, drawn perhaps by an opposite charge upon the rod ab. Then, according to the above definition, an electric current has passed from A to B. Very good reasons exist (see sect. 31, p. 38) for supposing that electric currents in liquids (electrolytes) are of this nature. The current probably consists of a swarm of charged particles of matter (ions) actually moving through the liquid under the influence of an electric field, just as the charged pith ball moves through the air.

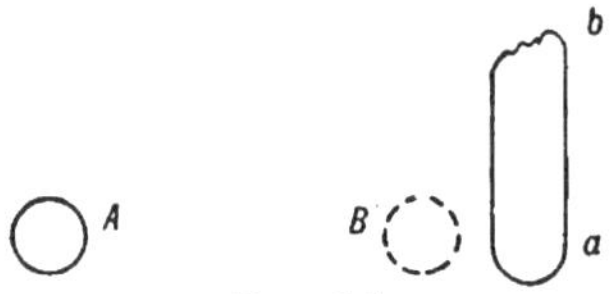

FIG. 15

But we have seen in section 3 that there is another way in which the charge upon A may arrive at B. Thus when an uncharged body is placed at B and then connected by a wire with the charged body A (Fig. 16), the instant the connection is made B is found to possess the same sort of attractive properties which were before possessed by A, while A is found to have lost some of its attractive power. The electric charge upon A (or part of it) has moved along the wire to B. According to the electron theory, in this case, also, a

FIG. 16

swarm of very minute charged particles has actually streamed through the metal from *A* toward *B*, or vice versa; but since no direct proof of the correctness of this view has yet been found, it must be regarded at present as merely an hypothesis. Nevertheless it is certain that a part of the charge which was at first upon *A* has actually passed in some way along the wire to *B*. Whether, then, a charge passes from *A* to *B* by the first method, in which matter is observed to move with the charge, or by the second method, in which minute particles are only assumed as the carriers of the charge, *it is the fact of its passage between A and B which we shall consider as constituting an electric current.*

That the properties possessed by a static charge which is made to move rapidly through space by mechanical means are in fact identical in all respects with the properties possessed by a charge which is moving along a wire (see sect. 25) was first proved in an elaborate investigation made in 1876 by the American physicist Henry A. Rowland. The correctness of this conclusion has been confirmed by many subsequent investigations.

25. Magnetic effect of an electric current. The new property possessed by a charge in motion, but not possessed by a charge at rest, is the property of exerting magnetic influences. That a charge at rest does not produce any magnetic effect may be clearly demonstrated as follows. If a charged body is brought east or west of, and near to, a magnet supported upon a point, the needle will indeed at first swing about toward the charged body as though the latter exerted a magnetic effect upon it. That this is in reality, however, an effect due merely to electrostatic induction may be convincingly shown by inserting a sheet of copper, zinc, or aluminum between the magnet and the charge, when the former will be found to swing back at once to its north-and-south position. For the sheet cuts off all electrostatic influences in accordance with the principle of electric screening (sect. 13); but it has no influence at all upon magnetic forces, as may be shown by inserting it between a magnet and the suspended needle.

Now if a strong static charge, for example that upon a Leyden jar, is passed through a wire which is wound one or two hundred times around a small glass tube containing an unmagnetized knitting needle, the needle will be found to have been quite strongly magnetized. If the sign of the charge is reversed and the knob of the jar touched to the same end of the wire as at first, a second needle will in general be found to have been oppositely magnetized. This certainly shows that a charge in motion produces some sort of a magnetic effect.

Again, if the positive terminal of a static machine is connected to one terminal of a coil of wire of many turns in the middle of which hangs a magnetic needle, the needle will be deflected when the machine is in operation; and if the terminals of the machine are reversed, the direction of deflection of the needle will be reversed also.

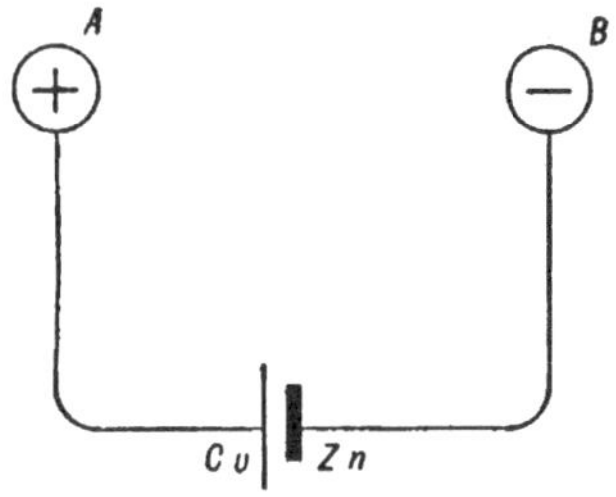

FIG. 17

A more convenient way, however, of showing that a charge in motion produces a magnetic effect is to make use of a galvanic cell.

Such a cell consists essentially of two dissimilar conducting solids immersed in any conducting liquid (the liquid cannot be a molten metal). It may be looked upon as a self-acting static machine. For if the bodies *A* and *B* be connected to the two plates of such a cell (see Fig. 17), they are found to be statically charged, just like the poles of a static machine, the one connected to the copper or carbon being positive, and the one connected to the zinc, negative. (It requires, however, a delicate electroscope to prove the existence of these static charges upon the terminals of a galvanic cell.) If now *A* and *B* are connected by a wire, the positive charge upon *A* discharges to *B* (or if it is preferred so to consider it, *B* loses its negative charge to *A*); but the chemical action which is set up in the cell recharges *A* (or *B*) as fast as it is discharged. Hence the galvanic cell, like the continuously turned static machine, produces in the wire a continuous current when its terminals are connected by a conductor.

If the wire connecting the terminals of such a cell is held over and parallel to a magnetized needle, the latter will be strongly deflected (Fig. 18). If the terminals of the cell are interchanged the deflection will be reversed. This experiment was first performed by the Danish physicist Oersted in 1819. It constitutes one of the most important discoveries which has ever been made in the history of science, for it established for the first time a connection of some sort between electricity and magnetism and paved the way for the marvelous electrical developments of the nineteenth century.

The preceding experiments have shown that reversing the direction in which the charge (or charges) moves past the needle reverses the magnetic effect observable near the path of the charge. According to the two-fluid theory the current consists in the motion of a positive charge in one direction and a negative charge in the other direction. According to Franklin's one-fluid theory it consists of a motion only of positive electricity in one direction.

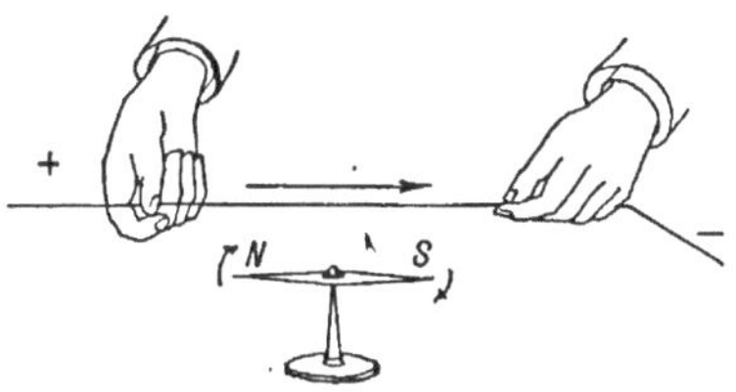

FIG. 18

According to the electron theory it consists of a motion only of negative electrons in the opposite direction. It is wholly immaterial which of these theories is correct so long as we understand the conventions which are in common use regarding direction of current. *According to universal convention the direction of a current is the direction from positive toward negative*, i.e. in the case of a galvanic cell from copper, or carbon (in the external circuit), toward zinc. Thus, according to this convention, when *B* was negatively charged and *A* uncharged, joining them with a wire caused a momentary electric current to flow from *A* to *B*, not from *B* to *A*. This definition makes it unnecessary to introduce the terms "positive current" and "negative current," although these terms are sometimes employed.

26. Form of magnetic field about a conductor. The form and direction of the magnetic field which surrounds a current may be obtained by mapping out the field with a compass needle in the manner described in Experiment 1, (A). Thus if the black center of Figure 19 represents the cross section of a conductor which passes vertically through the paper and through which the current flows from the reader toward the plane of the paper, as indicated by the cross in the middle of the conductor,* then the map of the magnetic field in the plane of the paper is that shown in the figure. The lines of force are circles about the current as a center, and the direction of these circles, i.e. the direction in which an isolated N pole would rotate about the current, is from left to right, as the observer looks in the direction of the current.

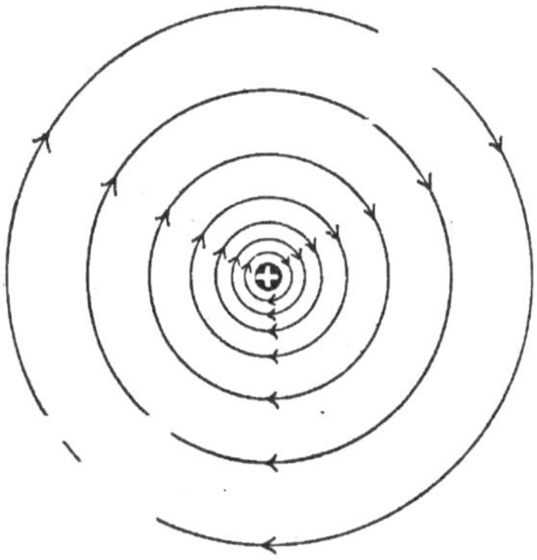

FIG. 19

This rule is often known as *the right-handed-screw rule*, for the relation between the direction of rotation of the lines and the direction of motion of the current is the same as that existing between the rotary and forward motions in a right-handed screw. The analogy stops here, however, for the lines of force are not spirals about the current. The plane of their direction is always perpendicular to that of the current.

It is to be expected from the above rule that an isolated pole would continue to move indefinitely in a circular path around a conductor carrying a current. That this is indeed the case may be strikingly shown by the following experiment. A vertical

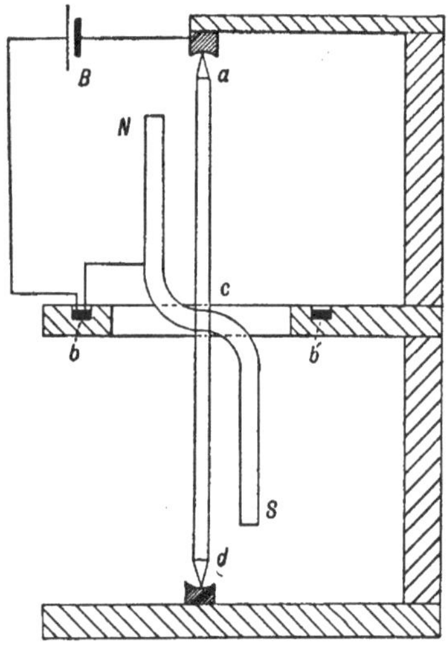

FIG. 20

* In general a dot in the middle of the cross section of a conductor is used to represent the head of an approaching arrow, i.e. a current flowing toward the reader. A cross in the conductor represents the tail of a retreating arrow, i.e. a current flowing away from the reader.

conductor ac (Fig. 20), free to turn about its own axis, carries a bar magnet NS. A circular mercury cup bb', shown in cross section, admits of a continuous connection (over the path bc) of the terminals a and c with a battery B. A current is now allowed to flow from a to c. This current produces no field at the point S and therefore may be considered as acting on an isolated pole N. Under its action the pole N will actually be found to rotate indefinitely in the direction determined by the right-handed-screw rule. If the direction of the current is reversed, the direction of the rotation is found to be reversed also. As is evident from the figure, connection with the battery may be made at b and d and the experiment performed with the S instead of the N pole.

27. The unit of current strength. Since an electric current is defined as an electric charge in motion, it is natural to measure the *strength*, or *intensity*, of current flowing through a conductor by the number of units of charge which pass through any cross section of the conductor per second. This is indeed the definition of current strength in the so-called *electrostatic system*. Thus if Q represents the number of units of charge (electrostatically measured, sect. 2, p. 2) which pass through the conductor in t seconds, then the mean current intensity I is given by the equation

$$I = \frac{Q}{t}. \tag{1}$$

The current flowing at any instant is the *rate* of passage of quantity, i.e. if we let dQ represent the quantity of electricity which passes a given cross section in the short element of time dt, then the rigorous definition of current is given by

$$I = \frac{dQ}{dt}. \tag{2}$$

A wire then carries a unit of current when a charge passes through it at the rate of unit quantity per second.

But Oersted was experimenting with a galvanic cell when he discovered the magnetic effect of a current, and neither he nor his contemporaries realized that a current was simply a charge in motion, and that the magnetic effect could be produced just as well by means of the electricity developed by means of a static machine. Hence another unit of current strength was chosen, and one which had no apparent connection with the unit of electrostatic quantity defined on page 3. The size of the magnetic effect was arbitrarily taken as the measure of current strength. But this effect was found to vary both with the length of the conductor and with the distance from the conductor to the point at which it was measured. Hence it was necessary to fix upon a wire of specified length and to measure the magnetic effect at a specified distance from it. A wire 1 cm. long was chosen, and in order to arrange to have all parts of this wire at the same distance from some point at which the magnetic effect was to be measured, it was decided to bend it into an arc of 1 cm. radius and to measure the magnetic effect at the center of this arc. Thus a current of unit strength was said to be flowing in the wire when a length of this wire equal to 1 cm. and bent into an arc of 1 cm. radius created at the center of this arc a magnetic field of unit strength. Otherwise stated, *unit current is that current unit length of which placed everywhere at unit distance from unit magnetic pole acts upon it with a force of 1 dyne.* Thus if the arc AB is 1 cm. and if r is 1 cm. (see Fig. 21), then I is one unit when the magnetic field strength at O is 1 dyne. Similarly I is 10 when the field strength at O is 10 dynes, etc.

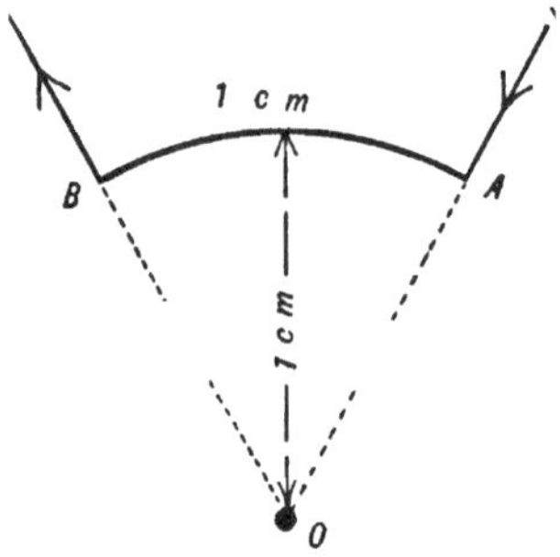

FIG. 21

Thus, in this so-called *electro-magnetic* system, it is the unit of current strength which is first defined and which is therefore the fundamental unit. The unit of quantity in this system is defined as the quantity conveyed per second through every cross section of a conductor which carries a current of unit strength. Thus, just as $I = Q/t$ was the equation which defined *current* in the electrostatic system, Q having been first defined, so the same equation written in the form $It = Q$ defines *quantity* in the electro-magnetic system, I having been first defined. Thus 100 units of quantity pass in 10 seconds through a conductor which carries a constant current whose strength is 10 units.

28. Ratio of the electrostatic and the electro-magnetic units of quantity, or of current. It is of interest to inquire which of these two units, the electrostatic or the electro-magnetic, is the larger. In order to answer this question experimentally, it is evident that it would only be necessary to collect a quantity of electricity which had been measured in electrostatic units by an observation of its attraction upon some known charge, and then to discharge this quantity in a known time through the arc of Figure 21, and measure the magnetic effect thus produced. Such a measurement shows that the electro-magnetic unit is enormously larger than the electrostatic unit, and further reveals the surprising fact that the ratio of the electro-magnetic to the electrostatic unit is equal to the velocity of light expressed in centimeters per second, namely 3×10^{10}. This fact played an important rôle in the discovery of the intimate relation which exists between electricity and light.

29. Practical units. No name has been given to the absolute units of current and quantity defined above. For purposes of convenience it has been decided to take as the commercial units one tenth of the absolute units of both current and quantity. Thus the commercial unit of current is 10^{-1} absolute electro-magnetic units of current. It is named an *ampere* in honor of the French physicist André Marie Ampère (1775–1836). Similarly the commercial unit of quantity is 10^{-1} absolute electro-magnetic units of quantity. It is named a *coulomb* in honor of the French physicist Charles Augustin Coulomb (1736–1806).

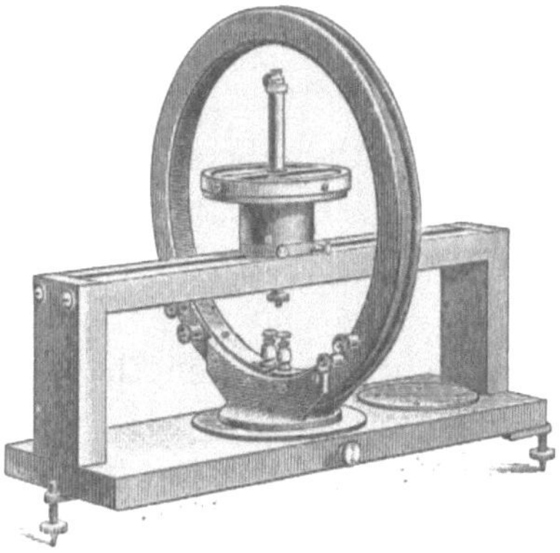
FIG. 22

30. Tangent galvanometer. It is evident from the definition of current strength that the direct method of measuring I must consist in measuring the strength of magnetic field produced by a known length of the current at a known distance from it. This is most easily accomplished by passing the current through a circular loop of wire of known radius and measuring the field strength at the center of the loop. If the current were of unit strength, then from the definition of current strength it would follow that every centimeter of length of the conductor carrying the current would produce at a point which is one centimeter distant from all parts of that centimeter of length a field strength of one dyne. Since all forces which emanate from point sources vary inversely as the squares of the distances from the sources, it is evident that the forces emanating from points on the wire vary inversely as the squares of the distances from those points. Hence at the center of the loop, a point which is r centimeters distant from all points on the wire, the field strength due to each centimeter of length must be $1/r^2$ and the field strength due to the $2\pi r$ centimeters of length must therefore be $\frac{2\pi r}{r^2}$. If the current has I units of strength, and if there are N loops instead of one, the field strength F at the center must be, therefore,

$$F = \frac{2\pi NI}{r} \text{ gausses.} \tag{3}$$

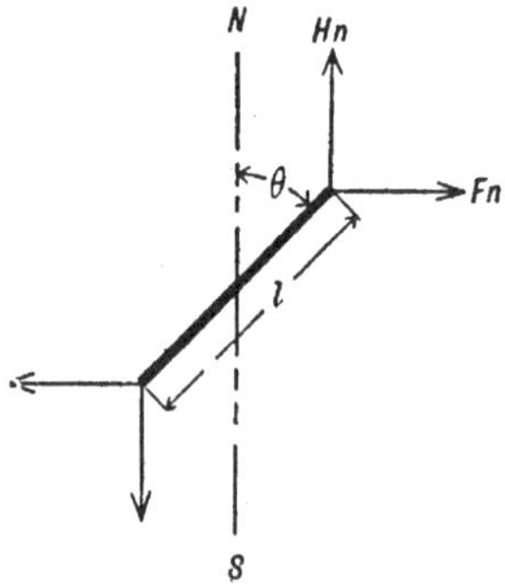

FIG. 23

In order to measure this field the plane of the coil (see Fig. 22) is set in the earth's magnetic meridian and a small suspended magnetic needle is placed at the center. In this position the field F due to the current tends to turn the needle into the east-and-west direction, while the earth's magnetic field tends to keep it in the north-and-south direction. The equation of equilibrium is, therefore (see Fig. 23 and sect. 19),

$$Hnl \sin\theta = Fnl \cos\theta, \tag{4}$$

in which l is the length of the small magnet.

Hence from (3)

$$H \tan\theta = \frac{2\pi NI}{r}, \quad \text{or} \quad I = \frac{rH}{2\pi N}\tan\theta. \tag{5}$$

Thus if the galvanometer is set up at some point for which H is known, the current strength I is determined in absolute units from a measurement of r, N, and $\tan\theta$.

31. Electrolysis. It is found that all of the liquids (save molten metals) which are conductors of electricity are solutions of chemical compounds, — salts, acids, or bases, — and that the passage of an electric current through such liquids is uniformly accompanied by the passage out of the solution of the components of the dissolved substance. Thus when a current is passed through a solution of copper sulphate ($CuSO_4$), metallic copper (Cu) is deposited upon the plate by which the current leaves the solution (the negative electrode or *cathode*), while SO_4 radicals collect about the positive electrode or *anode*, where their presence may be detected by the acid character which they impart to this portion of the solution. The systematic study of this phenomenon was first made by Faraday in 1832. He called the operation of separating the constituents of a compound in solution by means of a current *electrolysis*. Any liquid which was capable of being decomposed in this way he called an *electrolyte*. The constituents of the dissolved substance which appeared at the electrodes he called *ions*. The results of his investigation may be summarized thus.

(1) So long as the deposit of but one particular kind of ion, for example silver, was being studied, it was found that *the amount of the deposit by weight was proportional solely to the quantity of electricity which had passed through the solution.* It was independent of the nature of the solvent, of the nature of the silver compound in the solution ($AgCl$, $AgNO_3$, AgI, etc.), of the concentration of the solution, and of the strength of the current, save as this depended upon quantity through the relation $Q = It$.

The obvious interpretation of this result is that a given atom, or radical, in whatever compound it is found, is always associated with a certain definite quantity of electricity. In other words, *electricity, like matter, consists of discrete parts.* The charge associated with an atom of hydrogen was first called by Helmholtz an atom of electricity.

(2) When the investigation was extended to different ions of the same valency (combining power with respect to hydrogen), such, for example, as hydrogen, silver, iodine, or bromine, it was

found that *the amounts, by weight, deposited* (or liberated in the case of gases) *by the passage of equal quantities of electricity were exactly proportional to the atomic (or ionic) weights of the ions.* Thus an atom of silver weighs about 108 times as much as an atom of hydrogen, and the passage of one coulomb of electricity through a solution containing a silver salt was found to deposit 108 times as many grams of silver as there were grams of hydrogen liberated by the passage of one coulomb of electricity through a solution of hydrochloric acid (HCl) in water.

This evidently means that an atom of silver carries exactly the same charge as an atom of hydrogen, or, in general, that *all atoms having the same valency carry the same charges of electricity.*

(3) When substances of different valencies were compared, as for example hydrogen and copper, it was found that *the deposits by weight produced by the passage of a given quantity of electricity were proportional to the atomic weights of the ions divided by their valencies.* Thus the copper atom has a valency 2 and a weight of 63.2 as compared with the atom of hydrogen. The deposit of copper due to the passage of a given quantity of electricity was not 63.2 times the deposit of hydrogen, but instead exactly $\frac{63.2}{2}$, or 36.6 times this deposit. This signifies that a copper atom carries twice as large a charge as a hydrogen atom, i.e. that it carries two of Helmholtz's atoms of electricity, or, in general, that *the chemical valencies 1, 2, 3, 4, 5 correspond to electrical charges upon the ions in the ratios 1, 2, 3, 4, 5.*

These discoveries certainly indicate that chemical attractions are of electrical origin. They constitute the chief ground for the statement made in section 24, page 28, that the passage of a current through a liquid consists in the movement through the liquid of swarms of electrically charged particles (the ions).

The weight of a substance in grams deposited by the passage of one absolute electro-magnetic unit of quantity of electricity is called the *electro-chemical equivalent* of that substance. Thus the electro-chemical equivalent of silver (atomic weight 107.7, valency 1) is .01118, that of hydrogen (atomic weight 1, valency 1) is .0001038, that of copper (atomic weight 63.2, valency 2) is .00328.

32. Voltameter. This phenomenon of electrolysis is applied to the measurement of currents in an instrument known as the *voltameter.* The name "voltameter" is applied to any arrangement of apparatus whereby the weight or the volume of the ions caused to pass out of an electrolytic solution is used to determine the total quantity of electricity which has passed through the circuit in which the instrument is placed. Because of the accuracy with which weighings can be made, and because of the fact that, by taking the time long enough, a very small current can be made to deposit a considerable weight of metal, the voltameter, rather than the tangent galvanometer, is the instrument which is almost universally used for standardizing current-measuring instruments. However, it is to be remembered that the electro-chemical equivalents had first to be determined by reference to the tangent galvanometer, or some similar instrument which measures directly the strength of the magnetic field at a known distance from the current, and hence that the fundamental current-measuring instrument is not the voltameter, but the galvanometer.

33. Ammeter. Neither the tangent galvanometer nor the voltameter are convenient for ordinary commercial or laboratory work. Instruments called *ammeters* (see Fig. 24) are universally used for measuring current strength. These instruments are merely galvanometers, of a type to be described later, which have been empirically calibrated by passing a given current simultaneously through the ammeter and through an electrolytic solution (voltameter).

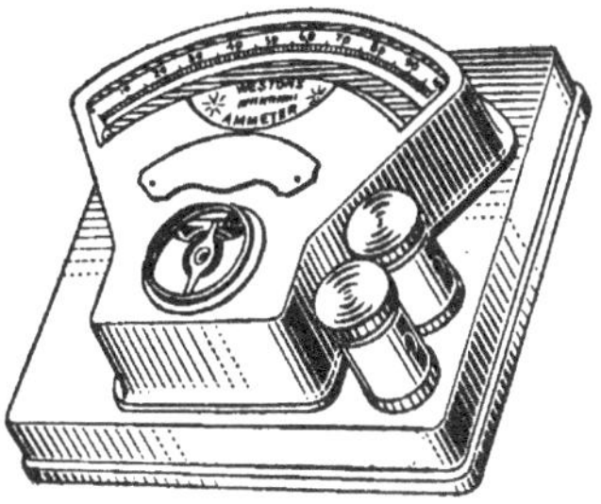

FIG. 24

EXPERIMENT 3

Object. To measure a current by means of (1) two copper voltameters, (2) a silver voltameter, (3) a hydrogen voltameter, (4) a tangent galvanometer, and (5) an ammeter; otherwise stated, to test Faraday's laws and to calibrate an ammeter.

Directions. I. *The solutions.* Make up two solutions of pure copper sulphate, one containing, say, 22 g., and the other 32 g. of crystals to 100 g. of water. Add to each solution about 1 per cent of strong sulphuric acid.

Make up for the silver voltameter a neutral silver nitrate ($AgNO_3$) solution by dissolving about 18 g. of crystals in 100 g. of water. Also make up for the hydrogen voltameter a solution of 50 or 60 g. of sulphuric acid to a liter of water.

II. *Connections and adjustments.* Set the coil of the tangent galvanometer in the magnetic meridian, i.e. parallel to the needle of a good compass. Then by means of the leveling screws adjust the instrument until the suspending fiber hangs over the center of the circular scale. Finally by means of the torsion head bring the suspended needle into the

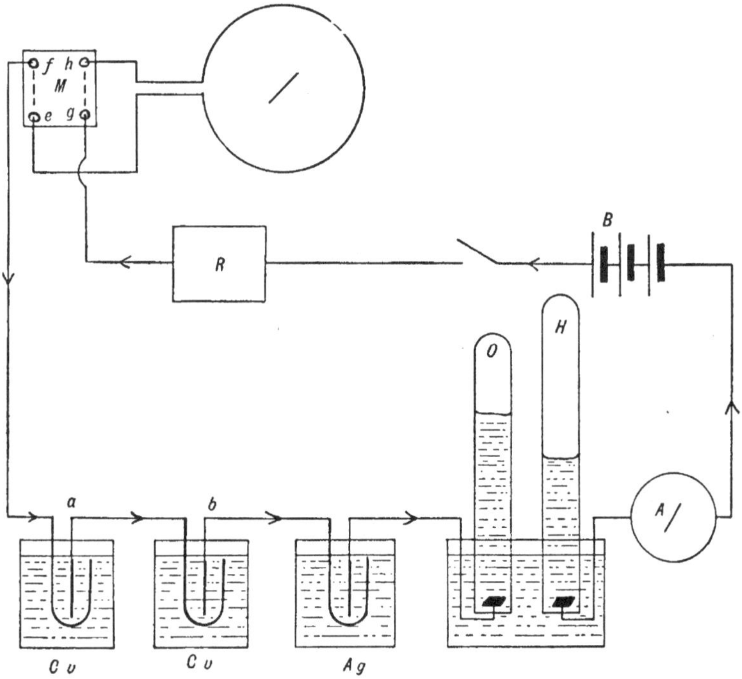

FIG. 25

magnetic meridian. This should be accomplished when each end of the aluminum index, which is attached rigidly to the needle at right angles to its length, is exactly above one of the two zeros of the scale.

Connect as in the diagram (Fig. 25), taking especial care to avoid loose contacts. The battery B may consist of three or four storage cells or six or eight fresh dry cells. The order in which various instruments are placed in the circuit is wholly unimportant, but the tangent galvanometer G should be not less than ten feet away from the milliammeter A in order that the magnet in the latter may not influence the needle of the former. Furthermore, since the deposit of a metal or of hydrogen is always upon the plate toward which the current flows in the solution (the cathode), care in getting the direction of the current through the voltameters as indicated by the arrows is of the utmost importance. The light lines in *B* represent the + terminals, i.e. the carbons of dry cells. The + terminals of storage cells are usually marked. If they are not marked they can easily be determined by connecting the two terminals with a few feet of about No. 30 German silver wire and noting the direction in which a compass needle held near the wire is deflected (see right-handed-screw rule, p. 32). *R* is a cheap, variable German silver resistance provided with a strong spring clamp for a sliding contact (Fig. 38, p. 53). *M* is a mercury commutator (Fig. 26), so made that connecting the mercury contacts *ef* and *gh* causes the current to traverse the tangent galvanometer *G* in one direction, while connecting the mercury cups *eg* and *fh* causes the current to pass through the galvanometer in the opposite direction. The negative electrodes *a* and *b* are pure copper plates of at least 50 sq. cm. area per ampere of current and well rounded at the corners; the positive electrodes are either bent copper plates of slightly greater width, as in Figure 25, or else two separate plates electrically connected as shown in Figure 27. The silver voltameter has similar electrodes of pure silver, the gain plate having an area of about 300 sq. cm. per ampere. In the instrument shown in Figure 27 the plates slip conveniently into spring clips attached to the lower side of the cover. The electrodes entering the graduated tubes *O* and *H* consist of pieces of platinum foil attached to platinum wires. Each wire may be sealed into the end of a glass tube bent into a U-shape, so that the platinum foil may be well within the graduated tubes; or the instrument may be constructed in the manner shown in Figure 28.

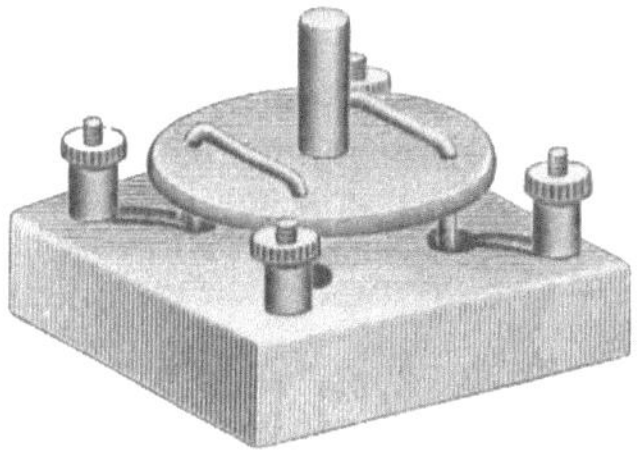

FIG. 26

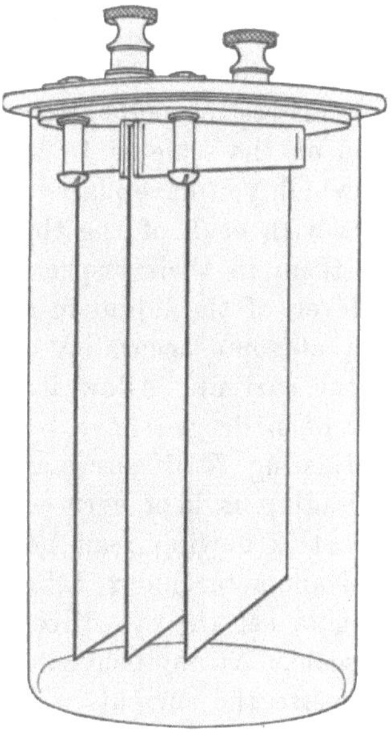

FIG. 27

If the plates have been in recent use proceed as follows.* Place the plates in their respective clips and lower them into the solutions. Fill the

* If the gain plates are not already clean they should be thoroughly polished with glass or sandpaper (not with emery).

tubes O and H with the sulphuric acid solution. Give the resistance R its maximum value. Close the circuit and then adjust R until the deflection of the tangent galvanometer needle is about 45°. (The time occupied by the experiment will be of convenient length if the galvanometer is so wound as to make this correspond to a current of from 200 to 250 milliamperes.) Allow the current to flow for four or five minutes so as to form a fresh deposit upon the plates. Then remove the gain plates, rinse first in distilled water, then in alcohol, and finally dry over a radiator or, if time is limited, by igniting, in an alcohol or Bunsen flame, the alcohol which cannot be shaken off. Do not touch the deposit at any time with the fingers.

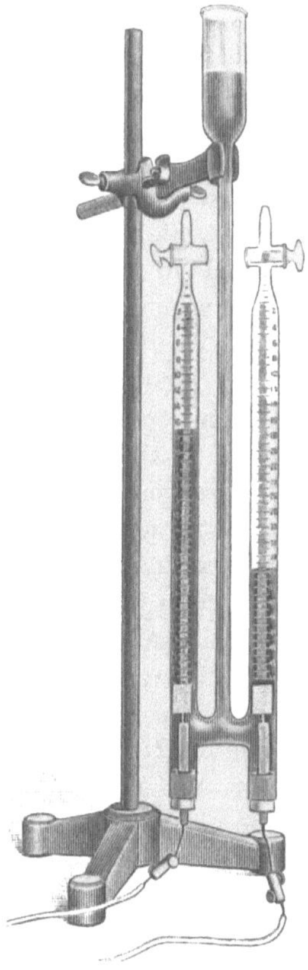

FIG. 28

III. *Weighings and readings.* Weigh the plates as follows. By means of a wire hook hang the plate to be weighed from one balance arm and add weights to the other pan until the resting point of the pointer, as determined by the method of oscillations,* is within one or two divisions of the middle of the scale over which the end of the pointer moves. Record this resting point, counting the middle division on the scale as 10, and record also the weight which you have placed on the other pan. Do this with each of the three cathodes, then replace them in their respective solutions.† Read the levels of the liquid in the tubes O and H. Then, at some accurately observed time, turn on the current. Allow it to run until 40 or 50 cc. of hydrogen have been formed, continually adjusting R, if necessary, so that the ammeter reading is kept very constant. While the current is flowing read both ends of the tangent galvanometer index, taking great pains in so doing to set the eye directly over the center of the scale. At any convenient time during the run reverse the current in the tangent galvanometer and again read both ends of the index. In reversing leave the current open long enough to allow the needle to make one half swing. Otherwise the needle may swing completely around. This interval of open circuit, which may well be measured with a stop watch, is to be deducted

* See "Mechanics, Molecular Physics, and Heat," Experiment 15.

† The plates should not be placed in the solutions more than a minute or two before the deposition is begun, for the copper plates tend to dissolve slowly in the solutions.

from the total observed time in order to obtain the actual time of duration of the experiment. At an accurately observed time break the circuit; remove the plates, rinse all three carefully in distilled water, then in alcohol, being exceedingly careful not to shake off any of the loosely adhering silver granules; dry over a radiator or by igniting the film of alcohol which clings to the surface, and weigh again as follows. Placing each plate on the same balance pan as before, record the weight which will bring the pointer to within two or three divisions of its first resting point. Record also the new resting point. Add 2 mg. and record the resting point for this added weight. From these last two resting points determine the sensibility of the balance, i.e. the number of scale divisions corresponding to an addition of one mg. to the pan. Using this sensibility determine the weight, to one tenth mg., which would bring the resting point of the second weighing into coincidence with that found in the first weighing. Thus suppose that, before the deposition, with a weight on the right pan of 35.965 g. the resting point was 9.5, and that after the deposition a weight of 36.195 g. produced a resting point of 8.6, while the addition of 2 mg. to the left side changed the resting point to 10.7. The sensibility is then $\frac{10.7 - 8.6}{2} = 1.05$, and the number of milligrams corresponding to the difference in resting points between 9.5 and 8.6 is $\frac{9.5 - 8.6}{1.05} = .9$ mg. Hence the weight, after the deposit, which would have been necessary to produce a resting point of 9.5 was $36.195 - .0009 = 36.1941$. Hence the weight of the deposit was $36.1941 - 35.965 = .2291$ g.

IV. *Calculation of current.* The calculation of the quantity of electricity which has accompanied the deposits of silver and copper is easily made from these deposits and the electro-chemical equivalents given on page 38. The current is then obtained from equation (1), page 33. The calculation of the current from the amount of hydrogen collected can be made as follows.

Let V represent the observed volume of hydrogen, let V_0 represent this volume reduced to 0°C. and 76 cm. pressure. The quantity of electricity which has passed (measured in absolute units) is equal to $\frac{V_0}{1.156}$, since 1.156 cc. is the volume at 0°C. and 760 mm. pressure occupied by .0001038 g. (the electro-chemical equivalent) of hydrogen. V_0 is obtained from V by the equation representing the combination of the laws of Boyle and Charles, namely

$$\frac{V_0}{V} = \frac{P}{760} \frac{T_0}{T}, \tag{6}$$

in which T is the absolute temperature of the room and P the pressure which the hydrogen in the tube exerts. If B represents the existing barometer height in millimeters, h the height, also in millimeters, of the final level of the water in the tube *above* that in the outer vessel (in the instrument of Fig. 28 this will of course be negative), and p the pressure of saturated water vapor (expressed in millimeters of mercury at the temperature of the room),* then evidently

$$P = B - \frac{h}{12} - p, \qquad (7)$$

the number 12 being taken as the approximate ratio between the density of mercury and that of the H_2SO_4 solution.

In the calculation of the current from the readings of the tangent galvanometer a sufficiently accurate correction for the torsion of the fiber may be made without determining T_0. Thus, while no current is flowing through the galvanometer, twist the torsion head through the same angle θ through which the needle was deflected when the current was flowing. Add to θ the angle through which the needle is now deflected in order to obtain a corrected value of θ for substitution in equation (5). The number of turns of the galvanometer will be given by the instructor. The value of H has been determined in Experiment 2, (**C**). Measure the mean diameter of the coil. Calculate the current from the data at hand.

EXAMPLE

After the current had been allowed to flow for several minutes the plates were rinsed, dried, and weighed. The current was then allowed to flow for 24 minutes and 30 seconds, less 7 seconds required for the reversal through the tangent galvanometer, i.e. the current flowed for 1463 seconds. The record of the weighing was as follows. With the first copper plate the weight used was 20.190 g. and the resting point 11.75. After the deposit 20.290 g. gave a resting point of 11.53. Adding 2 mg. gave a new resting point of 9.00. Hence the sensibility was 1.26 and the correct second weight 20.2898 g. Hence the gain was .0998 g. and the current .2070 ampere. With the second copper plate the first weight was 19.695 g. at 12.75 and the second 19.795 at 11.75. As the mass was practically that of the first plate weighed, it was unnecessary to redetermine the sensibility; hence the second weight was 19.7942 g., the gain .0992 g., and the current .2058 ampere.

Similarly with the silver plate the first weight was 20.540 g. at 11.56 and the second 20.878 g. at 11.30. Hence the final weight was 20.8778 g., the gain .3378 g., and the current .2063 ampere.

The initial volume of the hydrogen was 7.24 cc. and the final 47.04 cc. Hence the gain was 39.8 cc. The barometer reading was 744 mm. at

* See table No. 1 in the Appendix.

22°C. This reduced to 0°C. gave $B = 741.4$; h was 25 mm.; p from the table was 19.6 mm. Hence $P = 741.4 - 2.1 - 19.6 = 719.7$. Hence

$$V = \frac{719.7}{760} \cdot \frac{273}{273 + 22} \cdot 39.8 = 34.8,$$

and the value of the current as found from this was .2062 ampere.

The readings on the tangent galvanometer were, for the east end, 47.6, for one direction of the current, and 47.7 for the reversed direction. For the west end these readings were 47.5 and 47.6, respectively, giving a mean of 47.6°. Twisting the torsion head through 48° caused a deflection of the needle of .5°. Hence $\theta = 48.1°$. The average radius of the coil was 15 cm., and there were 24 turns. H from Experiment 2 was .1856. Hence the current was .2053 ampere.

The ammeter reading was kept constant throughout the experiment at .207 ampere.

The average of all the determinations of current was .2059 ampere, and the greatest deviation of any single determination was .3 per cent. The error in the ammeter at this point was approximately 1 milliampere.

CHAPTER IV

THE MEASUREMENT OF POTENTIAL DIFFERENCE

34. Potential difference between points in the neighborhood of charged bodies. The gravitational P.D. between two points has been defined as the amount of work required to carry unit mass between the two points against the force of the existing gravitational field. An exactly analogous definition was given for the electrical P.D. between two points, viz. the number of ergs of work required to carry unit positive charge between the two points against the force of the existing electrical field. This definition holds under all circumstances, whether static or current phenomena are under consideration. Thus the P.D. between the points *A* and *B* (Fig. 29) near a charge *P* is simply the number of ergs of work required to carry unit + charge against the repulsion of *P* from *B* to *A*; or, inversely, the work which the electric field does in carrying unit + charge from *A* to *B*. Or again, if the points *A* and *B* represent the localities of + and − charges respectively (see Fig. 30), the P.D. between *A* and *B* is the number of ergs of work required to carry unit + charge from *B* to *A*.

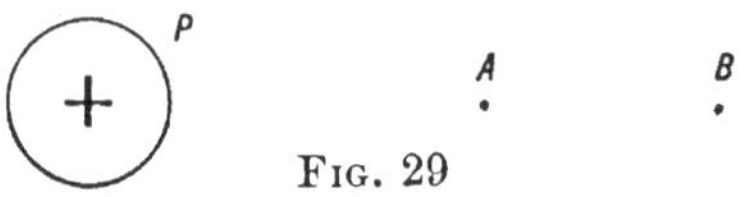

FIG. 29

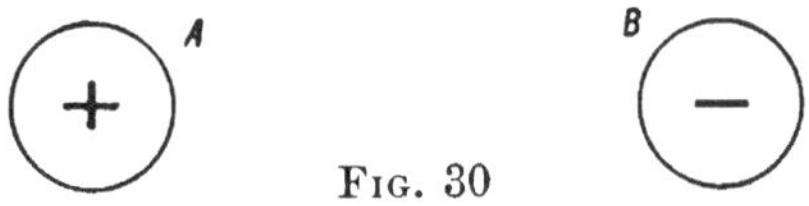

FIG. 30

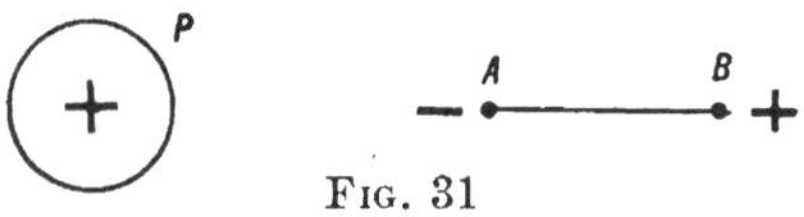

FIG. 31

If the points *A* and *B* in Figure 29 are connected by a conductor of any sort, as in Figure 31, we have seen (sect. 13) that a current at once flows through the conductor, and for convenience we have taken the direction of this current the same as the direction of the electric force, i.e. from *A*, the point of higher potential, to *B*, the point of lower potential. We have seen further that in this case the current can be only momentary, for since *AB* is an insulated conductor, the passage of a current from *A* to *B* means the appearance of a + charge at *B*, and of a − charge at *A*, since + and − charges always appear simultaneously and in equal amount. But since the accumulation of a + charge at *B* and a − charge at *A* means the creation of a field of force between *A* and *B* whose direction is opposite to that of the field due to *P*, we saw that a current can flow through an isolated conductor in obedience to the force of an outside electric field only until the new field between *A* and *B*, created by the charges accumulating at these points, is sufficiently strong to neutralize the initial field, i.e. the field due to *P*; that is, strong enough to reduce to zero the force which was initially causing a current to flow from *A* to *B*. This force gone, there can of course be no longer any P.D. between *A* and *B*; for the existence of a P.D. depends upon the existence of an electric force (sects. 9 and 11). It was thus seen that as soon as a current ceases to flow through a conductor in an electric field, all of the points of the conductor must have the same potential. Similarly, in Figure 30, the result of connecting *A* and *B* by a conductor is to cause a current to flow between these points until the field strength existing between them is reduced to zero, i.e. until *A* and *B* have the same potential.

35. Potential difference between points on the terminals of an electric generator. But if *A* and *B* are the terminals of a static machine, or of a galvanic cell, or of a dynamo, or of any generator *E* (Fig. 32) of electricity which is capable of building up a field of given strength between *A* and *B*, i.e. of maintaining, on open circuit, a given P.D. between them, then, when these points are connected by a conductor, the first result must be, as above, that a current flows from *A* to *B*, thus tending to discharge these bodies, and to cause the field between them to collapse, i.e. to cause their P.D. to disappear. But the generator *E* tends instantly to recharge *A* and *B*, and if this recharging takes place at a rate which is very rapid in comparison with the rate at which the

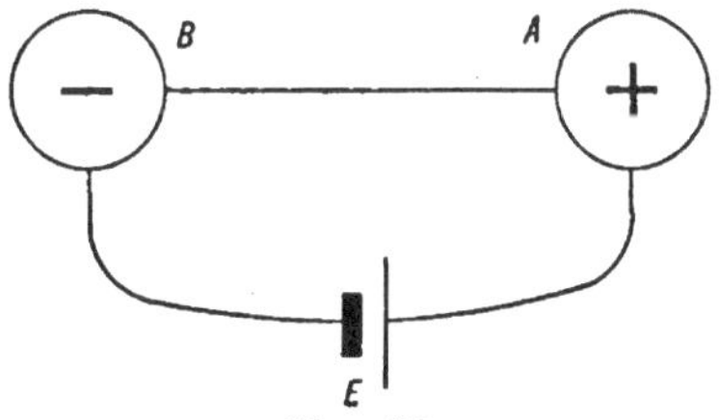

FIG. 32

connecting conductor discharges them, it is evident that, in spite of the presence of the conductor, a mean strength of field may be kept up between A and B which is almost as great as that kept up when the connecting conductor was absent. On the other hand, if the conductor discharges A and B very rapidly in comparison with the rate at which the generator is able to recharge them, the strength of field maintained between them must be small; in other words, A and B will have nearly the same potential. Thus, connecting the points A and B by a very good conductor may lower their P.D. almost to zero, — never, of course, quite to zero so long as a current is flowing from A to B, — while connecting them with a very poor conductor may lower their P.D. an inappreciable amount. Connections of intermediate conductivity must of course produce intermediate lowerings of the P.D. between A and B. *It is evident, then, that in all cases a P.D. exists between two points, whether on a conductor or off from it, only by virtue of the existence of a static field which makes it necessary to do work in order to carry a charge from one point to the other against the existing electrical force.*

36. Work done independent of path. If A and B are the terminals of a cell or a dynamo, and if these terminals are connected by a long wire $ACDB$ (see Fig. 33), the assertion that a certain P.D., say 10 absolute units, exists between A and B means simply that it would require 10 ergs of work to carry a unit + charge straight across through the air against the force of the electric field which exists between A and B. It also means that it would require just the same work to carry a unit charge from B to A by any other path whatever; for the work done is wholly independent of the path (see sect. 12, p. 11). Or, conversely, it means that the field would do 10 ergs of work in driving a pith ball charged with 1 + unit from A over to B; or, since work is independent of path, that it would do 10 ergs of work in carrying one unit of charge from A to B through the wire $ACDB$.

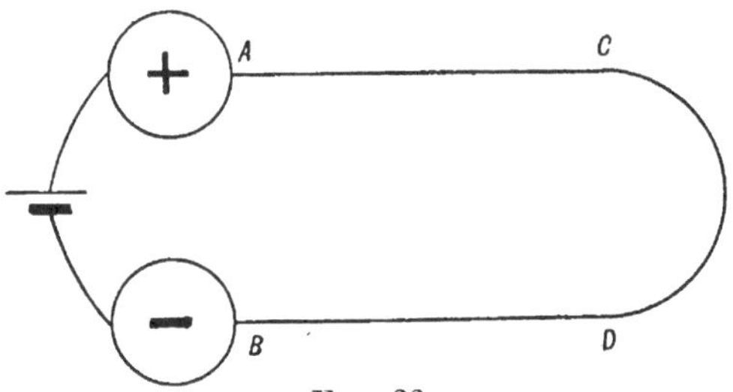

FIG. 33

37. How the work appears. If the pith ball were driven through the air from *A* to *B*, the work done by the field would be expended in overcoming the friction of the air and in imparting kinetic energy to the ball. This kinetic energy would all be given up, and a corresponding heat energy would appear, when the pith ball struck *B*. Hence all of the work done in moving the + unit charge from *A* to *B* would appear as heat. Similarly if the unit charge goes from *A* to *B* through the wire *ACDB*, the 10 ergs of work expended by the field in carrying it in this way from *A* to *B* must in general appear as heat. If any chemical changes took place in the wire because of this passage of current, this conclusion would not hold, nor would it hold if an electric motor were interposed anywhere in the wire *ACDB*. In these cases the work done by the field would be equal to the heat developed plus the chemical or mechanical work done. But if the heat is the only observable effect produced, then it follows from the principle of the conservation of energy that the work done by the electric field in moving a unit charge from *A* to *B* must be equal to the heat developed in the wire *ACDB* when this heat has been reduced to mechanical units. According to the electron theory the heating is in this case due to the frictional resistance which the wire offers to the passage of the electrons through it.

38. Measurement of P.D. It is evident then from the definition of P.D. that any measurement of P.D. must consist in measuring the work done in carrying unit charge between the two points whose P.D. is sought. This work may be obtained by measuring the strength of the field in dynes, and the distance between the points whose P.D. is sought

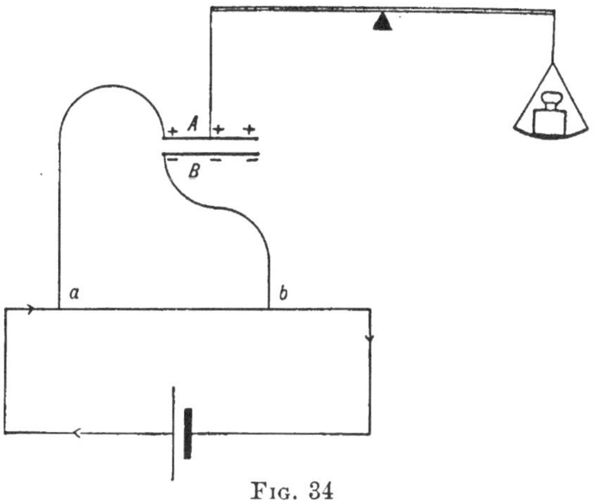

FIG. 34

in centimeters. These are the actual measurements which are made in the use of the so-called *absolute electrometer.* This is an instrument which consists essentially of two plates, A and B (Fig. 34), which are connected by wires to the points a and b between which the P.D. is sought. These plates are so arranged that the force of attraction between them may be measured. From the force of attraction of the plates the strength of the field between A and B is easily obtained. The product of this field strength by the distance between the plates gives the work necessary to carry unit charge from B across to A. Since all points on a conductor in which no current flows have the same potential (sect. 13, p. 12), it follows that the potential at A is the same as that at a, and that at B the same as that at b, i.e. the P.D. between A and B must also be the P.D. between a and b. Commercial electrostatic voltmeters, which are coming more and more into use, and which are in many respects the most satisfactory of all instruments for measuring P.D., differ but little in principle from the absolute electrometer. They consist of two fixed plates A and B (Fig. 35), which may be connected to the two points whose P.D. is sought, and which carry between them a gold leaf g, or some other movable system, the deflections of which may be taken as a measure of the field strength between the plates. The instrument is empirically calibrated so that given deflections of the movable system correspond to given differences in potential between A and B, and therefore between the points to which A and B are attached.

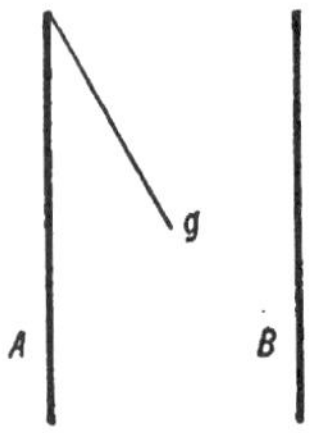

FIG. 35

Another method of making an absolute measurement of P.D. consists in measuring the number of calories of heat developed by the passage of a known quantity of electricity between the points whose P.D. is sought. Thus suppose that a constant current of strength I is made to flow through a platinum wire AB (Fig. 36) immersed in a calorimeter, and that a number of

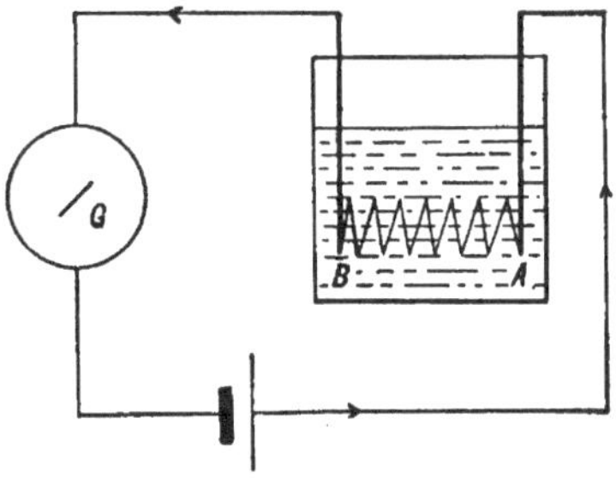

FIG. 36

calories of heat M are thus communicated to the water in t seconds. Then the amount of work done by the current is MJ ergs, where J is the mechanical equivalent of heat, i.e. the number of ergs equivalent to one calorie, namely 4.19×10^7. But since the P.D. between A and B is the number of ergs of work done in carrying unit quantity between these points, the work done by the passage of $Q(=It)$ units of quantity may be written $PD \times Q = PD \times It$. Equating these two expressions, we obtain

$$PD\ It = MJ, \qquad \text{or} \qquad PD = \frac{MJ}{It}. \tag{1}$$

Now the absolute electro-magnetic unit of P.D. is the P.D. which exists between two points when it requires 1 erg of work to carry 1 unit of quantity (measured in the electro-magnetic system) between the two points. The above equation evidently gives the P.D. in such units, provided M is measured in calories, t in seconds, and I in absolute electro-magnetic units of current. This absolute unit of P.D. is so extremely small, however, that it has been decided to use as the unit in practical work a P.D. which is 10^8 absolute units. This unit is called the volt. Hence *the P.D. between two points is 1 volt when it requires 10^8 ergs of work to carry one electro-magnetic unit of charge between these points.*

39. Voltmeters. Neither the calorimetric method nor that which makes use of the absolute electrometer is convenient for the rapid measurement of P.D.; but either one of them may be used for calibrating a so-called high-resistance galvanometer in such a way that P.D. may be read off upon it directly. The operation of finding the P.D. between any two points between which a current is flowing will then consist simply in touching the galvanometer terminals to the two points whose P.D. is sought, and observing the deflection produced.

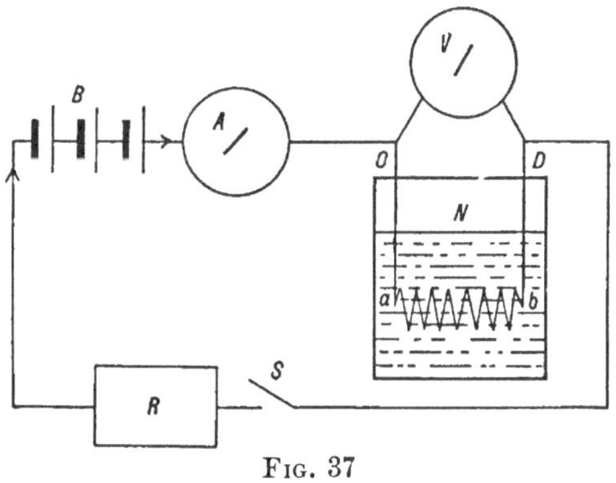

FIG. 37

Thus suppose a given P.D. is maintained between the terminals *C* and *D* of a platinum wire immersed in the calorimeter (Fig. 37), and suppose this P.D. has been found by measuring the rate at which heat is developed in the calorimeter. If now, while the current is still flowing, a galvanometer be connected across *C* and *D*, the conductivity of the galvanometer, added to that of the wire *CD*, will in general discharge the terminals *C* and *D* faster than they were being discharged when they were connected by the platinum wire alone. Hence the introduction of the galvanometer will in general cause a fall in the P.D. between *C* and *D*. But if this galvanometer carries a quantity ef electricity per second which is wholly negligible in comparison with that carried by *CD*, then the P.D. will remain essentially unchanged by the introduction of the galvanometer. The deflection of the galvanometer may then be marked and labeled with the number of units of P.D. which exist between *C* and *D*, as measured by the heat developed in the calorimeter. The P.D. between *C* and *D* may then be increased by the use, say, of a different generator, and another deflection labeled with a new P.D., as measured by the calorimeter. In this way the galvanometer may be empirically calibrated throughout its whole range of deflection, so that its readings always indicate the P.D. existing between the points to which its terminals are touched.

It is evident that such a voltmeter could not be used for determining the P.D. between two isolated static charges, for the P.D. between these charges would instantly become zero as soon as the galvanometer terminals were touched to them. It could be used for determining the P.D. between the terminals of a dynamo, or of a galvanic cell on open circuit, provided it carried so little current as not to lower appreciably this P.D. by its introduction between them. It could be used for determining the P.D. between two points on a conductor which is already carrying a current, provided it did not alter this P.D. by its introduction in shunt with the conductor, i.e. provided it carried a current which is negligible in comparison with that carried by the conductor. Of course after the galvanometer has once been calibrated for P.D. its readings represent under all circumstances the P.D. existing between its own terminals, whether the current which it carries is negligible or not; but if its current is not negligible, the P.D. between the points to which it is touched is not the same when it is in contact with these points as when it is not in contact with them. Thus a voltmeter is merely a galvanometer which has been calibrated empirically by reference to some absolute measurement of P.D.

(Because of the uncertainties of calorimetric measurement, the method given above for calibrating a voltmeter is not the one which is most commonly used (see Chap. V)).

EXPERIMENT 4

Object. To test the calibration of a voltmeter by an absolute determination of P.D.; otherwise stated, to verify the relation $PD \times Q = MJ$.

Directions. Connect as in the diagram of Figure 37. B is a generator capable of producing a constant P.D. of at least 20 volts. (Either 55 or 110 volts will serve the purpose admirably.) A is an ammeter with a range of, say, 0 to 10 amperes. ab is a platinum wire (No. 30 to No. 34) which is capable of being raised to a red heat by a current of from 3 to 6 amperes. S is a switch which is to be left open until the instructor has seen that all connections have been properly made, and which is in no case to be closed unless the platinum wire is immersed in water. R is a variable German silver resistance (Fig. 38) capable of reducing the current from 110 volt mains to from 3 to 6 amperes. N is a calorimeter the inner vessel of which has a capacity of about 300 cc. (Fig. 39 shows cut of the complete calorimetric outfit.) Figure 40 shows the construction of the voltmeter V which is to be calibrated.

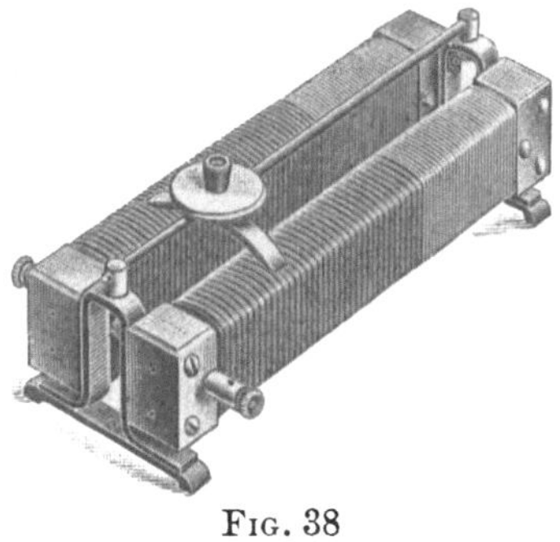

FIG. 38

Fill the calorimeter to within about 1 cm. of the top with water the temperature of which is from 10°C. to 15°C. below the temperature of the room, provided this temperature is not below the dew-point. If dew collects on the vessel the initial temperature should be taken a degree or two above that at which this dew is seen to appear. Take a preliminary observation as follows. Stir thoroughly, read the temperature, then throw on the current, and keep it on for just one minute, the ammeter reading being adjusted to $3\frac{1}{2}$ or 4 amperes. If the final temperature is not more than two or three degrees above room temperature, the conditions need not be changed. Otherwise alter the current until a run of a minute produces about the final temperature indicated above. Then fill the calorimeter again with cold water, stir it thoroughly, insert the platinum coil, and, while stirring continuously, take four or five successive readings of the temperature of the water, estimating the thermometer to tenths of the smallest division. As soon as possible after the last reading throw on the current, and keep it on for just a minute, stirring vigorously all the time and adjusting R, if necessary, so as to hold the current constant. Stir for at least a minute after the current has been turned off, then take the final temperature very carefully. If the experiment is performed in this way, no correction for radiation will in general be necessary.

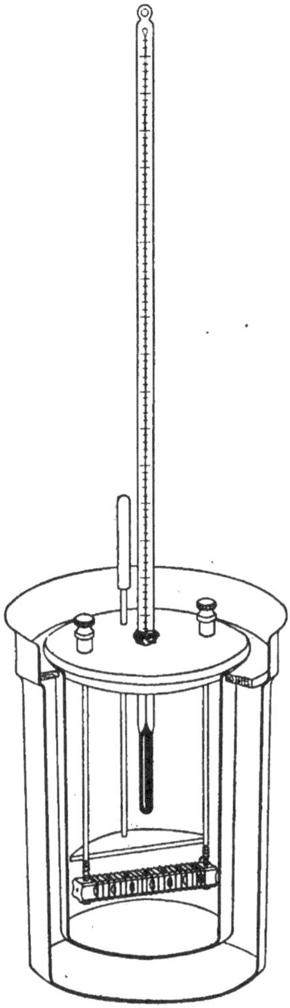

FIG. 39

But if the duration of the heating is several minutes, or if the final temperature is 8°C. or more above room temperature, the final temperature corrected for radiation may be found as follows. Observe the initial temperature precisely as indicated above; take the time at which the heating begins, the time at which it ends; then, beginning 1 minute after this latter time, record the temperature at minute intervals for at least 5 minutes, the stirring being continuous from the instant of throwing on the current. Next plot times, measured from this instant, as abscissas, and temperatures as ordinates, in the manner indicated in Figure 41. The full line drawn through these points shows the rate of cooling by radiation, and this line, extended back so as to cut the vertical line drawn through the time at which the current was discontinued, gives the temperature of the water at the instant of cessation of the current. This quantity could not have been observed directly because it is impossible, however vigorous the stirring, to keep all parts of the water at the same temperature during the heating. Next find from the curve the loss in temperature per minute per degree above room temperature. This quantity, multiplied by the time in minutes during which the current was flowing, and by the number of degrees that the average temperature during this time (as found by taking the mean of initial and final temperatures) was above the room temperature, gives the number of degrees to be added to the final temperature in order to obtain the final temperature corrected for radiation. If the average temperature during the heating is below room temperature, obviously heat is gained, and the correction, found exactly as indicated above, is to be subtracted.

As soon as possible after the observations on temperature, weigh on a rough balance the inner vessel of the calorimeter and the contained water,

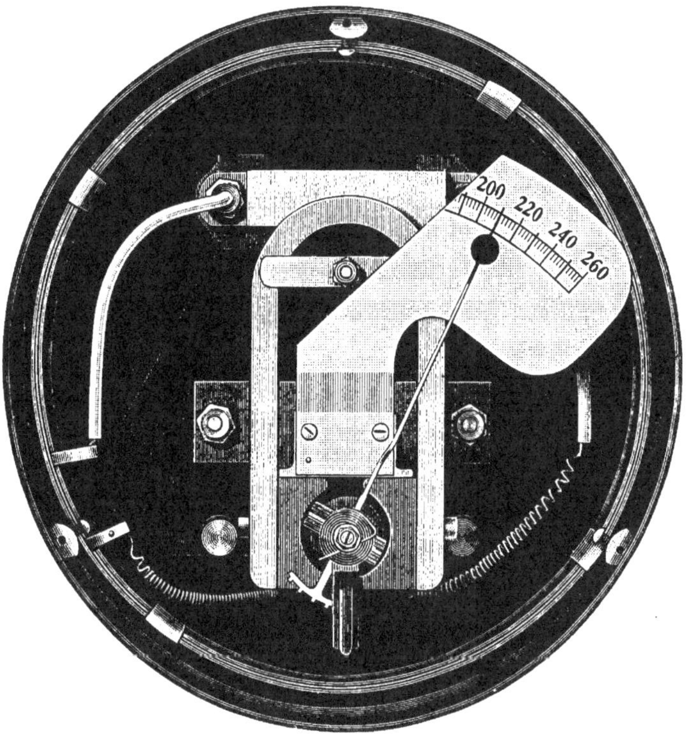

Fig. 40

then the empty calorimeter, then the stirrer. Estimate the volume of the electrodes in cubic centimeters from rough measurement of their dimensions.

The water equivalent of the inner vessel of the calorimeter, that is, the number of grams of water which would be raised 1° C. by the number of calories which is required to raise the calorimeter 1° C. is evidently the weight of the calorimeter × the specific heat of brass (.094). The water equivalent of the electrodes is their volume × their density

$(= 8.4) \times .094$. Similarly the water equivalent of the stirrer is obtained from its weight and specific heat. You can probably neglect the water equivalent of the thermometer.* The total weight of water plus these water equivalents, multiplied by the observed rise in temperature, evidently gives the number of calories of heat developed during the experiment. One calorie is equivalent to 427 gram meters of work or to 4.19×10^7 ergs. If I is the mean current in amperes, the heat developed, when reduced to ergs, represents the work in absolute units which has been done in transferring $Q(= It/10)$ absolute units of electricity from C to D. Hence the P.D., or the work expended in transferring one unit, is at once obtained. This is reduced to volts by dividing by 10^8. †

EXAMPLE

The room temperature was found to be 17.2°C. Initial temperature of water was 15.4°C. Final temperature (see Fig. 41) was 29.0°C. Mean temperature of water during heating was $\frac{29 + 15.4}{2} = 22.2$°C. Mean rate of cooling at end of experiment (eleven degrees above room temperature) was .15° per minute. ∴ Mean rate of cooling at one degree above room temperature would be .014° per minute. ∴ Rate of cooling at 22.2° was $.014° \times (22.2 - 17.2) = .07°$ per minute. ∴ Final temperature corrected for radiation was 29.07°C. ∴ Rise in temperature was 13.67°C. Weight of water and calorimeter was 282.9 g. Weight of calorimeter was 54.9 g. Weight of stirrer, 7.87 g. ∴ Weight of water was 228.0 g. Water equivalent of calorimeter was 5.16 g. Water equivalent of stirrer, .74 g. Volume of electrodes, .64 cc. Water equivalent of electrodes, .51 g. Water

* To calculate the water equivalent of the thermometer note that the heat capacity of mercury per unit of volume, viz. $13.6 (=\text{sp. gr.}) \times .033 (=\text{sp. ht.}) = .47$, is nearly the same as that of glass, viz. $2.5 (= \text{sp. gr.}) \times .19 (= \text{sp. ht.}) = .45$. Hence to calculate the heat capacity of a thermometer, estimate the volume immersed, and multiply by .46.

† The calculation may be freed from the use of powers of 10 as follows. The practical unit of work, the joule, is 10^7 ergs. One calorie is therefore 4.19 joules. If, then, M represents the number of calories developed, $4.19\,M$ is the heat developed expressed in joules. If PD and I are expressed in volts and amperes respectively, then the work done is $(PD)\ 10^8\ I\ 10^{-1} t$ ergs $= (PD)\ It$ joules. This gives $(PD)\ It = 4.19\ M$, where PD and I are expressed in practical units. A coulomb is defined as the quantity carried by a current of 1 ampere in 1 second. In general, then, the product of volts and coulombs gives joules. Also, since the practical unit of power, the watt, is defined as 1 joule per second, the product of volts and amperes gives the *rate of work* or *power* in watts.

equivalent of thermometer was .32 g. ∴ Total water equivalent was 234.7 g. Duration of heating was 60 seconds. Mean current I, 4.08 amperes. Therefore P.D. was 54.90 volts. The reading of the voltmeter was 54.8, which

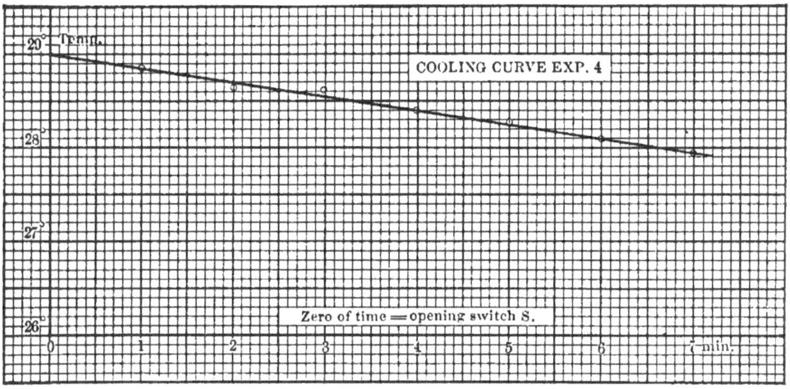

FIG. 41

differs from the calculated P.D. by but .2%. The voltmeter was therefore correct within the limits of observational error.

CHAPTER V

THE MEASUREMENT OF RESISTANCE

40. Ohm's law. In 1829 the Berlin physicist G. S. Ohm announced the famous law now called after its discoverer *Ohm's law.* This law asserts that when a continuous current is flowing through a given conductor, the temperature of which is kept constant, the ratio of the P.D. existing between the terminals of the conductor and the current carried by the conductor is a constant, no matter what the value of the current may be. Symbolically,

$$\frac{PD}{I} = \text{constant}, \qquad \text{or} \qquad \frac{PD}{I} = R. \tag{1}$$

It is customary, as above, to denote this constant, the value of which depends only upon the nature of the conductor, by the letter R, and to name it the *electrical resistance* of the conductor. The name is natural, because when the value of R is large, the amount of electricity which flows per second through the conductor as the result of a given P.D. is small, and vice versa. Ohm's law is to be regarded as an empirical law, which may be verified directly by varying the value of the P.D. between the terminals of a given conductor and measuring the corresponding currents. The most convincing evidence for the rigorous correctness of the law is found, however, in the agreement which exists between precise observations, like those made with a Wheatstone's bridge, and calculations which depend upon the assumption of this law. So far as the most delicate measurements have been able to show, Ohm's law is not, like Boyle's law, an approximation, but is, rather, a law of rigorous exactness.

41. Units of resistance. *The absolute unit of resistance is defined as the resistance of a conductor which conveys one absolute electro-magnetic unit of current when its terminals are maintained at a P.D. of one absolute electro-magnetic unit.* The practical unit of resistance is the resistance of a conductor which carries one practical unit of current (1 ampere) when its terminals are maintained at a P.D. of one practical unit (1 volt). This practical unit of resistance is called the *ohm.* Thus

$$\frac{\text{volts}}{\text{amperes}} = \text{ohms}.$$

Since 1 volt $= 10^8$ absolute units of P.D., and 1 ampere $= 10^{-1}$ absolute units of current, it is evident that 1 ohm $= 10^9$ absolute units of resistance.

It appears at once from the definition of R, namely PD/I, that the resistance of a conductor is determined once for all when an absolute measurement has been made both of the P.D. existing at any time between its terminals and of the corresponding current. Thus the resistance of the conductor ab in the preceding experiment is known, at the mean temperature of the calorimeter, from the measurements there made upon PD and I. When such an absolute measurement of the resistance of a wire has been once made, it is very easy to preserve the wire so that no change takes place in its character, and it is also easy to reproduce the temperature at which its resistance was determined. Hence with the aid of this wire and an ammeter standardized by comparison with a silver voltameter, it is easy to standardize any voltmeter; for according to Ohm's law the P.D. between the terminals of the wire will always be the product of the resistance of the wire times the current flowing through it, so that, R being known, we have only to vary the current I in a known way and take in every case the product RI in order to obtain PD. This is a method actually used in standardizing laboratories for calibrating voltmeters.

When one standard resistance has been found by any absolute method, other standards can easily be obtained by the method of comparison to be described in section 43. Many elaborate investigations have been undertaken for determining accurately the length of a mercury column 1 sq. mm. in cross section which will carry exactly 1 ampere of current when its terminals are maintained at a P.D. of 1 volt, that is, the length of a mercury column 1 sq. mm. in area which has 1 ohm of resistance according to the above definition. The methods employed for measuring the P.D. have not generally been calorimetric methods like the above, but have involved electro-magnetic principles to be considered later. The mean of the results of the most eminent observers gives 1 ohm as the resistance at 0° C. of 106.3 cm. of mercury 1 sq. mm. in cross section. Hence, at the Electrical Congress in Chicago in 1893, the resistance of a mercury column 1 sq. mm. in area and 106.3 cm. long was adopted as the "international ohm." Standards of resistance are made by comparison with the resistance of such a mercury column.

42. Laws of resistance. Experiment shows that R varies with the length, the area, the temperature, and the material of the conductor. It is found to be *directly proportional to the length and inversely proportional to the sectional area.* For metallic conductors it is found to increase with the temperature, the coefficient of increase per degree being nearly the same for all pure metals.

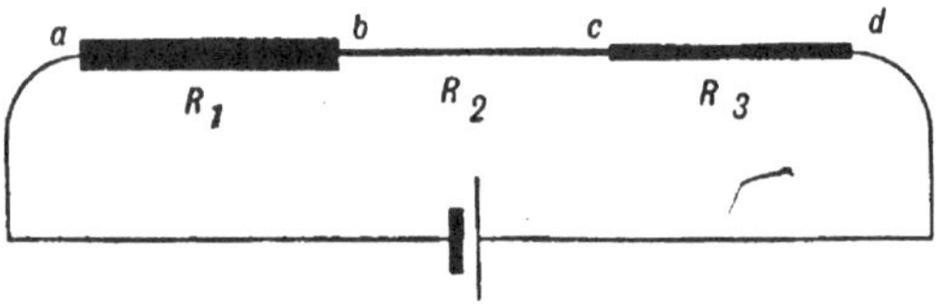

FIG. 42

If the separate resistances of conductors connected in series are R_1, R_2, and R_3 (see Fig. 42), the joint resistance of the three, that is, the total resistance R between the points a and d is easily proved to be

$$R = R_1 + R_2 + R_3. \tag{2}$$

For, if PD_1 represents the potential difference between a and b, PD_2 that between b and c, PD_3 that between c and d, and PD that between a and d, then, since the current which is flowing in all parts of the circuit must be the same, we have

$$\frac{PD}{I} = R,\ \frac{PD_1}{I} = R_1,\ \frac{PD_2}{I} = R_2,\ \frac{PD_3}{I} = R_3. \tag{3}$$

But
$$PD = PD_1 + PD_2 + PD_3. \tag{4}$$

Hence
$$\frac{PD}{I} = \frac{PD_1}{I} + \frac{PD_2}{I} + \frac{PD_3}{I}. \tag{5}$$

$$\therefore\ R = R_1 + R_2 + R_3. \tag{6}$$

If the conductors are connected in parallel (see Fig. 43), it can easily be shown that the joint resistance R, i.e. the total resistance between the points a and b, is given by the equation

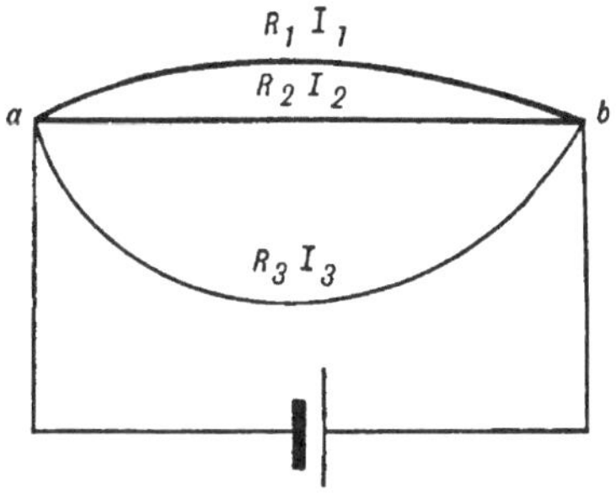

FIG. 43

$$\frac{1}{R}=\frac{1}{R_1}+\frac{1}{R_2}+\frac{1}{R_3}.$$

For, if PD represents the potential difference between the points a and b, then, by Ohm's law,

$$\frac{PD}{R}=I,\ \frac{PD}{R_1}=I_1,\ \frac{PD}{R_2}=I_2,\ \frac{PD}{R_3}=I_3. \tag{7}$$

But the total current I between a and b must be the sum of I_1, I_2, and I_3.

Hence
$$\frac{PD}{R}=\frac{PD}{R_1}+\frac{PD}{R_2}+\frac{PD}{R_3}. \tag{8}$$

$$\therefore \frac{1}{R}=\frac{1}{R_1}+\frac{1}{R_2}+\frac{1}{R_3}. \tag{9}$$

43. Theory of the Wheatstone bridge. Since at the present time standards of resistance are obtained not from the absolute measurement of P.D. and current, but from comparison with standards already existing, a very accurate method of comparing resistances is of extreme importance. *Wheatstone's bridge* is an instrument by means of which such accurate comparisons can be made. It consists of a divided circuit abc and adc (Fig. 44) composed of three known resistances P, Q, R, and of the unknown

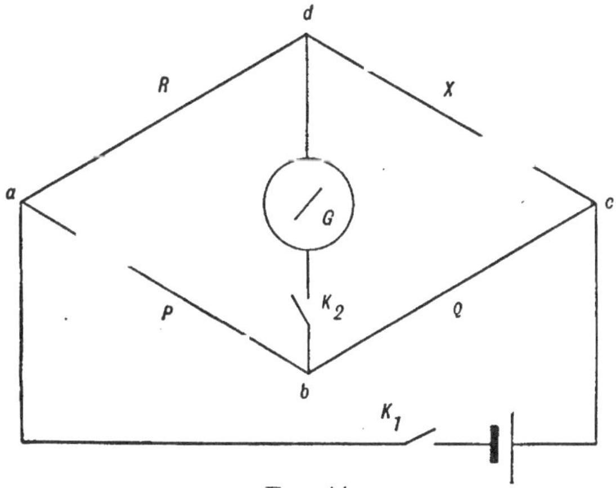

FIG. 44

resistance X, the value of which is to be found. The resistances P and Q are in general the parts ab and bc respectively of a long German silver wire connecting a and c.

As soon as the current has become constant in the circuit, a definite P.D. must exist between the points a and c. Hence the total fall of potential along the branch abc must be the same as the fall of potential along the branch adc. It is evident, therefore, that there must exist somewhere on abc a point which has exactly the same potential as the point d on adc. This point can easily be found by connecting one terminal of a galvanometer at d, and sliding the other terminal along abc until the point b of no deflection is reached. This must be the point which has the same potential as the point d. When the bridge has been put into this condition of balance, let PD_1 be the difference of potential between a and d, PD_2 that between d and c, PD_3 that between a and b, and PD_4 that between b and c. Also let I_1 and I_2 be the currents flowing in the upper and lower branches respectively. Then, by Ohm's law,

$$\frac{PD_1}{R} = I_1 \qquad \text{and} \qquad \frac{PD_2}{X} = I_1.$$

Hence
$$\frac{PD_1}{R} = \frac{PD_2}{X}, \qquad \text{or} \qquad \frac{R}{X} = \frac{PD_1}{PD_2}.$$

Similarly
$$\frac{P}{Q} = \frac{PD_3}{PD_4}.$$

But since b and d are points of equal potential, $\dfrac{PD_1}{PD_2} = \dfrac{PD_3}{PD_4}$.

Hence, finally,
$$\frac{R}{X} = \frac{P}{Q}, \qquad \text{or} \qquad X = \frac{QR}{P}.$$

As soon, then, as a condition of balance has been found, X is obtainable at once in terms of the known resistances P, Q, and R.

In the slide-wire form of bridge (Fig. 45) abc is a uniform wire. Since in this case resistance is proportional to length, P/Q is simply the ratio of the lengths P and Q. The heavy connecting strips of brass have negligible resistances.

44. The mirror D'Arsonval galvanometer. The accuracy with which the point b (Fig. 45) can be located, and hence the accuracy with which X can be found, evidently depends upon the sensitiveness

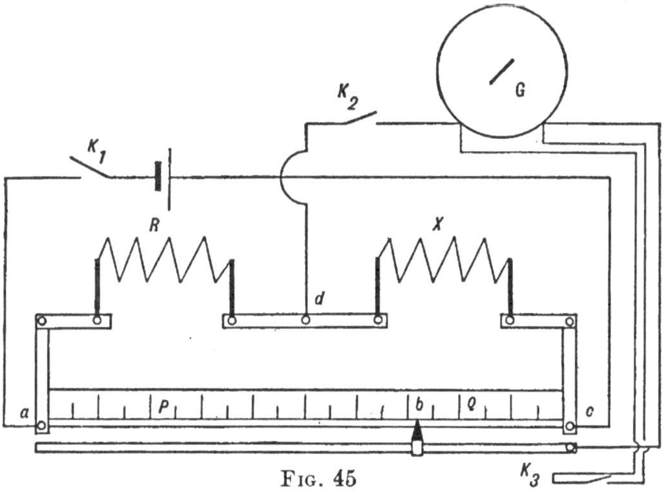

FIG. 45

of the galvanometer G. There are two general types of galvanometer. The first, like the tangent galvanometer (Fig. 22, p. 35), consists of a needle suspended in the middle of a coil of wire and hanging parallel to the plane of the coil. When a current flows through the coil the needle tends to set itself at right angles to the plane of the coil. This form of galvanometer becomes very sensitive when the coil is brought very close to the needle and the magnetic field in which the needle hangs is made very weak (see sect. 57). But a weak field means that the galvanometer is very susceptible to external magnetic changes. For this reason this so-called *Thomson* form of galvanometer is much less convenient for general laboratory work than is the *D'Arsonval,* or *moving-coil,* form. This form (see Fig. 46, and also Fig. 60, p. 80)

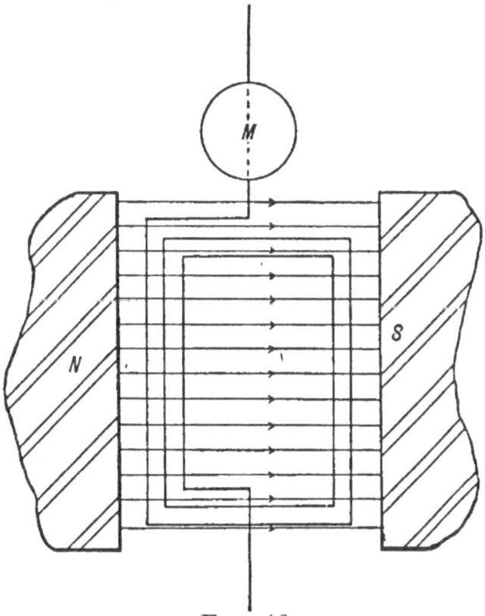

FIG. 46

differs from the Thomson in that the current is made to pass through a movable coil suspended in the field of a fixed magnet, instead of passing through a fixed coil at the center of which a movable magnet is suspended. The plane of the coil is initially parallel to the lines of magnetic force. As soon as the current traverses the coil it tends to rotate so as to place its plane at right angles to the direction of the magnetic lines. A mirror M is attached to the coil so that, with the aid of a telescope and scale placed at a distance, an extremely minute deflection of the coil can be detected. Since in this galvanometer the magnetic field in which the coil is suspended is very powerful, slight changes in the external magnetic conditions have no observable influence upon the galvanometer readings.

EXPERIMENT 5

Object. To test as accurately as possible, by means of Wheatstone's bridge, the laws of series and parallel connections.

Directions. I. *Setting up the galvanometer.* After setting the coil approximately parallel to the direction of the magnetic field, and leveling the galvanometer until there is perfect freedom in the swing of the coil, shut out all air currents by covering the galvanometer with a box provided with a glass window in front. Then, setting the scale and telescope at a distance of about a meter from the galvanometer, focus the telescope upon the reflected image of the scale, as follows. Locate first the image of your eye in the galvanometer mirror and set the telescope pointing approximately along the line of sight. Then, sighting along the telescope, — *not through it*, — adjust the scale vertically until its reflected image is seen in the mirror. Next focus the eyepiece of the telescope until the cross hairs appear most distinct. This is accomplished by sliding the eyepiece alone in or out of the telescope tube. Now, leaving the eyepiece as adjusted, focus the telescope until the image of the scale, as seen in the telescope, coincides with the image of the cross hairs. This coincidence is obtained by the method of parallax as follows. The head is moved slightly from side to side while the scale is viewed through the eyepiece. If there is no apparent motion of the cross hairs with respect to the scale, the image of the scale formed by the objective is exactly coincident with the cross hairs, and observations may be begun. If the cross hairs appear to move to the right across the scale when the head moves to the right, the cross hairs are farther from the eye than is the image of the scale, so that the draw tube of the telescope should be pulled back slightly toward the eye.

If the cross hairs appear to move over the scale in a direction opposite to that of the motion of the head, the draw tube should be pushed in until all apparent motion vanishes.

II. *Connections.* Following Figure 45 connect the unknown resistance x_1 to the bridge in the place of X. Also connect a known resistance at R. Both of these connections should be made of heavy strips of copper of negligible resistance. The resistance R may be either a single standardized coil, or a resistance box. Figure 47 shows the manner in which the coils in such a box are connected. When a plug is taken out the current is obliged to pass through the resistance coil which lies beneath it in the box. When the plug is put in this particular resistance is cut out. K_1 and K_2 are keys which may be combined, if desired, into a double key of the form shown in Figure 48, but in this case the battery circuit should be arranged to close through the upper contact points, the galvanometer circuit through the lower contact points. This is because the equation for the bridge was deduced from Ohm's law, which holds only for constant currents, not for variable currents such as exist at the instant at which the current of a battery is closed (see Chap. XIV). K_3 is a damping key which merely short circuits the galvanometer. It is used for bringing the coil quickly to rest (see Chap. XII).

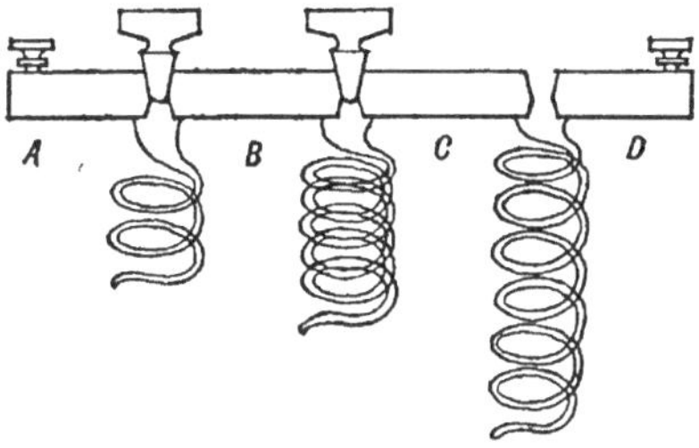

FIG. 47

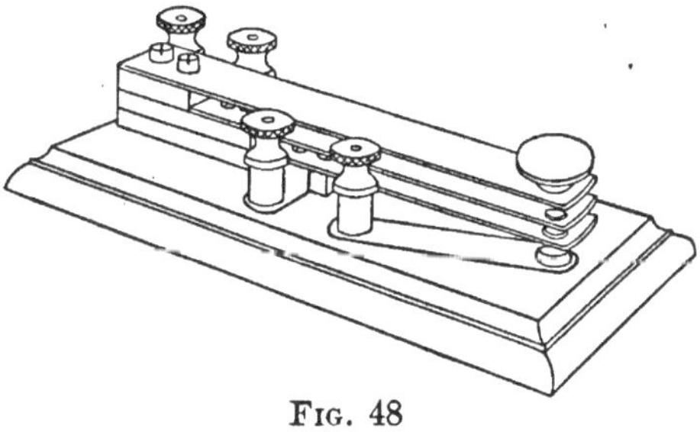
FIG. 48

Determine first the value of x_1 as explained in the next paragraph. Then in a similar manner determine x_2. From these values calculate the combined resistance of x_1 and x_2, first, when connected in series (x_s), and second, when connected in parallel (x_p). Then, using x_1 and x_2 in series as X, determine the value by observation and compare with the calculated value. Make a similar determination and comparison for the parallel connection.

III. *Readings.* If the known resistance R is adjustable, i.e. a resistance box, remove some plug, close the short-circuiting key K_3, and, setting the slider about the middle of the bridge, press first the battery key K_1 and then the galvanometer key K_2. Notice the direction of the galvanometer deflection. Now try in the same way some widely different resistance for R. If the deflections are in the same direction, both of these values of R lie on the same side of the value necessary for a balance. If the deflections are in opposite directions, the value sought lies between those used. Select some value of R which will bring the position of the slider near the middle of the bridge and complete the adjustment by moving the slider. If the known resistance is not adjustable, try two different positions of the slider to get an idea of the necessary position. The short-circuiting key K_3 was only closed at first in order to prevent violent throws. As soon as a first approximation toward a balance has been made, it should be opened and the accurate setting made with it in this condition. After each deflection the coil may be brought quickly to rest by closing the key K_3.

In order to find the position of the slider for which there is no deflection, find two points which correspond to the smallest observable deflections to the right and to the left, K_3 being open. If these points are more than .2 or .3 mm. apart, the galvanometer is not sufficiently sensitive. The point midway between these two is the point of balance. Since imperfect contacts in the resistance box may lead to erroneous results, it may be well to try the point of balance when the resistance R is made up of different combinations of the resistance coils of the box. If a large variation results, clean the plugs by wiping them with a cloth wet with benzine.* *Do not push down hard on the plugs; merely twist them until you are sure that they make good contacts.* Be sure that all other contacts are firm.

In order to minimize the error due to lack of uniformity in the bridge wire and also to eliminate thermo-electric currents, † take the mean of two determinations of X, making the second determination as follows. Interchange the positions of R and X and obtain values P' and Q' for the lengths on the bridge wire.

EXAMPLE

The resistance R was given a value of 20 ohms and remained unchanged throughout the experiment. For x_1 the point of balance on the bridge wire was found at $P = 30.41$ cm. and $Q = 19.59$ cm. When R and x_1 were interchanged P was found to be 30.49, and Q, 19.51. The average value of x_1 was, therefore, $\frac{19.55}{30.45} \times 20$, or 12.84 ohms. Similarly x_2 was found to be 21.07 ohms. Also x_s was found to be 34.00 ohms, a value which agreed to within .3 per cent with the calculated value of 33.91 ohms, obtained by adding x_1 and x_2. The value found for x_p was 7.93 ohms, which agreed well with the calculated value of 7.94, found from the law for parallel connections by taking the reciprocal of the sum of the reciprocals of x_1 and x_2.

* The chief source of error in all work with resistance boxes lies in imperfect contacts at the plugs. Hence it is of great importance to have the plugs scrupulously clean, and to twist each plug in the plug seat at the time of its insertion.

† Thermo-electric currents are discussed in Chapters VI and X. This interchange of R and X eliminates these currents only partially.

CHAPTER VI

TEMPERATURE COEFFICIENT OF RESISTANCE; SPECIFIC RESISTANCE

45. Disadvantages in the slide-wire form of Wheatstone's bridge. For accurate work the form of Wheatstone's bridge described in the preceding experiment is open to four serious objections. First, the bridge wire may not be uniform in diameter and resistance throughout its length. The exchange in position of R and X tends to minimize the error thus introduced in assuming P and Q proportional to their lengths; but since, except for a point of balance at the center of the wire, a certain portion of the wire is included in both positions in the larger of the two quantities P or Q, the error mentioned is not entirely eliminated. In general, however, the error due to this cause is small.

A second and more serious objection is this. Since the wire is in general not more than a meter in length and must, for mechanical reasons, be of considerable diameter, its resistance is small. The fall in potential which takes place along the wire is consequently small, so that a small change in the position of the slide causes but a small variation in the P.D. between the terminals of the galvanometer and consequently but a small deflection.

A third objection is that thermo-electric differences of potential are introduced which cannot rigorously be eliminated by the interchange of R and X. A thermo-electric difference in potential exists between the two junctions of a metal with a second metal if these junctions are at different temperatures (see Chap. X). The source of this error in the wire bridge is largely the result of the heating by the hand of the junction of the brass slider and the German silver wire. Between this junction and one of the other junctions of the wire with the brass strips of the bridge there may exist a P.D., causing a deflection of the galvanometer even when P, Q, X, and R are in balance. The point of balance actually observed is where this P.D. is counteracted by a small P.D. in the opposite direction, due to having the resistances slightly out of balance. Since the heating due to the presence of the hand is not constant, the assumption of a constant thermal current is not rigorous, and the interchange of R and X does not entirely eliminate the error. Another place where this error may be introduced is at the brass-platinum contact of the key in the galvanometer circuit. Since the platinum contact is thin, and consequently cannot differ appreciably in temperature at its two junctions with the brass of the key, the error here introduced is small. On the assumption of a constant error it would be eliminated entirely by reversing the connections of the battery, since this reverses the directions of all the differences of potential in the bridge except those due to the thermal effects.

A fourth objection lies in the fact that only resistances of about the same order of magnitude may be compared. For evidently the observational error introduced in setting the slider or reading its position causes a minimum error in the determination of the resistance when P and Q are the same, i.e. when R and X are equal. If R and X are widely different, this error may become quite large.

46. The post-office-box form of Wheatstone's bridge. A form of bridge known from its original use by the telegraph department of the British post office as the post-office-box bridge overcomes the first, second, and fourth of these objections, and also reduces largely the errors due to thermo-electromotive forces. It possesses also the advantage of compactness, since the resistances P, Q, and R, as well as the keys for the battery and galvanometer circuits, are contained in the same box.

The essential difference between the box and slide-wire forms consists in the substitution for ab and bc (see Figs. 44 and 45) of two series of accurately determined resistance coils. This is shown diagrammatically in Figure 49. The branches ab and bc consist of 1000-, 100-, 10-, and 1- ohm coils. The branch ad is merely the ordinary set of coils to be found in a resistance box. The ratio of

P to Q is limited to values of 1000, 100, 10, 1, $\frac{1}{10}$, $\frac{1}{100}$, and $\frac{1}{1000}$. In the use of this form a fixed ratio is chosen for P/Q, and the resistance R is varied until a balance is obtained. For example,

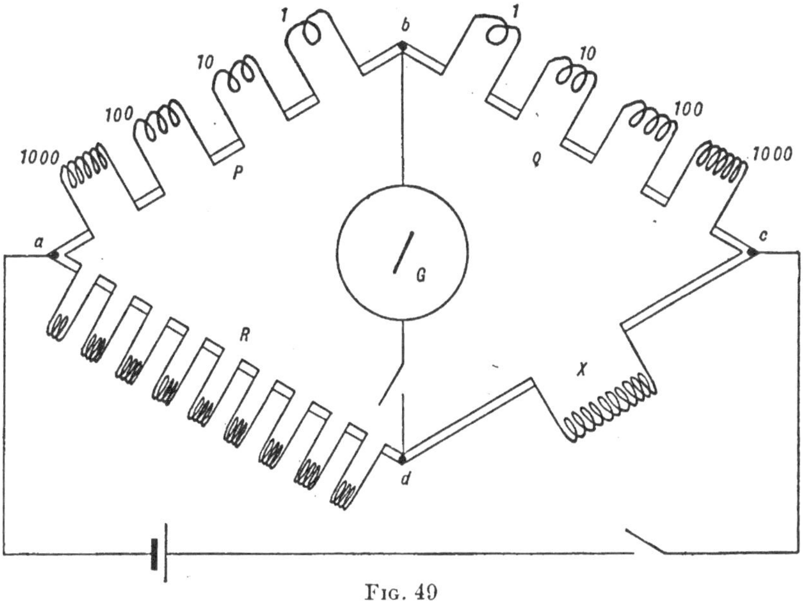

FIG. 49

in measuring a resistance X, known to lie between 3 and 4 ohms, the ratio P/Q was made 1000, and R for a balance was found to be 3682 ohms, thus giving $X = 3.682$ ohms.

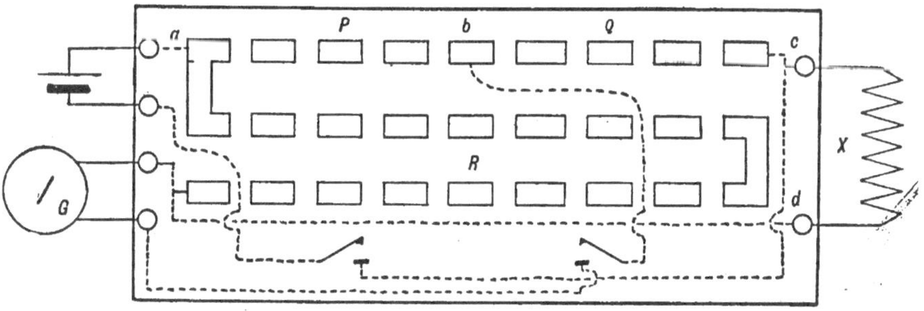

FIG. 50

In practice the boxes are not made in the form shown in Figure 49, which has been introduced as a transition between Figure 44 and Figure 50, in which is shown the ordinary scheme of connections. The dotted lines indicate connections made within the box. Figures 51 *a* and 51 *b* show, in diagram and in perspective, a box similar to that represented diagrammatically in Figure 50,

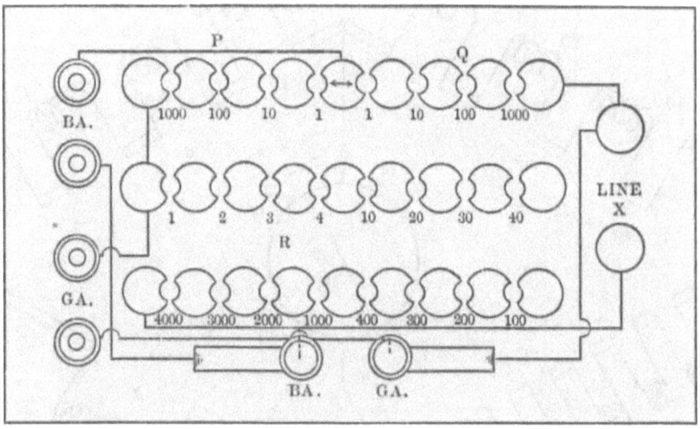

FIG. 51 *a*

the only difference being that in Figure 51 the battery is connected between the points *b* and *d* of Figure 49, and the galvanometer between the points *a* and *c*. This has no effect upon

FIG. 51 *b*

the equation of balance, for in considering the diagram of the Wheatstone bridge it is evident that if the battery and galvanometer are interchanged, the condition for equilibrium becomes

$P/R = Q/X$. But this is true if $P/Q = R/X$. That is, the galvanometer and battery are interchangeable in position without affecting the relation $P/Q = R/X$.*

47. Temperature coefficient of resistance. As was stated on page 60, the resistance of a conductor depends upon its temperature. The temperature coefficient of resistance of a conductor is defined as *the ratio between the change in resistance per degree change in temperature and the resistance at 0° C.* It will be seen that the definition is altogether analogous to that given for the coefficient of expansion of a gas. In symbols, if R_t and R_0 represent the values of the resistance of any conductor at t°C. and 0°C. respectively, then the temperature coefficient of resistance α is defined by the equation

$$\alpha = \frac{R_t - R_0}{R_0 t}. \tag{1}$$

This coefficient is positive for the metals, and has a value of approximately .0038 for all pure metals. It is negative for carbon; that is, the hot resistance of an electric-light carbon is less than its cold resistance. It will be seen for the above equation that the resistance R_t of a conductor at any temperature may be written

$$R_t = R_0(1 + \alpha t). \tag{2}$$

48. Specific resistance. The specific resistance of a substance at any temperature is defined as the resistance between two opposite faces of a centimeter cube of the substance. Since the resistance of a conductor varies directly as its length, and inversely as

* The reason that post-office-box bridges are commonly connected as in Figure 51, rather than as in Figure 50, may be seen from the following rule taken from Maxwell's "Electricity and Magnetism," Art. 349. The deduction of the rule cannot be taken up in a text of this scope. "*The rule, therefore, for obtaining the greatest galvanometer deflection in any given system is as follows. Of the two resistances, that of the battery and that of the galvanometer, connect the greater resistance so as to join the two greatest to the two least of the four other resistances.*" Since a galvanometer will practically always have a higher resistance than an ordinary battery, and since, in the use of the post-office box (Fig. 49), the resistance R is almost always larger than X, and, therefore, P larger than Q, it will be seen that, in accordance with the rule, the galvanometer should be connected across ac, and the battery across bd, if the highest sensibility is to be obtained.

its cross-sectional area, the resistance R of any conductor, at the temperature for which its specific resistance r is known, may be written

$$R = \frac{rl}{A}, \quad \text{or} \quad r = \frac{RA}{l}, \tag{3}$$

in which l is the length in centimeters, and A the average area of cross section expressed in square centimeters. Knowing, then, the resistance R at a given temperature and the dimensions of a conductor, its specific resistance at that temperature may be readily calculated from (3), and if its temperature coefficient is also known, its specific resistance at 0°C. may then be obtained from (2) by throwing it into the form

$$r_0 = \frac{r}{1 + at}. \tag{4}$$

EXPERIMENT 6

(A) Object. To determine the temperature coefficient of a commercial copper conductor.

Directions. I. *Connections.* The copper wire of which it is desired to find the temperature coefficient of resistance is wound on a wooden frame and supported by a wooden top within a thin, brass tube closed at the lower end. Heavy copper wires pass from the coil through the cover (see Fig. 52). An opening in the cover admits a thermometer supported so as not to touch the metal. This tube is immersed in a large vessel of water to which the heat is applied. The resistance of the wire is determined by means of a post-office box, first at the temperature of the room, then at a series of higher temperatures. This data is plotted as a curve on coördinate paper, using temperatures for abscissas and resistances as ordinates (see Fig. 53). The curve, which is a straight line, may be prolonged backwards, and its intercept on the axis of resistances (i.e. on the

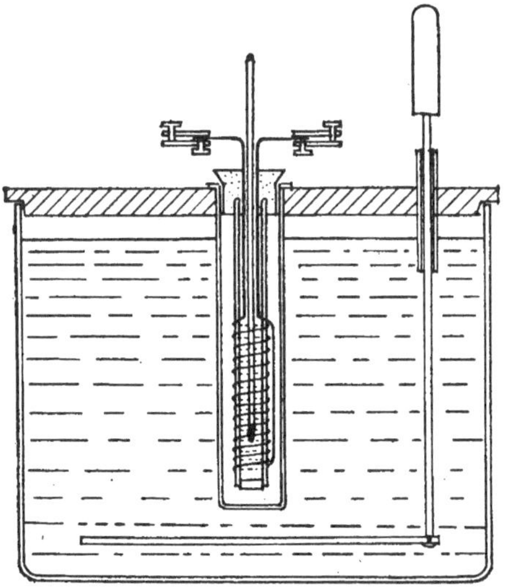

FIG. 52

ordinate corresponding to the temperature 0°C.) taken as R_0. The slope of this line, namely $\frac{(R_t - R_0)}{t}$, divided by R_0 gives α.

II. *Observations.* Measure the resistance of the coil, including the connecting wires, at the temperature of the room, as follows. Make $P/Q\ \frac{10}{10}$,

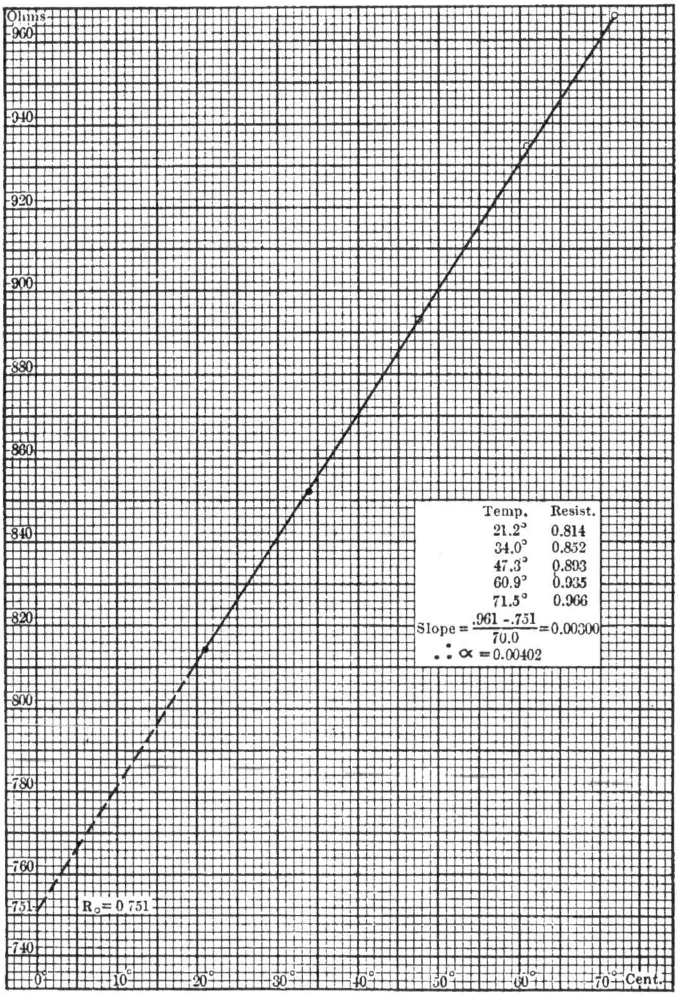

FIG. 53

and vary R until a balance is obtained to within 1 ohm. (For example, 3 ohms too small, 4 ohms too large.) It will be best to provide the galvanometer with a damping key as in Experiment 5, and to keep the latter closed during this operation. Now make $P/Q\ \frac{100}{1}$. R will be between 300 and 400 ohms. Again find the value to within 1 ohm. Suppose R is found to lie between 368 and 369. Then make $P/Q \frac{1000}{1}$ and find R to the nearest ohm. If R lies between 3682 and 3683, take the number which gave the smallest deflection, e.g. 3683. Then $X = 3.683$. If the galvanometer is sensitive, the accuracy can be pushed at least one place farther by the method given below in (**B**), but with the accuracy attainable in the temperature readings of this experiment, it is scarcely advisable to attempt greater refinements in the resistance measurements.

Keeping $P/Q = \frac{1000}{1}$, raise the temperature of the water about 10° above that of the room. Stir thoroughly, and by regulating the flame of the Bunsen burner keep the temperature of the water as constant as possible. Since the air surrounding the coil is a poor conductor of heat, the coil will be slow in rising to the temperature of the water. Follow its rise by varying the resistance in the post-office box and take a reading of this resistance as soon as it becomes constant. Note the temperature of the thermometer in the inner tube.

Next begin to plot your readings as follows. Plot temperatures along the X axis of the coördinate sheet, using, for example, 1 division to represent 2°C., or choosing some other scale which will make 70 or 80 degrees occupy nearly the full width of the page. Plot resistances along the Y axis, using any convenient scale. The method of selecting a scale which will make the fullest use of the sheet may be seen from the example presented in Figure 53. The first two observed temperatures were 21.2°C. and 34.0°C. The increase in resistance corresponding to this increase of 12.8° in temperature was .038 ohm. In raising the temperature from 0°C. to 70°C. it was estimated that there should be an increase of about 5.5 times this amount, or approximately .210 ohm. Now there were on the particular sheet shown in the figure 110 available spaces along the Y axis. Dividing the approximate number of thousandths of an ohm, namely 210, through which the resistance was expected to be raised, by the number of available divisions, gave 1.95 as the number of thousandths of an ohm to be represented by 1 space, if the sheet was to be completely utilized. Since this number was very inconvenient both for plotting the graph and for reading the values from it, the nearest whole number above 1.95, namely 2, was taken as the number of thousandths of an ohm to be represented by 1 division. The general rule for this choice would be as follows. *Select the nearest whole number larger than the quotient found as above except where this number would be 3, 6, 7, or 9, in which case it would be more convenient to choose 4, 5, 8, or 10 respectively.*

Do not call the intersection of the two axes zero resistance, but make it instead some convenient number which is about 10 per cent less than the resistance of the coil at the temperature of the room (see Fig. 53).

Having plotted the first two points, raise the temperature by about ten-degree steps, and plot each point as soon as it is obtained. If any point does not lie on, or very near to, the right line joining the first two points, repeat the observation before raising the temperature farther. If results continue to be erratic, wipe all the plugs with a clean cloth moistened with benzine and look to your contacts. For example, turn each plug until it catches. Do not, however, push down hard upon the plugs, as this may spring the lugs and thus injure the box. After carrying the temperature to 70°C. or 80°C., disconnect the connecting wires from the heater, connect their terminals to each other, and measure the resistance of these wires.

III. *Calculations.* Draw a straight line which will come as near as possible to all of the plotted points. Produce the curve thus found until it cuts the Y axis (the line of zero temperature), and take this point of intersection as the resistance of the coil at 0°C. Subtract from this value the resistance of the connecting wires to find R_0. Find α by dividing the slope of the curve by R_0 (see Fig. 53).

(B) Object. To find the specific resistance of copper at 0°C.

Directions. In order to determine a specific resistance the dimensions of the wire must be accurately measured. This condition necessitates a large wire, and hence one of small resistance. But neither the slide-wire nor the post-office form of bridge is suitable, without modification, for the measurement of a resistance which is so small as to be comparable with the resistances of the contacts or of the connecting strips. These are often several thousandths of an ohm. A very satisfactory method of eliminating all errors due to contacts, connecting wires, or thermal electromotive forces, and of measuring with considerable accuracy a resistance of a few thousandths of an ohm is as follows.

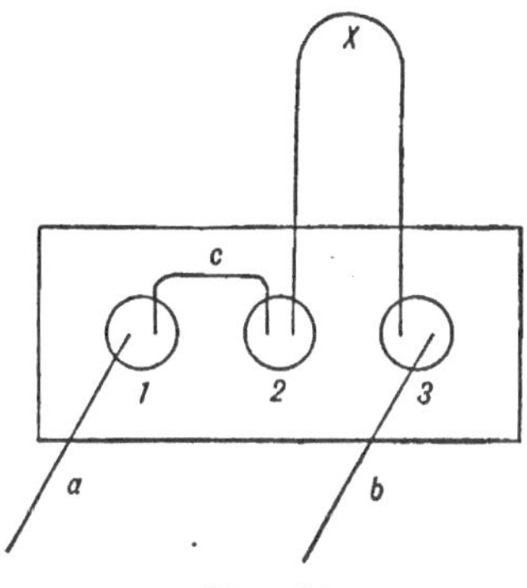

FIG. 54

Let large connecting wires a and b from the post-office box terminate in two holes *1* and *3* (Fig. 54) bored in a wooden block and filled with mercury. Let an auxiliary wire c connect holes *1* and *2*, and the wire of unknown resistance X complete the circuit by connecting holes *2* and *3*. Let the surfaces of contact of all the wires with the mercury be well amalgamated by dipping the ends of the wires into nitric acid, then into mercury, then rubbing dry with filter paper. Let P/Q be made $\frac{1000}{1}$ and let R be varied until a change of 1 ohm causes a change in the direction of the deflection. Then take the permanent deflection produced by closing K_2 (Fig. 45) when R has the lower value, say 6 ohms. Then add 1 ohm to R and take the permanent deflection in the other direction. If, for example, the first deflection were 4.35 mm. and the second 62.70 mm., then the value of R for a perfect balance would have been $6 + \frac{4.35}{4.35 + 2.70 + 6} = 6.065$ ohms. Now, with as little delay as possible, remove X, transfer *b* from *3* to *2* (see Fig. 55), and again find precisely as before the value of R which would produce a balance. If this is found to be, for example, 4.107 ohms, then obviously $X = \frac{6.065 - 4.107}{1000} = .001958$ ohm. Repeat the observations and see how nearly the *difference* between the two values of R can be checked. Changes may occur in the individual values of R because of temperature changes, but the differences should remain very nearly constant. Next take the mean of some ten or more measurements of the diameter of the wire, using the micrometer caliper. Measure with a tape the distance between the two points on the wire to which it was immersed in the mercury. From these data and the temperature coefficient as found in **(A)** compute the specific resistance of commercial copper at 0° C. by the aid of equations (3) and (4).

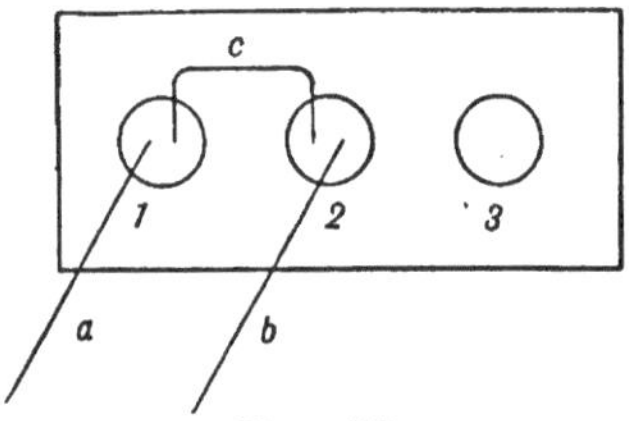

FIG. 55

EXAMPLE

(A) The curve plotted as above (Fig. 53) was the record of this experiment. Tabulated in one corner are found the various values of the resistance and temperature, the value of R_0, the value of the slope of the line $\frac{(R_t - R_0)}{t}$, and the temperature coefficient α.

(B) Total resistance = .00595 ohm ; temperature = 26° C. ; resistance of connecting wires and contacts = .00403 ohm ; length of wire = 137 cm. ; mean radius = .205 cm. ; therefore specific resistance at 0° C. = .00000167 ohm = 1.67×10^{-6} ohms = 1670 absolute units.

CHAPTER VII

GALVANOMETER CONSTANT OF A MOVING-COIL GALVANOMETER

49. Galvanometer constant. In the formula for the current flowing through a tangent galvanometer, namely

$$I = \frac{Hr \tan \theta}{2 \pi N},$$

it is seen that in general the current is not proportional to the deflection, although it would be so if deflections were always kept so small that $\tan \theta$ were approximately equal to θ. In galvanometers in which the current *is* proportional to the deflection, the reduction factor by which the angle of deflection must be multiplied in order to give the current is known as the *galvanometer constant.* Thus if I represents the current which produces in such a galvanometer a deflection of θ radians, then the definition of the galvanometer constant K is given by the equation

$$I = K\theta, \qquad \text{or} \qquad K = \frac{I}{\theta}. \tag{1}$$

That is, *the galvanometer constant is defined as the constant ratio between the current and the deflection produced by it.* This ratio for small angles is still called in many cases the galvanometer constant, even though the instrument be one with which the ratio may not remain constant for large angles. It is the object of this discussion to find the expression for this constant in the case of a moving-coil (i.e. a D'Arsonval) galvanometer. To do this it will be necessary first to reconsider the definition of unit current given in section 27, page 34.

50. Restatement of definition of unit current. Unit current has been defined as a current which, when flowing in a conductor of unit length bent into the arc of a circle of unit radius, exerts unit force on a unit magnetic pole lying in the plane of the conductor and at the center of the arc. Thus the unit pole at O (Fig. 56) is urged from the plane of the paper toward the reader with a force of 1 dyne if the current in the wire has unit strength.

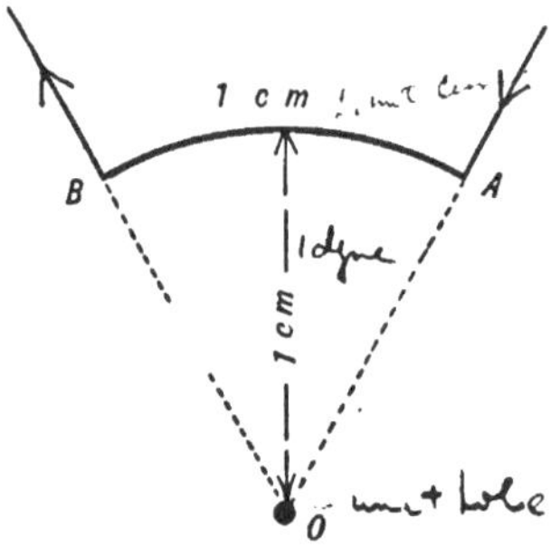

FIG. 56

Since action and reaction are equal and opposite in direction, the conductor is urged from the reader toward the paper with an equal force. Now the magnetic field due to the isolated unit pole has a strength at all points on the wire of unity, and a direction at right angles to the wire, since its lines of force are straight lines radiating from it in all directions. Instead, then, of stating the definition of unit current in terms of the force which the current exerts on the magnet, we may state it in terms of the force which the magnetic field exerts on the conductor. Thus *unit current is that current unit length of which, when flowing in unit field at right angles to the direction of the field, experiences unit force.* In symbols, the force F, in dynes, exerted on l cm. of length of a conductor carrying I electro-magnetic units of current, by a field of strength $\mathcal{H}$, at right angles to l is, by definition of current,

$$F = Il\,\mathcal{H}. \tag{2}$$

51. The motor rule. The relation between the directions of the magnetic field, the current in the conductor, and the force exerted on the conductor may be seen from the figure to be stated in the following rule, known, from its especial application to the direct-current motor, as the "motor rule." *Extend the thumb, forefinger, and second finger of the left hand in directions at right angles to one another; let the forefinger point in the direction of the magnetic field, and the second finger in the direction of the current; the thumb will then point in the direction of the force acting on the conductor* (Fig. 57).

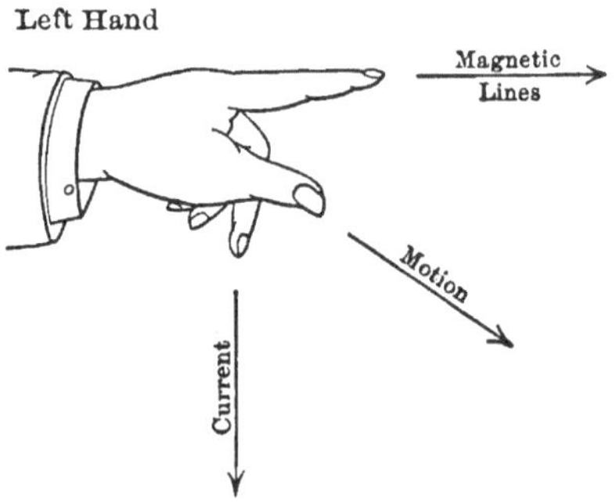

FIG. 57

52. Faraday's explanation of electro-magnetic forces. The fact that the force between a magnet and a current does not act along the line connecting the magnet and the conductor, but at right angles to this line, constitutes a striking difference between electromagnetic forces and the forces met with in the study of mechanics; for these uniformly act along the lines connecting the acting bodies. Further, the forces of mechanics, excepting gravitation, act only as the result of the transmission of stresses through matter. Because of the difficulty of the conception of action at a distance, that is, action assumed to take place without the intervention of a medium, and in order to admit of a more perfect visualization of the actions of magnets and electrical charges, Faraday conceived of these electrical and magnetic actions as having their seat in the lines of force; that is, he imagined, merely as an aid to

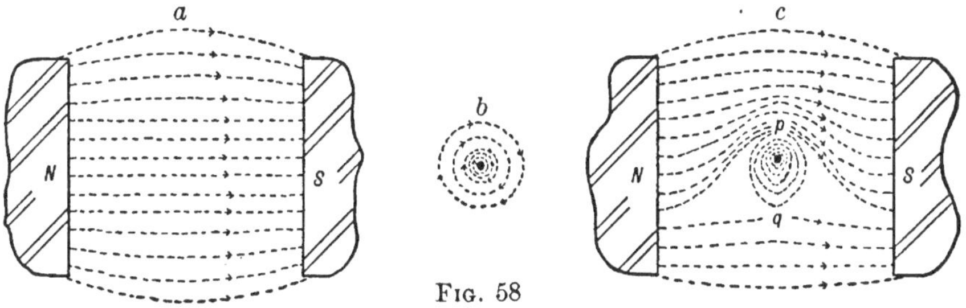

FIG. 58

thinking, that these lines of force constituted a sort of system of invisible, stretched, elastic bands endowed with the following properties: (1) a tension in the direction of their length, and (2) a repulsion at right angles to this direction, so as to cause them to act on one another as if a hydrostatic pressure existed at right angles to their direction. Thus in Figure 58, *a*, which shows the field existing between two opposite magnetic poles, the lines at the top and bottom were thought of as forced out by a repulsion, not balanced, as at the center, by an equal and opposite repulsion due to other lines.

Again in Figure 58, *b*, is shown the field surrounding a wire carrying a current toward the paper, while in Figure 58, *c*, is the resultant field due to the presence in the uniform field of *a* of the conductor of *b*. The resultant field at *p* is reënforced, while that at *q* is weakened. The lines are therefore crowded together more closely at *p* and less so at *q*. There results from these hypothetical

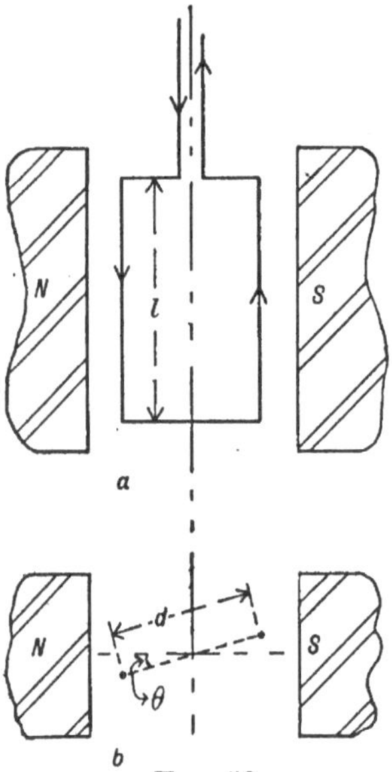

FIG. 59

characteristics of the lines an unbalanced repulsion forcing the conductor across the field toward the bottom of the page, in accordance with the demands of the motor rule. This mechanical picture of the action of electric and magnetic forces has so permeated the literature of electricity and magnetism that a familiarity with it is of importance to the student of the subject.

53. The equation of the moving-coil galvanometer. In Figure 59, *a*, is shown a diagram representing the magnets and one single loop of the coil of a moving-coil galvanometer (see also Fig. 60). Suppose that a current of I units is flowing through the coil and that the strength of the magnetic field between N and S is $\mathcal{H}$ units. Then the left side of the loop is urged toward the reader with a force of $Il\mathcal{H}$ dynes (sects. 50 and 51) and the right side is urged away from the reader with an equal force. Upon the horizontal wires no force acts, for they carry currents parallel to the direction of the field. If the distance between the two sides is d centimeters, the moment of force $\overline{Fh}$ acting to twist the loop about a vertical axis midway between its sides is given by

$$\overline{Fh} = Il\,d\mathcal{H}. \qquad (3)$$

Since ld is simply the area a of the loop, this equation may be written

$$\overline{Fh} = Ia\mathcal{H}. \qquad (4)$$

If there are n loops of average area a, the total twisting moment

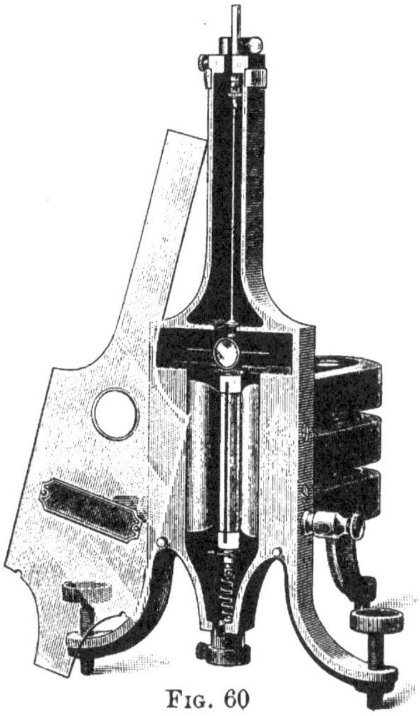
FIG. 60

is of course $Ian\mathcal{H}$; and if we call an the total area of the coil and represent it by A, we have

$$\overline{Fh} = IA\mathcal{H}. \tag{5}$$

This is the value of the couple which acts on the coil so long as the plane of its loops is parallel to the direction of the field $\mathcal{H}$. Under the influence of this moment of force it rotates until it is brought to rest by the restoring moment due to the torsion of the suspending fiber. Suppose that when equilibrium is established the coil has rotated through an angle θ (see Fig. 59, b). The couple arm will then have changed from d to $d \cos\theta$, so that the moment of force producing the deflection is now $IA\mathcal{H}\cos\theta$ instead of $IA\mathcal{H}$, while the restoring moment is $T_0\theta$, if we represent by T_0 the moment of torsion of the suspending fiber. Hence the equation of equilibrium of the moving-coil galvanometer is

$$IA\mathcal{H}\cos\theta = T_0\theta. \tag{6}$$

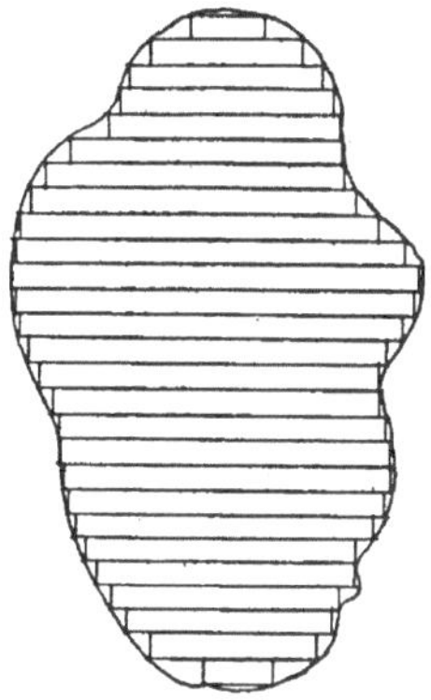

FIG. 61

In deducing this expression we have assumed a rectangular coil, but it can very readily be seen that the result is precisely the same whatever its shape. For any irregular coil may be considered to be made up of infinitesimal rectangular elements (see Fig. 61) and we have just seen above that the moment of force acting on a rectangular coil is $Ia\mathcal{H}\cos\theta$, where a is the area of the rectangular element. The total moment is therefore $\Sigma Ia\mathcal{H}\cos\theta$ or $AI\mathcal{H}\cos\theta$, where the total area of the coil, found by summing up the infinitesimal areas, is A.

Equation (6) shows that the current I is not proportional to the deflection θ, but rather to $\dfrac{\theta}{\cos\theta}\cdot$ If, however, θ is so small that $\cos\theta$ may, without appreciable error, be taken as unity, then equation (6) becomes

$$I = \frac{T_0\theta}{A\mathcal{H}},$$

and it is clear that the galvanometer constant K, i.e. I/θ (see eq. (1)), is given by

$$K = \frac{T_0}{A\mathcal{H}}\cdot \tag{7}$$

This expression makes it obvious why K is called the galvanometer constant, for it involves only the nature of the suspending fiber (T_0), the area of the coil (A), and the strength of the magnetic field ($\mathcal{H}$).

Keeping clearly in mind the restriction limiting the determination and use of K to small angles, it is possible to find K for any galvanometer by measuring the current which produces a small angular displacement and dividing the current by the displacement.

For some purposes it is more convenient to have this reduction factor expressed, not in terms of current per radian, but in terms of the current necessary to produce a deflection of 1 mm. on a scale at a distance of 1 m. The current in amperes necessary to do this is known as the *figure of merit* of the galvanometer. Since the mirror of a galvanometer actually turns through one half the angle through which the reflected ray is rotated, and since 1 mm. is .001 m., it is evident that the figure of merit k of a galvanometer is $K/2000$.

Again, instrument makers often express the "sensibility" of a galvanometer in terms of the number of megohms (million ohms) which would have to be placed in series with it in order to reduce the deflection to 1 mm. on a scale 1 m. distant, when a P.D. of 1 volt exists between the ends of this resistance. This is obviously the reciprocal of the figure of merit. Thus a galvanometer which has a constant K equal to .000004, I being measured in amperes, has a figure of merit of $.000000002 = 2 \times 10^{-9}$, and a sensibility of $1/2 \times 10^{9} = 500$ megohms.

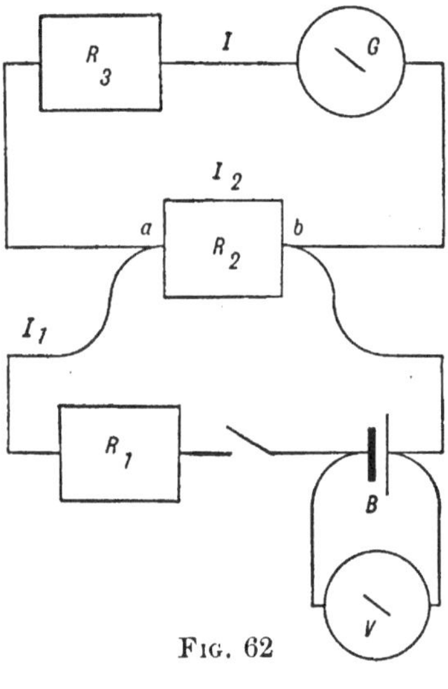

FIG. 62

54. Method of determining *K*. The scheme of connections which is used for determining K is shown in Figure 62. The current from a Daniell cell B is made small by inserting into the circuit a large resistance R_1. Only a small fraction of this current is passed through the galvanometer G, which is placed in one arm of a divided circuit made up on the one side of the small resistance R_2 and on the other side of the large resistance R_3 plus the galvanometer resistance G. The deflection corresponding to a given arrangement of the resistances is observed and the current passing through the galvanometer is calculated from the reading (P.D.) of the voltmeter and the values of the various resistances. K is then found by dividing the current by the deflection, the latter being expressed in radians.

The calculation of the current follows directly from Ohm's law. If, as is usually the case, the resistance of the branched circuit from a to b is negligible in comparison with R_1, then the current I_1 in the main circuit is given by

$$I_1 = \frac{PD}{R_1}. \tag{8}$$

Since the potential difference between a and b is the same for both branches, the currents, I through the galvanometer, and I_2 through R_2, will vary inversely as the resistances of their respective branches. That is

$$\frac{I}{I_2} = \frac{R_2}{R_3 + G}. \tag{9}$$

Since the sum of these two currents must be the current in the main circuit, we have

$$I + I_2 = I_1. \tag{10}$$

From (9) and (10) we obtain

$$I = I_1 \frac{R_2}{R_2 + R_3 + G}. \tag{11}$$

This shows that in general *the current which flows through one side of a shunt is obtained by multiplying the total current by the resistance of the other side divided by the resistances of both sides.*

From (8) and (11) we at once obtain I in terms of the various resistances and the voltmeter reading P.D.

If this current causes a deflection of d centimeters on a scale D centimeters from the mirror, the angle through which the mirror rotates (one half that through which the reflected beam rotates) is $d/2D$. Hence

$$K = \frac{2ID}{d}. \tag{12}$$

55. Resistance of the galvanometer. The resistance of the galvanometer can be found most accurately by connecting it as the unknown to a post-office box and measuring it by the Wheatstone-bridge method, using another galvanometer. But if R_3 is large (compared with G), G may be found very much more conveniently and with sufficient accuracy by connecting a Daniell cell B, a voltmeter V, a milliammeter A, and the galvanometer G as in Figure 63. The reading of V in volts, divided by that of A in amperes, gives the resistance in ohms of G and A together, and the resistance of A will, in general, be either negligible or known.

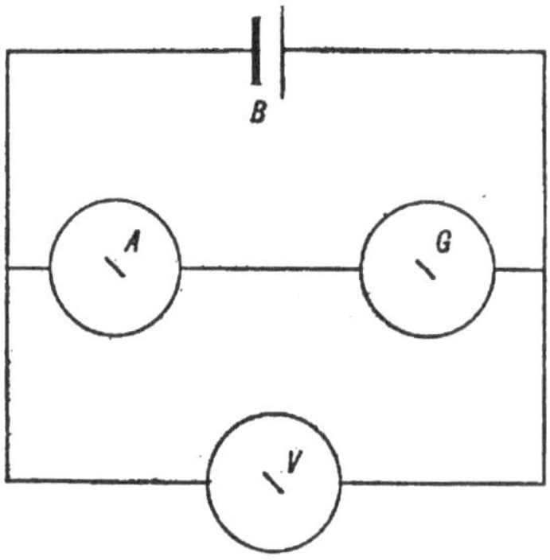

FIG. 63

56. Direct-reading moving-coil galvanometers. If the term $A\mathcal{H}\cos\theta$ in the expression $I = T_0\theta/A\mathcal{H}\cos\theta$ (see eq. (6), p. 81) can be made constant, the deflection will be proportional to the current for all values of θ. This condition means that the pole pieces must be so shaped that the field strength $\mathcal{H}$ acting on the vertical wires of the coil in any position shall be inversely proportional to the cosine of the angle through which it has been deflected. Instruments in which this condition has been met may be used for the measurement of widely different currents on a scale of uniform divisions. The common, commercial, direct-reading ammeter is an instrument of this sort.

The plan of an ammeter is shown in Figure 64. M is a permanent magnet of which aa' and bb' are the shaped pole pieces. E is a soft-iron cylindrical core which concentrates the field. The coil c is wound on a rectangular aluminum frame, and turns on pivots in jeweled bearings against the torsion of two flat spiral springs not shown in the figure. The pointer D is attached to the coil and moves over a scale S empirically calibrated, but having practically constant divisions. The terminals of the coil are attached to heavy wires ww' which connect with the binding posts BB'. Across these wires, in shunt with the coil, are a number of small resistances r which carry the greater part of the current, allowing only a fraction to pass through the movable coil.

The current flows through the coil of the ammeter in a direction such that it produces the magnetic field indicated by the small letters *n* and *s*. The coil therefore tends to rotate to the right. The angle θ is reckoned from a position of the coil parallel to the lines of force *NS*. Thus θ has a value of about 45° for the initial position of the coil shown in the figure. Owing to its rotation

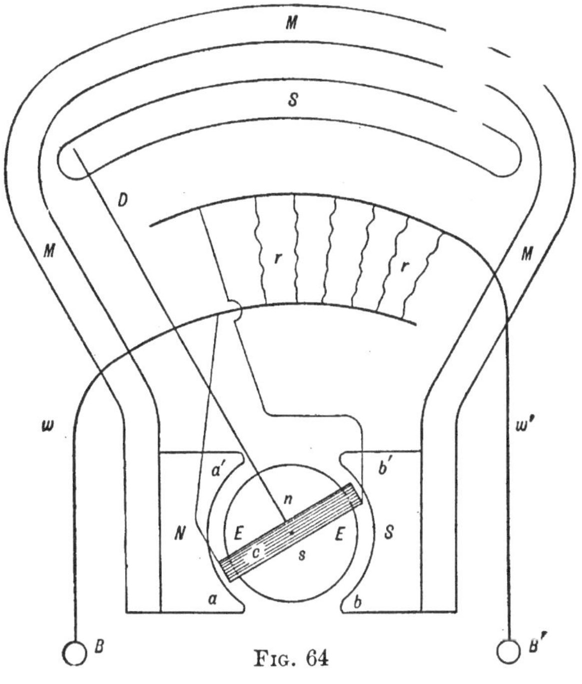

FIG. 64

from this position, θ decreases to zero when the coil is parallel to *NS*, and then increases to about 45° for a position parallel to *a'b*. This means that $\cos\theta$ increases to unity and then decreases. In order that $\mathcal{H}\cos\theta$ shall be constant $\mathcal{H}$ must be larger at the tips of the pole pieces *a*, *a'*, *b*, and *b'* than at the center. This is accomplished by giving the pole pieces the shape shown in the figure.

A current-measuring instrument of high resistance may be calibrated to read units of P.D., e.g. volts, as was explained on page 52. Such an instrument, of the movable-coil type, is the voltmeter.

It differs essentially from the ammeter of Figure 64 only in the matter of resistance. Instead of the small resistances r in shunt between the binding posts BB' there is a single large resistance in series with the movable coil c.

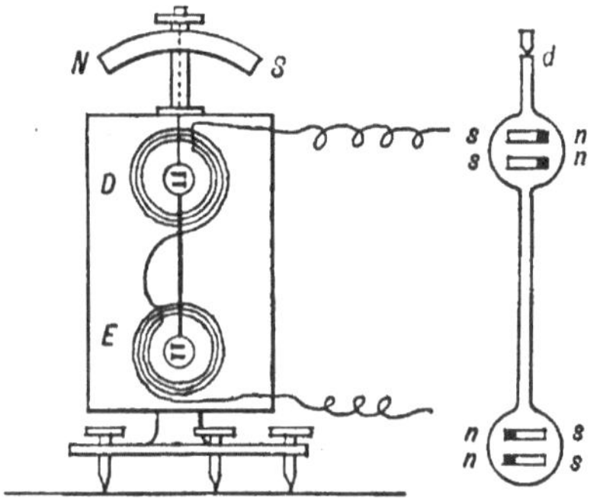

FIG. 65 *a*

57. Types of current-measuring instruments. Thus far two distinct types of instruments for measuring currents by the mutual actions of two magnetic fields have been described. In the first the current is measured by the deflection of a magnet which assumes a position of equilibrium under the action of an external field and the field due to a fixed coil through which the current to be measured is flowing. The tangent galvanometer is an instrument of this type, the external field being that of the earth.

The Thomson galvanometer described in connection with Figure 46 is also an instrument of this type, but in it the external field is in general the resultant field due to the earth and a bar magnet attached to the galvanometer in the manner shown in Figure 65 *a*. This magnet is used to oppose the earth's field at the galvanometer, and, by weakening the resultant field, to increase the sensitiveness of the instrument. In the most sensitive instruments of the Thomson type the effect of the earth's field is reduced practically to zero by the use of a so-called astatic system. This consists of two oppositely directed sets of small magnetic needles *ns* (Fig. 65 *a*) mounted in the same plane on mica disks which are attached to a light rigid frame *d*. The earth's field obviously tends to make one set rotate in one direction and the other in the opposite direction, so that, if the needles were exactly alike, the resultant

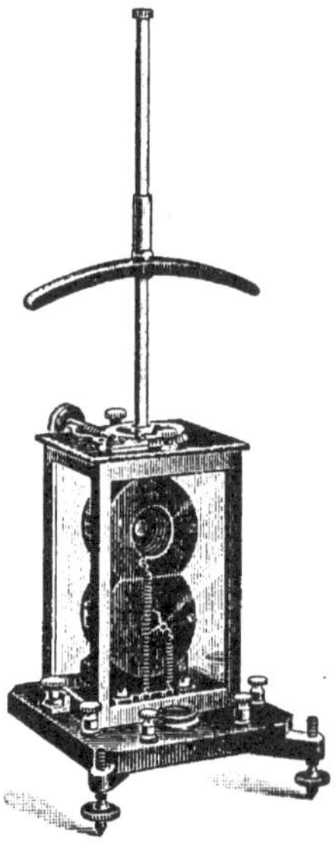
FIG. 65 *b*

effect would be zero. In order to avoid having the effect of the current on one set opposite to its effect on the other set, the two sets are hung in separate coils *D* and *E*, through which the current flows in opposite directions as shown in the figure. A commercial form of this instrument is shown in Figure 65 *b*.

The D'Arsonval galvanometer and its practical form, the ammeter, are the chief representatives of the second type of instruments, in which a movable coil rotates in a fixed magnetic field due to permanent magnets. While instruments of this type cannot be made nearly as sensitive as the Thomson galvanometer, they are much more satisfactory for ordinary work because of the fact that they are not influenced by the magnetic disturbances due to the electric currents which flow within or near to most buildings.

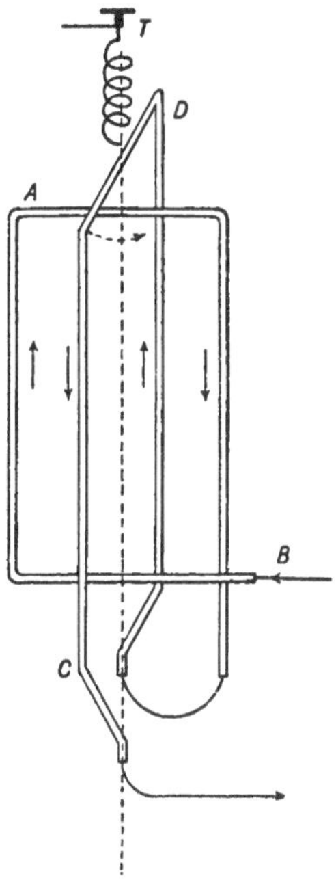

FIG. 66 *a*

There is a third form of current-measuring instrument in which the movable coil is placed in the magnetic field due to a fixed coil. The principle underlying many different varieties of instruments of this type may be seen from Figure 66 *a*. A fixed coil *AB* carrying the current to be measured sets up a magnetic field in which there is suspended by a helical spring *T* a movable coil *CD* connected in series to the fixed coil and also carrying the current to be measured. Since the magnetic field due to the fixed coil is proportional to the current, and since the reaction between the current flowing in the movable coil and this magnetic field is proportional to this current times the strength of field, it follows that the moment of force tending to rotate the movable coil is strictly proportional to the square of the current, provided that the relative positions of the coils are kept unaltered. This condition is satisfied by twisting the helical spring until the torque it exerts balances the moment acting on the coil. The angle θ through which the spring is twisted is proportional to this moment and consequently to the square of the current I. That is,

$$I \propto \sqrt{\theta}, \qquad \text{or} \qquad I = k\sqrt{\theta}. \tag{13}$$

The reduction factor k, by which the square root of the angle must be multiplied to give the value of the current, may be found from an observation of the current and the corresponding deflections. Thereafter the current may be found from the relation of equation (13). A commercial form of this "dynamometer" is shown in Figure 66 *b*.

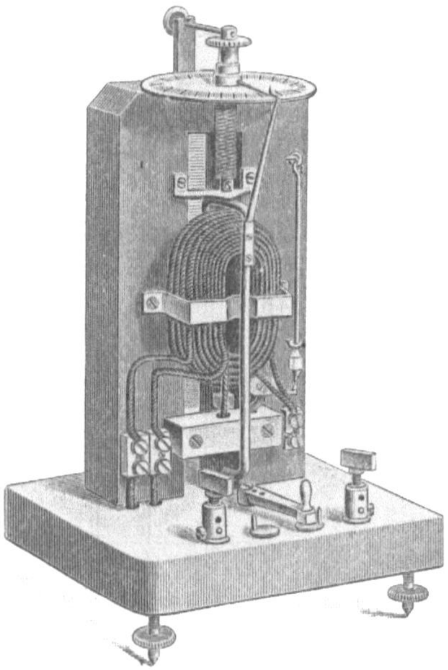

FIG. 66 *b*

The great majority of alternating-current ammeters and voltmeters are instruments of the dynamometer type. For, in view of the fact that any change in the direction of the current takes place simultaneously in both coils, the direction of the torque remains always the same; so that, if the alternations are sufficiently rapid, an alternating as well as a direct current will produce a constant deflection which is proportional to the mean square of the current strength. The dynamometer also has the advantage of having no permanent magnets whose magnetism may undergo slow changes with time. This is indeed the chief source of inaccuracy in calibrated instruments of the D'Arsonval type.

EXPERIMENT 7

Object. To find the constant K of a moving-coil galvanometer.

Directions. Find the resistance of the galvanometer* by either of the two methods stated above, as directed by the instructor. Connect the

* The galvanometer should be the ballistic one which it is intended to use in Experiment 8.

resistances, a voltmeter and a Daniell cell, as in Figure 62. See that the scale is normal to the line drawn from the zero reading to the mirror. Make $R_1 = 10{,}000$ ohms and $R_3 = 1000$ ohms. Make R_2 successively 1, 2, and 3 ohms, and observe the corresponding deflections. If these values of R_1, R_2, and R_3 do not cause deflections of from 3 to 10 cm. on a scale at 1 meter's distance, choose other values which will keep the deflections within about these limits. Since the chief source of error in this experiment lies in imperfect contacts, it is very important to make all connections carefully and to see that the plugs, particularly of the resistance box R_2, are well cleaned with benzine and carefully inserted.

Reverse the direction of the current through the galvanometer and repeat. Read the P.D. on the voltmeter at the time of taking the readings of the deflections.*

Calculate the value of K for each value of R_2, using in each case the mean deflection for d. Neglect in the calculation any figures the dropping of which from the result will not introduce into the value of K an error larger than that due to the observational error in reading the voltmeter. Express K in amperes per radian and also in absolute electro-magnetic units per radian. Record also the "figure of merit" and the "sensibility" of the galvanometer.

EXAMPLE

The galvanometer was found to have a resistance of 536.8 ohms. The resistance R_1 was made 20,000 ohms, and R_3 1000 ohms. When R_2 was 1 ohm the average deflection for direct and reversed currents was 3.10 cm. on a scale distant 143.4 cm. from the galvanometer mirror. As the voltmeter reading across the terminals of the Daniell cell used was 1.05 volts, this deflection corresponded to a current of .0000000342 ampere. Hence K was .00000315 ampere. Similar readings for $R_2 = 2$ and $R_2 = 3$ ohms gave 6.3 cm. and 9.48 cm. as values of the deflection, and .00000310 and .00000309 for the corresponding values of K. The average value of K was therefore .00000311 ampere, or .000000311 absolute unit of current. The figure of merit of the galvanometer was 1.55×10^{-9}, and the sensibility 645 megohms.

* If the voltmeter is in demand it may be freed from this experiment as follows. Observe carefully one deflection when the voltmeter is connected as in Figure 62, then disconnect the voltmeter. If the deflection changes, the P.D. at the terminals of the cell when the voltmeter is disconnected is the voltmeter reading multiplied by the ratio of the second deflection to the first. Thenceforth use the cell without the voltmeter, but use in the calculations this corrected value of the P.D.

CHAPTER VIII

THE ABSOLUTE MEASUREMENT OF CAPACITY

58. Definition of capacity. The difference of electrical potential between any conductor and the earth is commonly called simply the potential of the conductor and is designated by the letter V. In other words, the electrical potential of the earth is arbitrarily chosen as the zero from which the potentials of all other conductors are measured.

It follows from this convention, and from the definition of P.D. given on page 11, that the electrical potential in absolute units of any conductor A (Fig. 67) on which there is a charge of Q units of electricity is equal to the number of ergs of work required to carry unit charge of positive electricity from the earth E up to A against the force which the charge Q exerts upon this unit charge.

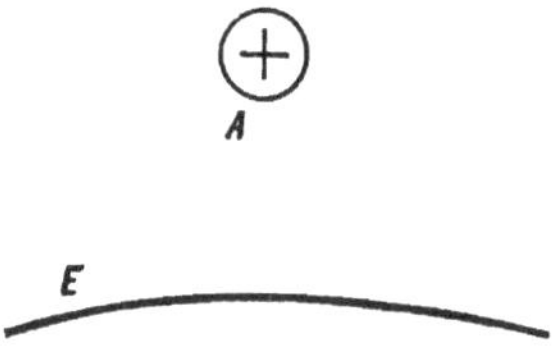

FIG. 67

Suppose now that the quantity Q were to be doubled. The field strength at all points between the earth and A would obviously be doubled also, and hence the work required to carry unit charge up to A would be doubled. In other words, under the conditions indicated, the ratio between the charge on A and the potential which this charge imparts to A is a *constant.*

This constant ratio between the charge on a conductor and its potential is called the *electrical capacity* of the conductor, and is denoted by the letter C. Thus for the defining equation of capacity we may write

$$\frac{\text{Charge}}{\text{Potential}} = \text{Capacity}, \quad \text{or} \quad \frac{Q}{V} = C. \tag{1}$$

We may state the definition in words as follows: *The capacity of any conductor is the charge which must be placed upon it in order to raise its potential one unit above the potential of the earth.*

59. Factors upon which capacity depends. A charge placed upon an isolated conducting sphere establishes at any point outside the sphere a field intensity of the same value as would be produced at that point were the whole charge concentrated at the center of the sphere. If, then, the size of the sphere *A* is increased while the charge upon it is kept constant, the unit charge which we are imagining to be brought up from the earth to *A* does not need to approach so near to the center of the charge *A* in order to be placed upon the sphere, and therefore does not require as great an amount of work. Indeed it may be shown, by a more extended analysis than will be attempted here, that the amount of work required varies inversely as the radius of the sphere. Since the amount of work required represents the potential of the sphere, it follows that the potential of an isolated conducting sphere carrying a given charge varies inversely as its radius, and hence that the capacity of a sphere varies directly as its radius.

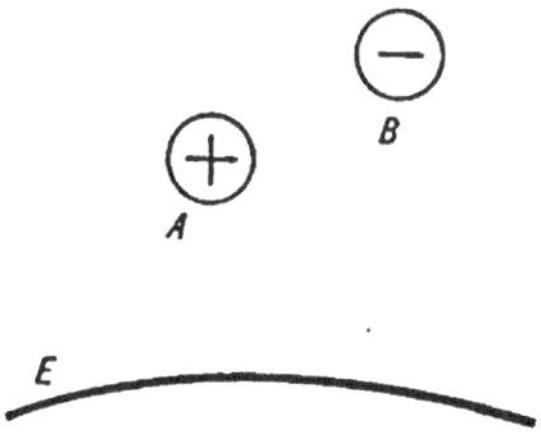

FIG. 68

A second and very important factor upon which the capacity of a conductor may depend is *the presence of neighboring conductors.* Thus suppose that a negatively charged body *B* is placed near *A*, as in Figure 68. It is clear that the work required to bring a unit positive charge from the earth to *A* will now be less than when *A* was isolated, for the attraction which the negative charge on *B* exerts upon the unit positive charge which we are imagining to be brought from the earth to *A* will partially neutralize the repulsion due to *A*. Hence the capacity of *A*, i.e. the charge required to raise it to a given potential, is much greater when *B* is near to it than when *B* is remote.

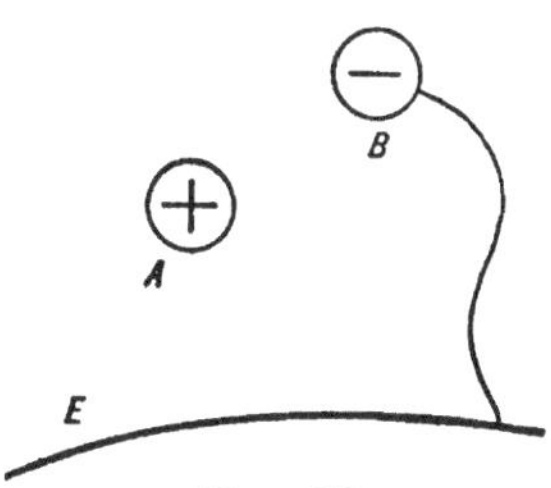

FIG. 69

If the body *B* is not initially charged, but is simply connected to earth by a conductor as in Figure 69 and then brought near to *A*, it will acquire a charge by electrostatic induction (see sect. 5, p. 4). That is, *A* will induce upon *B* a charge opposite in kind to that upon itself by attracting toward itself an unlike kind of electricity and by repelling a like kind along the wire toward the earth. The conductor B will therefore lower the potential and thus increase the capacity of A, just as well as though the negative charge had been imparted to it before it was brought near to A. In other words, *the capacity of a conductor may be very greatly increased by simply bringing near it another conductor which is connected to earth.* On account of this fact two conductors which are very close together and so arranged that one of them can be connected to earth are said to constitute an *electrical condenser.*

60. The condenser. It will be clear from the above discussion that if we wish to increase enormously the capacity of a conductor by the presence of an adjacent conductor, we have only to give to the two conductors such shapes that one can be brought almost into coincidence with the other. This condition is realized by making the conductors thin metallic sheets, or plates, which are separated from each other by a very thin layer of air, or other insulating material. Hence condensers usually take the form of two sheets of tin foil separated by thin sheets of mica, or glass, or paraffined paper.

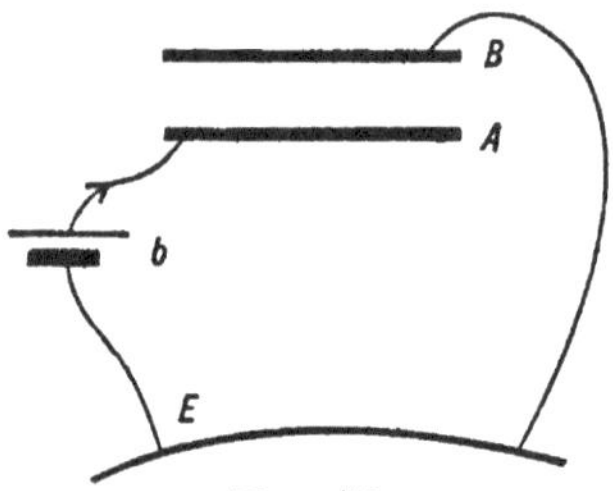

FIG. 70

To charge such a condenser we might connect one plate B to earth and join the other plate A to one terminal of any electrical generator b, for example a static machine or a galvanic cell. But if we wished to raise A to as high a potential as possible above the earth, we should also connect the other terminal of the generator to earth (Fig. 70), for then the full difference of potential which the generator is able to maintain would be built up between the earth and A. Furthermore, since B is also connected to earth, its potential is necessarily zero, i.e. the same as that of the earth, since all points on a conductor in a static condition are always at the same potential (p. 12). Hence the capacity of A in this condition would be simply the charge upon it divided by the P.D. between A and B, i.e. by the amount of work required to carry unit charge from B to A. But it is clear that we could produce the same P.D. between A and B by connecting the terminals of the generator directly to A and B without the intervention of the earth, for in either case the P.D. between A and B would be the total number of volts which the generator is able to maintain. Hence it is customary to charge a condenser by connecting it, as in Figure 71, with a battery b and pressing the key k_1. It is evident that the two opposite charges upon A and B are necessarily of the same size, since A and B are simply the terminals of the generator b, and since positive and negative electricities always appear in exactly like amount whatever be the means by which they are generated. We have, then, the charge Q upon either plate A or B divided by the P.D. between A and B as the capacity of the condenser. We can obtain the P.D. at once with a voltmeter attached to the terminals of the cell b. To obtain the charge Q upon either plate, we discharge the condenser through the galvanometer G. This is done by opening k_1 and closing k_2. We may conceive of the discharge as being accomplished either by the passage of the total charge on A around to B (this would accord with Franklin's one-fluid theory, which made negative the absence of positive), or by the passage of the total charge on B around to A (this would accord with the electron theory, which makes the negative the mobile kind of electricity), or by the passage of half of the charge on A around to B and half of that on B around to A (this would be in accordance with the old two-fluid hypothesis). In any case, however, the total quantity which would pass through the galvanometer would be the total charge Q on one of the plates.

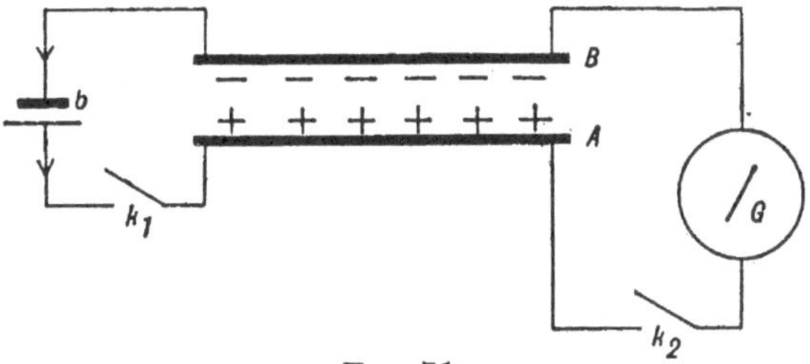

FIG. 71

61. To find Q from the throw θ of the galvanometer, the galvanometer constant K, and the period t of its half vibration. While the quantity Q passes through the coil of the D'Arsonval galvanometer, it constitutes an electrical current which reacts upon the magnetic field of the galvanometer so as to impart an impulse to the coil. If i represent the mean value of this current, then the mean moment of force $\overline{Fh}$ acting on the coil while this current is flowing is given by

$$\overline{Fh} = i\mathcal{H}A \quad \text{(eq. (5), p. 81).} \tag{2}$$

If τ represents the time during which this discharge takes place, then the total impulse acting on the coil, i.e. the moment of force times the time, is given by

$$\overline{Fh\tau} = i\tau\mathcal{H}A = Q\mathcal{H}A.. \tag{3}$$

Now, by Newton's second law, the *impulse*, or product of the force by the time during which it acts, is equal to the momentum imparted ($ft = mat = mv$). Similarly, in rotation, the moment of an impulse, or the impulse multiplied by its lever arm, is equal to the moment of momentum, or angular momentum, imparted.

If, then, I represents the moment of inertia of the coil and ω the angular velocity imparted to it by the impulse, we have

$$\overline{Fh\tau} = I\omega. \tag{4}$$

Furthermore, we have seen (p. 81) that the constant of the galvanometer K is equal to $T_0/A\mathcal{H}$. We get, then, by substitution of these new values of $\overline{Fh}\,\tau$ and $\mathcal{H}A$ in (3),

$$I\omega = Q\frac{T_0}{K} \quad \text{or} \quad Q = K\omega\frac{I}{T_0}. \tag{5}$$

Since we do not observe directly the initial angular velocity ω of the coil, but rather the angular distance θ through which it moves because of this velocity, we must find a way of expressing ω in terms of θ. This is most easily done by equating the initial kinetic energy of the coil, namely $\frac{1}{2}I\omega^2$, to the work done in moving this coil an angular distance θ against the torsional resistance of the suspension. The value of this resistance at the end of the swing, i.e. when the deflection is θ, is $T_0\theta$, in which T_0 is the moment of torsion of the suspension. Since the resistance of torsion is proportional to the displacement, the mean resistance of torsion while the coil is swinging through the angle θ is $\frac{1}{2} T_0\theta$. The work done against torsion is the resistance times the distance. We have, then,

$$\tfrac{1}{2} I\omega^2 = \tfrac{1}{2} T_0\theta^2 \qquad \text{or} \qquad \omega = \theta\sqrt{\frac{T_0}{I}}.$$

Substituting in (5) we obtain

$$Q = K\theta\sqrt{\frac{I}{T_0}}. \tag{6}$$

Now the half period t of any torsional system is given by

$$t = \pi\sqrt{\frac{I}{T_0}}.^{*}$$

Substituting this value of I/T_0 in (5) we obtain

$$Q = \frac{K\theta t}{\pi}. \tag{7}$$

This gives us Q in terms of the very easily observed quantities θ the throw, K the galvanometer constant, and t the half period of vibration of the suspended coil.

62. Correction of the formula for damping. In the above deduction of the relation between ω and θ we tacitly assumed that the vibration of the galvanometer coil takes place altogether without frictional losses of any sort; that is, we assumed an *undamped* swing. The damping which in fact always exists may be taken into consideration as follows: If the line oo' (Fig. 72) represents the position of rest of the galvanometer, and if the amplitudes θ_1, θ_2, θ_3, etc., of successive swings are represented by the distances of the successive turning points from the line oo', then it is found by experiment that the law which always holds,

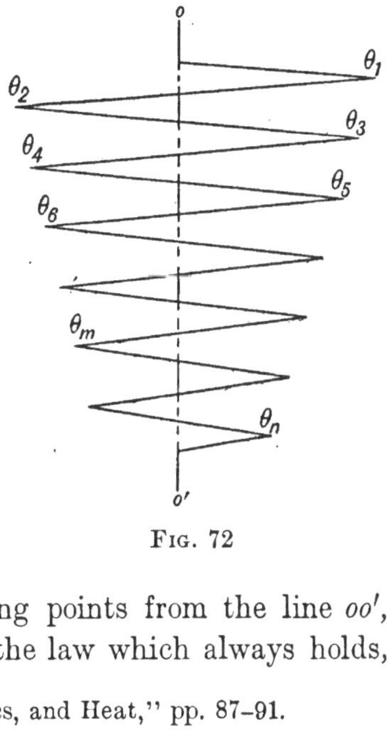

FIG. 72

* See "Mechanics, Molecular Physics, and Heat," pp. 87–91.

approximately at least, for any damped vibration is that the successive amplitudes bear to one another a constant ratio. If we call this ratio ρ, we have

$$\rho = \frac{\theta_1}{\theta_2} = \frac{\theta_2}{\theta_3} = \frac{\theta_3}{\theta_4}, \text{ etc.}$$

That is, $\theta_1 = \rho\theta_2, \qquad \theta_2 = \rho\theta_3$, etc.

$$\therefore \theta_1 = \rho^2\theta_3, \qquad \theta_2 = \rho^2\theta_4, \text{ etc.};$$

or, in general, $$\theta_m = \rho^{n-m}\theta_n. \tag{8}$$

This equation tells us that any amplitude θ_m may be obtained from any later amplitude θ_n by multiplying θ_n by the damping factor ρ raised to a power corresponding to the number of swings through which the damping acts between θ_m and θ_n.

In accordance with this rule the amplitude θ, which would have been attained in the first swing if there had been no damping, may be obtained from the actual amplitude θ_1 of the first swing by multiplying θ_1 by ρ raised to the power corresponding to the number of swings through which the coil has been damped between the instant at which it starts and the instant at which it reaches θ_1. Since this is one half swing, we have $\theta = \rho^{\frac{1}{2}}\theta_1$. The value of ρ may be found by observing any two amplitudes, say the first and the twenty-fifth, and then substituting in equation (8), which, for this case, becomes

$$\rho = \sqrt[24]{\frac{\theta_1}{\theta_{25}}}. \tag{9}$$

We obtain, then, as the final form of the equation for the determination of the quantity Q in terms of the throw of the galvanometer,

$$Q = \frac{Kt\theta_1\sqrt{\rho}}{\pi}. \tag{10}$$

If, then, V denotes the potential difference between the plates and C the capacity of the condenser, we have

$$C = \frac{Kt\theta_1\sqrt{\rho}}{\pi V}. \tag{11}$$

63. The ballistic galvanometer. It is evident that we cannot determine the capacity of a condenser by the above method unless we use a galvanometer which oscillates for a long time, that is, one which does not damp down rapidly; for t cannot be determined accurately unless a considerable number of swings can be observed. Furthermore, the damping law mentioned above obviously cannot hold for a very rapid rate of damping, for if the system should not make more than one or two swings, we evidently could not say that the ratio of successive swings was constant.

Now a galvanometer which is designed to reduce damping to a minimum is called a ballistic galvanometer. The only differences between a ballistic and a nonballistic D'Arsonval galvanometer lie in the absence from the former of all mechanical damping devices, and in the fact that the coil is not wound on a conducting frame, for such a frame causes electro-magnetic damping.*

64. Units of capacity. If Q and V are measured in absolute electro-magnetic units, then C will also be obtained in absolute units; but if Q is measured in coulombs and V in volts, then C will be obtained in practical units. The practical unit of capacity is named the *farad* in honor of Faraday. It is the capacity of a condenser which acquires a P.D. of 1 volt when it receives a charge of 1 coulomb. Thus

$$\text{farads} = \frac{\text{coulombs}}{\text{volts}} = \frac{10^{-1}}{10^{8}} = 10^{-9} \text{ absolute units.} \qquad (12)$$

The farad is so large a unit that the microfarad (=.000001 farad) is the unit which is now most commonly in use.

EXPERIMENT 8

Object. To make an absolute measurement of the quantity of electricity discharged by a condenser, charged to a known difference of potential, and hence to determine the capacity of the condenser.

Directions. I. Set up in the manner indicated in Figure 73 a standard condenser C (between .1 and 1. microfarad), a Daniell cell B, a voltmeter

* The conducting frame rotating in the magnetic field of the galvanometer would have induced in it a current in such a direction as to oppose its motion. See Chapter XII.

V,* a ballistic D'Arsonval galvanometer G, a commutator c, a damping key k_2, and a discharge key k_1. This last instrument is merely a special form of two-point key (see Fig. 74). It is best to connect it so that when the tongue is pressed down the battery will charge the condenser through the lower contact points and when the tongue is released the condenser will discharge through the galvanometer by way of the upper contact points. If a commutator is not available, commutate by interchanging the battery terminals at d and e (see Fig. 73).

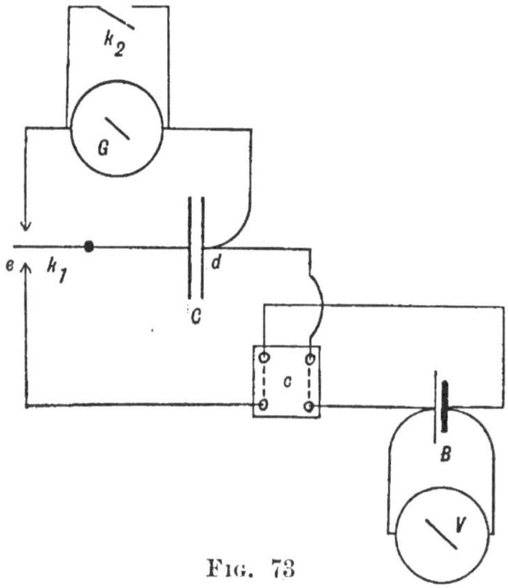

FIG. 73

II. See that the scale upon which you observe deflections is set normal to the line of sight, then bring the galvanometer quite to rest by closing k_2. If a swing of a few millimeters occurs, because of thermal effects, when k_2 is opened, it will be easy with a little practice to press k_2 temporarily at such instants as to check the slight remaining swing and bring the coil to rest even when k_2 remains open. Record this point as the true zero, then charge the condenser by depressing the tongue. Discharge by releasing it and observe the throw in millimeters.† Reverse the commutator and repeat. The mean of six throws, three to right and three to

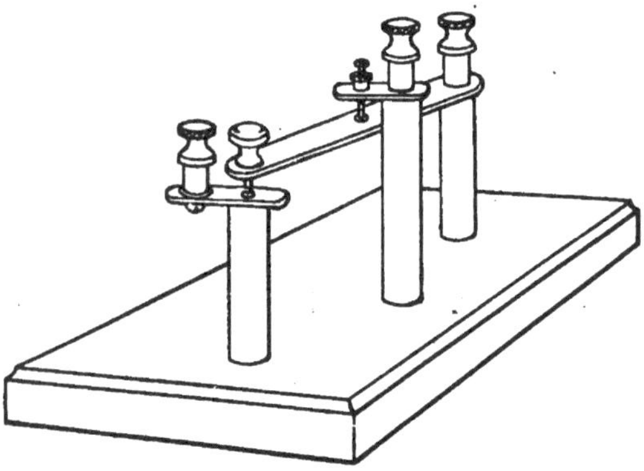

FIG. 74

* As in Experiment 7, the voltmeter need not be kept attached to the battery throughout the experiment, provided, in calculating capacity (equation 11), the voltmeter reading is multiplied by the ratio between two throws taken, one with the voltmeter disconnected, the other with it connected.

† If it is found too tedious to bring the coil exactly to rest when k_2 is open, the throw can be taken quite accurately, even when there are one or two millimeters of swing, in the following way. Observe the middle point about which

left, divided by twice the distance from the scale to the mirror, should be taken as the correct value of θ in radians. Since the zero may be slightly variable, a new zero reading should be taken for every throw. Further, if the first observations of the deflections do not agree closely, record no readings until successive readings in a given direction do agree.*

III. To obtain t, discharge the condenser again, and with a stop watch take the time of at least 20 half swings, timing your counts at the instants of passage of the coil through its zero position, not through its end positions.

IV. To obtain ρ bring the galvanometer again to rest, charge and discharge the condenser, note carefully the first throw, and, calling this first turning point 1, the next 2, etc., *count* the turning points (the readings corresponding to them need not be taken) until the amplitude has been reduced to about one third its first value, then note carefully that amplitude. Calculate ρ from equation (9).

V. Take K from Experiment 7, or from a value furnished by the instructor. Calculate in microfarads the value of the capacity of the condenser used, and compare with the value marked upon the condenser.

EXAMPLE

The condenser used was charged to a potential of $V = 1.06$ volts as given by the voltmeter reading. When discharged through the galvanometer it caused a mean deflection of 15.58 cm. on a scale at a distance of 143.4 cm. Reversing the direction of charging, an average throw of 15.56 cm. was observed. Hence $\theta = .05430$ radians. The time of 20 half vibrations was found to average 125.8 seconds, hence $t = 6.29$ seconds. The damping factor ρ was found by observing $\theta_2 = 15.7$ and $\theta_{22} = 4.5$. Hence $\rho = \sqrt[20]{\frac{15.7}{4.5}} = 1.064$. $\therefore\ \rho^{\frac{1}{2}} = 1.031$. The galvanometer constant K as found in Experiment 7 was .00000311 ampere. Hence

$$C = \frac{.00000311 \times 6.29 \times .05430 \times 1.031}{\pi 1.06}$$

$$= .000000329 \text{ farad} = .3293 \text{ microfarad.}$$

The condenser used was marked $\frac{1}{3}$ microfarad by the manufacturer, hence the per cent of difference was 1.2. (This error is not larger than the uncertainty in the calibration of most voltmeters.)

the oscillation takes place and call this the zero reading. Then discharge the condenser at the instant at which the coil is passing through one of its positions of rest at the ends of its swing. The deflection will then be the difference between the zero and the extreme reading produced by the discharge.

* The reason for this disagreement is that frequently the suspending wire acquires a "set" or tendency to a greater deflection in one direction than in the other. Taking several throws in the same direction will in general, however, result in constancy for that direction of throw.

CHAPTER IX

COMPARISON OF CAPACITIES, THE DETERMINATION OF DIELECTRIC CONSTANTS, AND THE RATIO OF THE ELECTROSTATIC AND ELECTRO-MAGNETIC UNITS

65. Comparison of capacities by means of a ballistic galvanometer. From the definition of capacity (in symbols, $C = Q/PD$) it follows that the quantities of electricity acquired by two separate condensers when charged to the same P.D. vary directly as their capacities. Since the throws of a ballistic galvanometer are proportional to the quantities of electricity passing through it, the ratio between the capacities of two condensers may be found by charging them to the same P.D. and noting the ratio of the throws they cause when discharged through the same ballistic galvanometer. If one of the condensers is a standard of known capacity, the value of the capacity of the second condenser may be found from this ratio.

66. The bridge method of comparing capacities. A comparison of capacities may also be made by a method which is analogous to the Wheatstone-bridge method of comparing resistances, i.e. a method in which a condition of balance is indicated by no deflection of the galvanometer (a so-called "zero method"). The two condensers to be compared are connected as in Figure 75. C_1 and C_2 represent their capacities. R and S are two resistance boxes. The key K permits of the charging or discharging of the condensers. A galvanometer is connected between a and c. When the galvanometer shows no deflection on charging or discharging, there exists a condition of balance for which, as is shown in the next paragraph, the following relation holds:

$$\frac{C_1}{C_2} = \frac{S}{R}.$$

Since by the assumption of no current through the galvanometer the points a and c are always at the same potential, it follows that during the whole time of discharge, for example, the P.D. between b and a is always equal to that between b and c. Now it follows from Ohm's law that if the P.D. between b and a is the same as that between b and c, the currents flowing in the branches R and S are inversely proportional to the resistances of R and S. Since, however, the time during which the condensers are discharging is the same for both, — for otherwise a current would have to

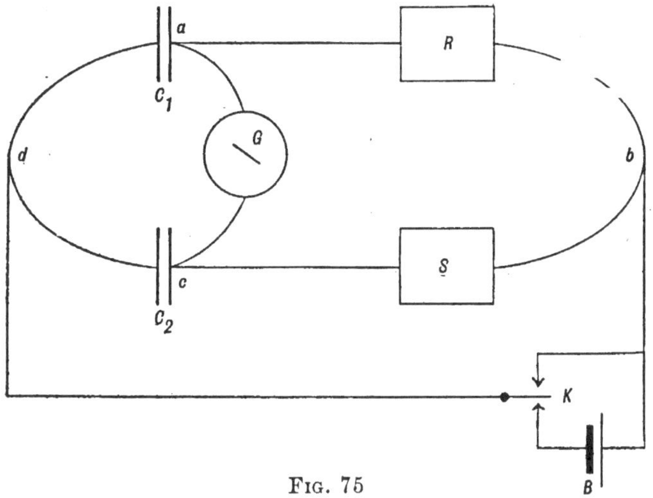

Fig. 75

flow through the galvanometer, — it follows that the quantities discharged through R and S by the condensers *1* and *2* are proportional to these currents, and hence inversely proportional to the resistances R and S. That is,

$$\frac{Q_1}{Q_2} = \frac{S}{R}. \qquad (1)$$

But since a and c are always at the same potential, the condensers must always be charged to the same P.D., and hence the quantities which they hold must be proportional to their respective capacities. Substituting for Q_1/Q_2 the equal ratio C_1/C_2 we obtain, as the condition which must exist for no galvanometer deflection,

$$\frac{C_1}{C_2} = \frac{S}{R}. \qquad (2)$$

67. The bridge method, using an induction coil. No essential difference is introduced if the charging and discharging of the condensers in the method just described is accomplished, not by the key, but by the use of an alternating current. To this end the points *b* and *d* in Figure 75 are connected to the secondary of an induction coil, to the primary of which is connected the battery. The effect of the automatic make and break of the current in the primary circuit is to set up in the secondary a current increasing and decreasing in alternate directions. This charges the condensers, permits their discharge as it decreases, and, recharging in an opposite direction, again permits their discharge. If the balance is not perfect, there will be a rapidly alternating P.D. between *a* and *c*. In the place of the galvanometer a telephone receiver is therefore used, and the test for a balance becomes the absence of a buzzing sound in the telephone.*

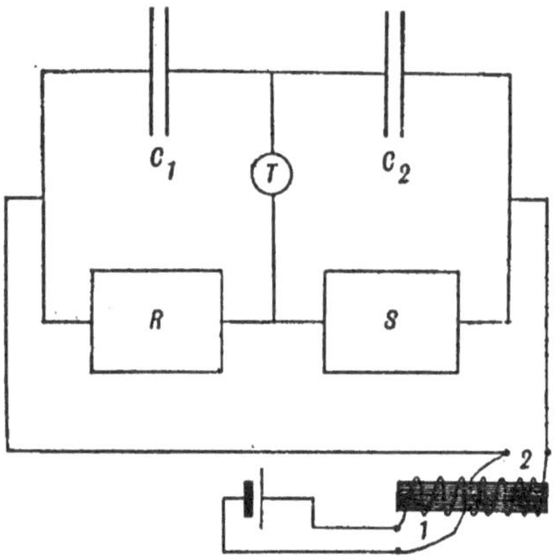

FIG. 76

68. The dielectric constant. When the capacities of condensers which differ from one another only in the nature of the medium between the plates are compared by either of the methods given above, it is found that they differ widely. Hence Faraday introduced the term "*specific inductive capacity*" *to denote the constant of the medium which is measured by the ratio between the capacity of a condenser which has the given medium between its plates and the capacity of the same condenser when air (or more strictly a vacuum) is between the plates.* Thus the specific inductive capacity of ordinary glass is from 2.3 to 2.8, that of hard rubber is from 2 to 3, that of paraffin 1.8 to 2.3, that of mica 2 to 4. This constant relates, of course, only to insulators or *dielectrics.* Hence it is now commonly called the *dielectric constant.*

* Precisely the same equations as were developed in section 66 hold, not only for the scheme of connections mentioned in this section, but also for that shown in Figure 76, where the telephone *T* and the induction coil have been intercharged.

69. Dielectric absorption. The results of a comparison of two condensers by the bridge method, described in section 67, will often not agree closely with the results of a comparison by the method of ballistic throws. Particularly is this true if the condensers compared are made of quite different insulating materials, e.g. mica and paraffin. The reason for such a disagreement is as follows.

The capacity of a condenser has been defined as the constant ratio between the charge given it and the P.D. to which it is charged. But when we measure a capacity by finding the ratio of the quantity discharged to the charging P.D., we tacitly assume the equality of the quantity of charge and that of discharge. As the result of a phenomenon known as dielectric absorption, such an equality does not always exist. The insulating medium between the condenser plates, called the dielectric, appears to absorb a portion of the charge. Thus if a discharged condenser be allowed to stand for a short time on open circuit, a second or even a third charge may be taken from it. This residual charge is practically negligible, however, in the case of the ordinary mica condenser. The amount of the absorption for any given condenser is some complicated function of the time consumed in charging and discharging. Methods of comparison, therefore, which employ markedly different times will not in general agree closely.

70. Laws for the combination of condensers. Using either of the methods of comparison described above, it is possible to test the laws for the combination of condensers. These laws follow at once from the defining equation of capacity.

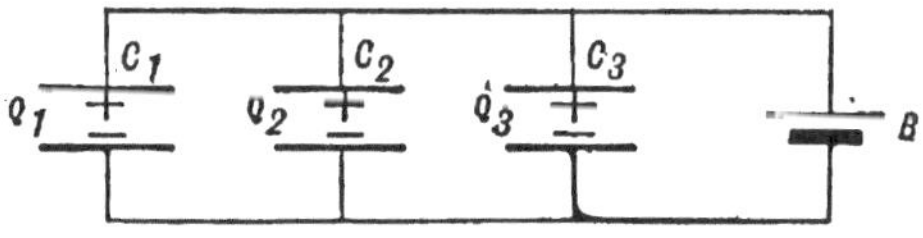

FIG. 77

Thus in Figure 77 are shown three condensers of capacities C_1, C_2, and C_3 respectively, connected in parallel. Obviously they are all charged to the same potential. By definition, the quantities they receive, Q_1, Q_2, and Q_3, are then expressed in terms of their common P.D. as follows:

$$Q_1 = C_1(PD), \quad Q_2 = C_2(PD), \quad Q_3 = C_3(PD). \tag{3}$$

The total charge Q which they receive is the sum of these separate charges. That is,

$$Q = Q_1 + Q_2 + Q_3. \tag{4}$$

Now if C represents their joint capacity, $Q = C(PD)$, and therefore, by substitution in (4) of the values of Q, Q_1, Q_2, and Q_3, we obtain

$$C = C_1 + C_2 + C_3. \tag{5}$$

That is, *for condensers in parallel the joint capacity is the sum of the several capacities.*

In Figure 78 are shown these same condensers connected in series. PD_1, PD_2, and PD_3 represent their respective differences of potential. In this case the total potential difference PD between the ends of the series is necessarily equal to the sum of the several P.D.'s. That is,

$$PD = PD_1 + PD_2 + PD_3. \tag{6}$$

The quantities of electricity held by all these condensers are the same. For obviously the + charge on condenser *3* equals the − charge on condenser *1*, since the generator B must develop equal amounts of positive and negative electricity. But since the + charge and the − charge on the plates of a condenser are equal, the − charge on condenser *1* equals the + charge on the opposite plate. And this + charge is the same as the − charge on the plate of condenser *2*, for the + charge on condenser *1* and the − charge on condenser *2* are simultaneously produced by electrostatic induction, and, according to the laws of induction as discovered by Faraday, + and − charges always appear in equal amount. From these considerations it is clear that

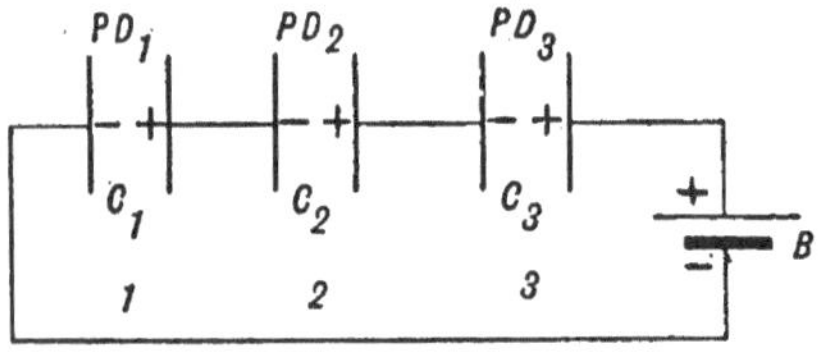

FIG. 78

$$Q = Q_1 = Q_2 = Q_3. \tag{7}$$

But, by definition,

$$Q = C(PD),\ Q_1 = C_1(PD_1),\ Q_2 = C_2(PD_2), \text{ and } Q_3 = C_3(PD)_3. \tag{8}$$

By substitution in (6) of the values of PD, PD_1, PD_2, and PD_3, given in (8), it follows, in consideration of (7), that

$$\frac{1}{C} = \frac{1}{C_1} + \frac{1}{C_2} + \frac{1}{C_3}. \tag{9}$$

That is, *for condensers in series the reciprocal of the joint capacity is the sum of the reciprocals of the several capacities.* Hence the law for capacities in series is similar to the law for resistances in parallel, and vice versa.

71. Standard condensers. Condensers carefully constructed of tin-foil plates and sheets of mica as the dielectric are arranged in boxes similar in form to resistance boxes. In Figure 79 is shown a standard form of subdivided condenser. The binding posts b and b' are connected to the brass strips M and N respectively. Between M and N, and capable of connection by plugs to either M or N, are other strips A, B, C, D, E, and F. The condensers are connected between these as indicated by the dotted connections. To connect between b and b' any single capacity, for example that of the condenser between E and F, we connect M to E by a plug

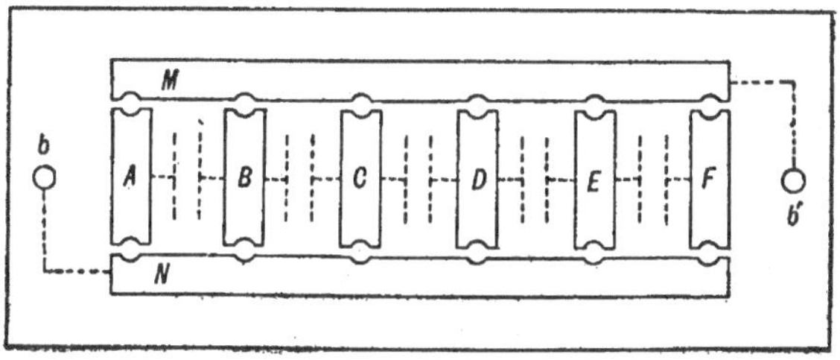

FIG. 79

and also N to F. If it is desired to have the sum of the capacities of the three condensers CD, DE, and EF between the binding posts, since capacities in parallel are added, we connect the condensers in parallel by plugging D and F to M, and also C and E to N.

Since the sum of the reciprocals of several capacities in series is the reciprocal of their joint capacity, to obtain a small fraction, we connect, for example, A to M and C to N, thus putting condensers AB and BC in series between b and b'.

72. The ratio of the electro-magnetic and the electrostatic units. As stated in section 28 (p. 35), the electro-magnetic unit of quantity is found to be equal to 3×10^{10} electrostatic units, and the fact that this number is precisely the same as the velocity of light in centimeters has been one of the chief factors in the establishment of the electro-magnetic theory of light.

This relation between the units is made the object of an experimental verification in the experiment following this chapter. It is not, however, a quantity of electricity which will be directly measured in the two systems, but rather the capacity of a condenser. From this latter measurement the former can be easily calculated, as will appear from the following considerations. If the electro-magnetic unit is in fact 3×10^{10} electrostatic units, it follows that when a given quantity of electricity is measured in electrostatic units we should get a number 3×10^{10} larger than when we measure the same quantity in electro-magnetic units. Furthermore, since the unit of P.D. in either system is the P.D. between two points when it requires 1 erg of work to carry unit quantity, measured in that system, between the points, it will be seen that a P.D. measured in electro-magnetic units should be 3×10^{10} times as large a quantity as the same P.D. measured in electrostatic units.

Now since $C = Q/PD$, it follows that

$$\frac{C \text{ electrostatic}}{C \text{ electro-magnetic}} = \frac{Q \text{ electrostatic}}{Q \text{ electro-magnetic}} \times \frac{PD \text{ electro-magnetic}}{PD \text{ electrostatic}}$$
$$= 9 \times 10^{20}.$$

That is, if the ratio of the electro-magnetic unit of quantity to the electrostatic unit of quantity is 3×10^{10}, the capacity of a condenser measured in electrostatic units should be 9×10^{20} greater than its capacity when measured in electro-magnetic units. In order, then, to determine the ratio of the units of quantity, we have only to measure in electro-magnetic units, by the method of Chapter VIII, the capacity of a condenser of simple geometric form for which the capacity in electrostatic units may be calculated according to relations which will be established in the next section. Since, however, the capacity of a condenser of form suitable to this calculation is very small, we shall determine its capacity in electro-magnetic units by comparison with a larger condenser, the capacity of which has already been measured in absolute electro-magnetic units by the method discussed in the preceding chapter.

73. The calculation of the capacity of a plate condenser in electrostatic units. We can compute C in electrostatic units at once, if we know the area of the condenser plates, the distance between them, and the dielectric constant of the medium. To see how this is done, let us first find the field strength in dynes which exists between two condenser plates in air when each of these plates is charged with Q electrostatic units. Let σ represent the amount of electricity upon each square centimeter of each plate. This quantity σ is known as the *density of charge.* Consider the force f which the charge upon any little element of surface ds (Fig. 80) exerts upon a unit charge at any point a between the plates. The number of units of charge upon ds is σds. If we represent the distance of this charge from a by r, then by the law of electrostatic force given in section 2, page 2, we have

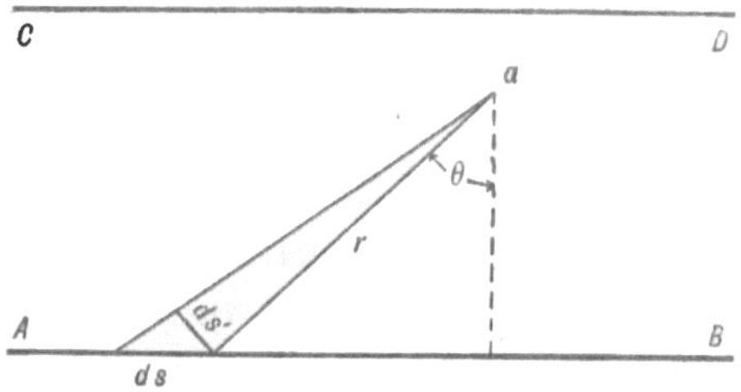

FIG. 80

$$f = \frac{\sigma ds}{r^2}. \tag{10}$$

Now the component of this force which is normal to AB is the only one with which we are concerned, since it is evident that, except at the very edges of the plates,* the resultant force which the whole charge on AB exerts upon a unit charge at a is directed along the normal from AB toward CD. If f' represents the value of the normal component of f, we have

$$f' = \frac{\sigma ds}{r^2} \cos\theta. \tag{11}$$

* The error introduced by assuming that the lines of electrostatic force between the plates are everywhere normal to the plates, even at the very edges, is of the same order of magnitude as the ratio of the distance between the plates to their diameter. In the experiment which follows this chapter the first distance will be, perhaps, .1 mm. and the last 20 cm. or 30 cm. Hence no appreciable error will be here introduced by assuming the lines to be everywhere normal to the plates.

But if ds' represents the projection of ds upon a plane normal to r and passing through ds (which it will be remembered is an element of infinitesimal area), then we have

$$ds \cos\theta = ds'. \tag{12}$$

Hence

$$f' = \frac{\sigma ds'}{r^2}. \tag{13}$$

Now the solid angle subtended by ds at the point a is by definition ds'/r^2. If then we call this solid angle u, we have

$$f' = \sigma u. \tag{14}$$

That is, *the force, normal to the plate, which each element of charge upon the plate exerts upon a unit charge at any point between the plates, is equal to the density of charge upon the element, multiplied by the solid angle subtended by the element at the point considered.* Now if the density of charge is the same at all points upon the plate, then the total force F' which all the little elements of charge on AB exert on the unit charge at a is σ times the sum of all the solid angles subtended by all the elements of surface on AB; i.e. it is σ times the total solid angle subtended at a by the whole plate AB. But since the distance between AB and CD is very small as compared with the dimensions of these plates, the total solid angle subtended at a by AB is practically the solid angle subtended by a hemisphere at its center. The solid angle about a point is by definition the surface of a sphere of radius r about that point as a center divided by r^2; that is, it is $\frac{4\pi r^2}{r^2} = 4\pi$. The solid angle subtended by a hemisphere at the center is therefore 2π. Hence we have

$$F' = 2\pi\sigma. \tag{15}$$

Since the charge upon the plate CD exerts a like force upon the unit charge at a, the total field strength F between the condenser plates is given by

$$F = 4\pi\sigma. \tag{16}$$

Hence, if d is the distance from CD to AB, the P.D. between the condenser plates, i.e. the work required to carry unit positive charge from one to the other, is $4\pi\sigma d$. But if A is the total area of one plate, the charge Q upon it is $A\sigma$. Hence the capacity of the condenser C, which is by definition Q/PD, is given by

$$C = \frac{A\sigma}{4\pi\sigma d} = \frac{A}{4\pi d}. \tag{17}$$

If we replace the air by a medium of dielectric constant K, we have, by definition of K,

$$C = \frac{AK}{4\pi d}. \tag{18}$$

EXPERIMENT 9

(A) Object. To test the laws for the combination of condensers in series and in parallel by the comparison of several such combinations with a standard condenser.

Directions. Using for C_1 a subdivided condenser and for C_2 a single standard condenser of say $\frac{1}{3}$-microfarad capacity, connect as indicated in Figure 75, except that a telephone receiver replaces the galvanometer G and a small induction coil replaces the discharge key **K**. R and S should be resistance boxes capable of a range of from 1 to 10,000 ohms. Give to R a fixed value and vary S until a balance is obtained. Approach the value of S which corresponds to a balance from the side of too large a resistance and also from the side of too small a resistance. Record both observations and use the average in the calculation. Arrange the plugs in the subdivided condenser box so as to make the following combinations of condensers: (1) series connections, .2 and .05 microfarad; (2) parallel connections .05, .05 and .2 microfarad.

Compare these combinations with the $\frac{1}{3}$-microfarad condenser. Calculate the joint capacity from the laws of combination of condensers and compare with the observed value.

(B) Object. To determine the dielectric constant of mica.

Directions. Place a thin smooth sheet of mica between the metal plates of the parallel-plate condenser shown in Figure 81. Determine the capacity of the condenser so formed by the method used in (A), being careful to select a value for C_2 which is as near as possible to that of the unknown (say, .05 or .1 microfarad). Cut off three small pieces (3 to 5 mm. square) from the corners of the sheet of mica and, separating the plates by these,

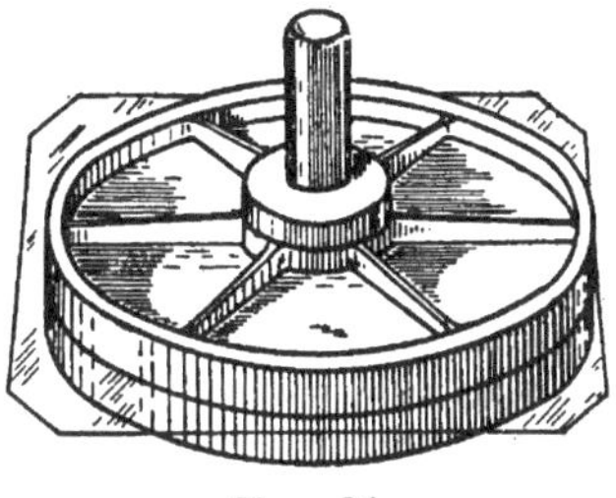

FIG. 81

form an air condenser of the same thickness as that of the mica.* Measure its capacity in the same manner and determine the dielectric constant from the ratio of the capacity with mica as the dielectric to the capacity with air as the dielectric.

(C) Object. To determine the ratio of the units of capacity, and hence of the units of quantity, in the electrostatic and the electro-magnetic systems.

Directions. The capacity in electro-magnetic units of a parallel-plate air condenser has already been determined (Exp. 9, (**B**)). It only remains to calculate its capacity in electrostatic units from the relation of equation (17). Measure carefully the thickness of the small mica strips used in Experiment 9, (**B**) to keep the plates separated. Take the average of at least ten observations for the value of d. Measure the radius of the condenser plates and from this data calculate the capacity. Determine the ratio of the E.S. to the E.M. value of this capacity, remembering that one microfarad is 10^{-15} absolute electro-magnetic units (see p. 97). Extract the square root of this ratio to find the ratio of the electro-magnetic unit of quantity to the electrostatic unit of quantity. Compare this value with the value of v (the velocity of light), namely, 3×10^{10}.

EXAMPLE

(A) The resistance S was kept constant at 1000 ohms. The value of the standard condenser was $\frac{1}{3}$ microfarad. The .05 and .05 microfarad in the multiple condenser box were connected in series. For a balance R was 74 ohms. Hence $C = \frac{74}{1000} \cdot \frac{1}{3} = .0247$. The calculated value found by the relation $\frac{1}{C} = \frac{1}{.05} + \frac{1}{.05}$, namely .0250, differed from the observed value by 1.2 per cent. Similar observations with .05 and .05 in parallel gave for a balance $R = 1000$, $S = 301$; that is, $C = .1003$ observed value and .1000 calculated value, a difference of .3 per cent.

(B) The average of thirty readings of the thickness of the mica was .0112 cm. The average of ten readings on the small strips used in separating the plates for the air condenser was .0119 cm. Using a .1 microfarad and making $R = 2000$ ohms, S was found to be 130 ohms when the condenser contained mica. Therefore $C_{mica} = .00650$ microfarad. Keeping $R = 2000$ ohms but using .05 microfarad, S was found to be 113 ohms when the condenser contained air. Therefore $C_{air} = .002825$ microfarad. The balance in both cases was so exact that a change of 1 ohm in S disturbed it. Since the mica and air condensers differed in thickness, it

* If the mica plate is not of uniform thickness, a correction must be made as is shown in the example following this experiment.

was necessary to increase the capacity of the air condenser to the value which it would have had if it had been of the same thickness as the mica. This value is evidently $\frac{.0119}{.0112} \times .002825$, or .00300. Hence the dielectric constant of mica was found to be $\frac{.00650}{.00300} = 2.17$.

(C) In Experiment 9, **(B)**, the capacity of the air condenser had already been found to be .002825 microfarad for a thickness of .0119 cm. The radius of the plates was 10.9 cm. Hence the area $A = \pi\, 10.9^2$, and $C = \frac{S}{4\,\pi d} = \frac{\pi\, 10.9^2}{4\,\pi .0119} = 2496$ electrostatic units. Hence $\frac{\text{the capacity } E.S.}{\text{the capacity } E.M.}$ $= \frac{2496}{.002825 \times 10^{-15}} = 8.84 \times 10^{20}$, and therefore the ratio of the electromagnetic unit of quantity to the electrostatic unit was found to be 2.973×10^{10}. This determination differs from 3×10^{10} by .9 per cent.

CHAPTER X

ELECTROMOTIVE FORCE AND INTERNAL RESISTANCE

74. Definition of electromotive force. Whenever two dissimilar substances are brought into contact one of them is found to take up a potential higher than the other. Thus, if copper and zinc are joined as in Figure 82, the zinc is found to acquire a slight positive charge, and the copper a corresponding negative charge. In order to assign a cause to this phenomenon, it is customary to assume that there exists at the surface of contact of copper and zinc an agent which we call an *electromotive force* (E.M.F.), which has the ability to drive positive electricity from the copper to the zinc; that is, in terms of the electron theory, the ability to drive negative electrons from zinc to copper. In obedience to this agent a current flows, according to our conventions (see p. 31), from the copper to the zinc until the action of the agent is brought to rest by the back pressure, that is, the P.D. created between the zinc and the copper because of the charges which they acquire. We then define an electromotive force as any agent which is able to cause the appearance of positive and negative electrical charges; that is, any agent which is able to set up a P.D. And we measure the strength of an E.M.F. by the P.D. which it is able to maintain. The relations of E.M.F. and P.D. are then simply the relations of cause and effect.*

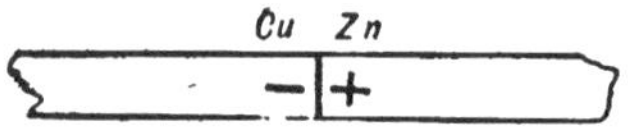

FIG. 82

* It is to be noted that in the above the phenomenon of contact E.M.F. has merely been described, and the term E.M.F. defined. None of the present theories as to the nature and cause of this phenomenon have as yet been generally accepted. Helmholtz imagined the phenomenon to be due to differences in the specific attraction of different kinds of matter for the two kinds of electricity. In terms of the electron theory this view reduces to a difference in the specific attraction of different substances for the free electrons. If, however, there is a larger number of free electrons in one substance than in another,

75. Thermo-electromotive forces. Perhaps the easiest way of showing the existence of electromotive forces of contact between metals is to place a sensitive galvanometer in a circuit consisting of two different metals, for example copper and zinc (Fig. 83), and then to heat one of the junctions. So long as the temperature of the two junctions *a* and *b* is the same, the E.M.F. from copper to zinc at *a* is necessarily exactly equal and opposite to the E.M.F. from copper to zinc at *b*, and hence no current can possibly flow through the circuit. But as soon as one junction is heated even a fraction of a degree above the other, a current is at once found to appear in the galvanometer. This is because the value of the E.M.F. of contact between any two substances changes with the temperature. Hence if one junction undergoes a change in temperature, while the other does not, the balance between the E.M.F.'s at the two junctions is destroyed and a current flows through the circuit which is proportional to the resultant E.M.F. of the circuit. If this current flows from *a* over to *b* through the zinc, and from *b* back to *a* through the copper, then there will of course be a continuous fall of potential from *a* to *b* in the zinc and from *b* to *a* in the copper. Furthermore, the sum of these falls in potential through the zinc and copper must of course be exactly equal, numerically, to the resultant E.M.F. in the circuit, for otherwise there would be either a cause (an E.M.F.) without an effect (a P.D.), or else an effect (a P.D.) without a cause (an E.M.F.).

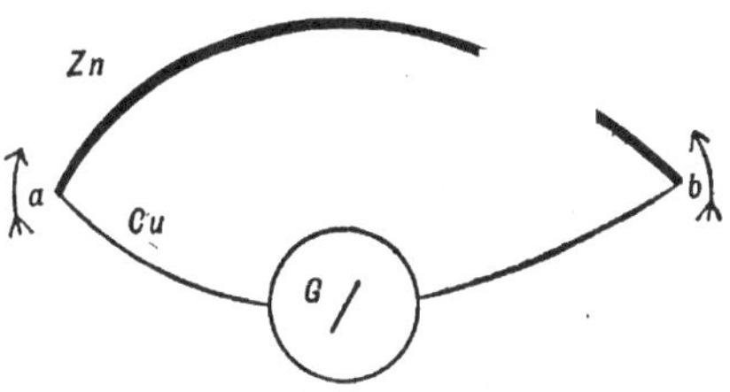

FIG. 83

76. The E.M.F. of a galvanic cell. In a galvanic cell (Fig. 84) there is an E.M.F. at every surface of contact of two dissimilar substances, and the total E.M.F. of the cell is simply the algebraic sum of all these E.M.F.'s of contact. It is measured by the total P.D. which the cell is able to maintain between its terminals *A*

there would, of course, be a tendency to diffuse more rapidly in one direction than in the other, and this fact would have to be taken into account, in connection with the specific attractions, in order to determine which of two substances would have an excess of negative electrons.

and B on open circuit. That is, the resultant E.M.F. of contact causes a positive charge to accumulate on A and a negative charge on B until the back pressure through the cell, set up by the accumulation of these charges, just balances the total E.M.F. which is producing it, and thus brings to rest the action of this E.M.F.

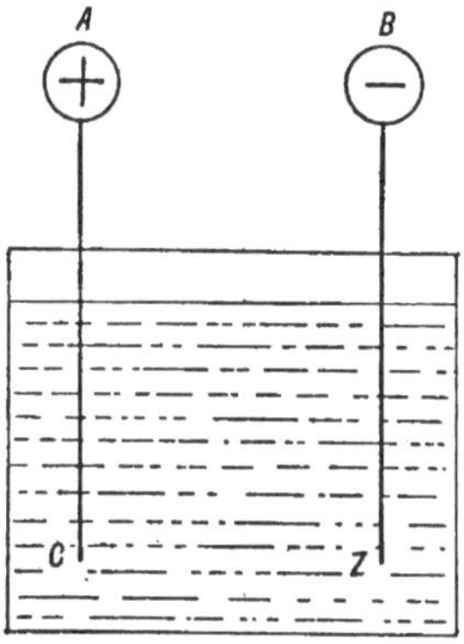

FIG. 84

The water analogy of the galvanic cell is presented in Figure 85. The weight W corresponds to the E.M.F. of the cell. It is the cause of the movement of the water from B to A. Furthermore, just as the weight W is brought to rest as soon as the back force against the paddle wheels, due to the difference in levels in A and B, becomes equal to the forward force which W is able to exert, so the actions at the surfaces of contact of the dissimilar substances in a galvanic cell are brought to rest by the P.D. which the E.M.F. of the cell sets up between its terminals.

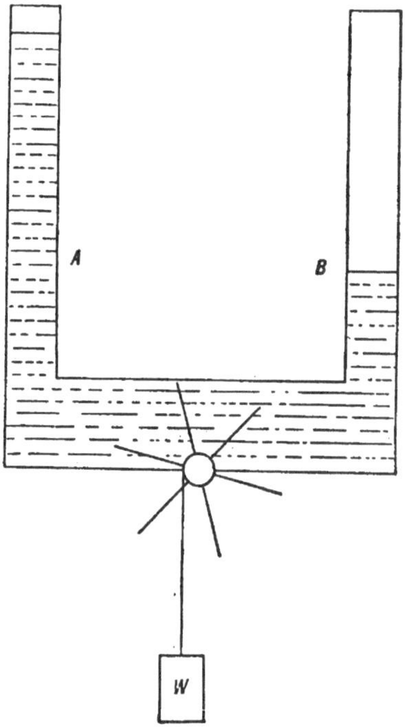

FIG. 85

77. Internal resistance of a galvanic cell. Connecting the terminals A and B of the galvanic cell by a conductor (Fig. 86) causes the P.D. between them at once to fall, just as connecting the tanks A and B by a pipe mn (Fig. 87) causes the difference in level between them to diminish. Further, just as the weight W at once begins to cause the wheel to rotate and thus to restore the lost difference in level, so the E.M.F. of the cell begins at once to act to restore the lost P.D. between the terminals of the cell as soon as they are connected by a conductor.

The fact that a current flows through the circuit composed of the conductor AB, the dissimilar metals Cu and Zn, and the liquid in the cell, means that there is a fall in potential from A to B in the external, and from B to A in the internal, portion of the circuit. The total fall in potential throughout all of the conductors of the circuit (Fig. 86) is equal to the resultant E.M.F. of contact of the dissimilar substances in the circuit, just as in the case of Figure 83 total P.D. equals resultant E.M.F. Similarly, in the case of the water analogy of Figure 87, the water flows through the external circuit mn, and through the internal circuit po; and the sum of the falls in pressure in the external and the internal portions of the circuit is equal to the difference in pressure produced by the wheel. In the case of a galvanic cell this E.M.F. is the result of a difference in the energy of combination of the zinc with the liquid (for example, sulphuric acid) and the energy of combination of the copper with the liquid. In the case of the thermal E.M.F.'s the resultant E.M.F. was the direct result of energy supplied to one junction in the form of heat.

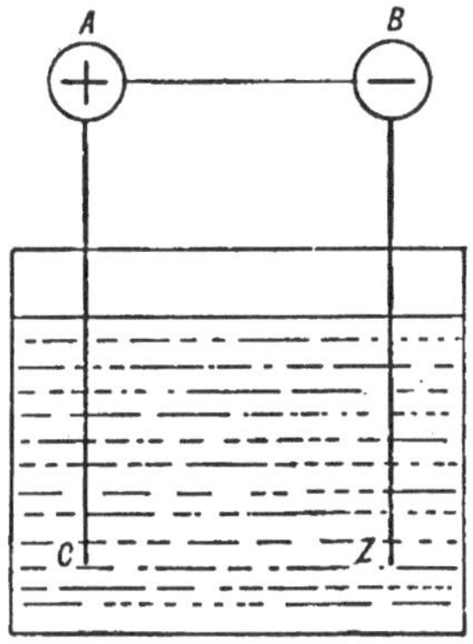

FIG. 86

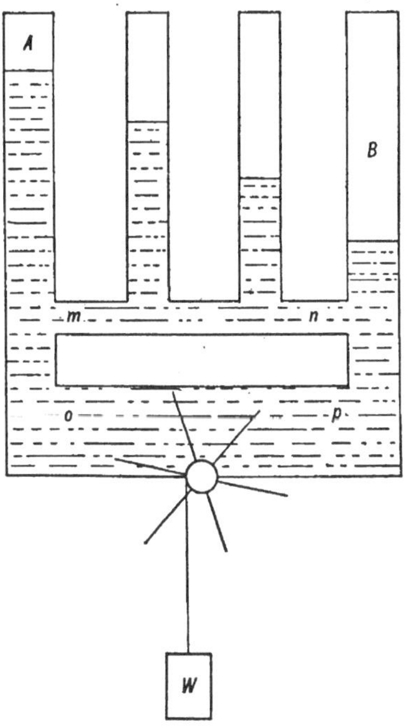

FIG. 87

How near to the original P.D. the E.M.F. of the cell is able to maintain the terminals A and B (Fig. 86) when they are connected by the conductor depends wholly upon the relative amounts of difficulty which the conductor, on the one hand, has in discharging A and B, and which the E.M.F., on the other hand, has in forcing new charges through the cell up to these terminals; that is, it depends upon the relative *resistances* of the *internal* and the *external* portions of the circuit (see p. 47). Similarly, the difference in water level maintained by W between A and B (Fig. 87) will depend wholly upon the relative capacities of the pipes mn and op for carrying water.

If the resistance of the conductor connecting A and B (Fig. 86) is very large in comparison with the resistance of the liquid between Z and C, then the P.D. maintained between A and B will be practically the same as upon open circuit. But in any case the total E.M.F. of the cell must be equal to the sums of the falls in potential in the external and internal portions. Thus if I denote the current furnished by the cell, R its external and r its internal resistance, and if PD_1 and PD_2 denote external and internal falls in potential respectively, then, by Ohm's law, $PD_1 = RI$, and $PD_2 = rI$; or, since $PD_1 + PD_2 = EMF$, we have

$$EMF = I(R + r). \tag{1}$$

Otherwise stated, while Ohm's law applied to any part of a circuit is $PD/I = R$, as applied to a complete circuit it is

$$\frac{EMF}{I} = R + r. \tag{2}$$

78. Polarizing and nonpolarizing cells. A cell which consists simply of two dissimilar metals, for example zinc and copper, immersed in a conducting liquid like sulphuric acid, will, in general, have neither a constant E.M.F. nor a constant internal resistance; for, since the positive ions which are already in the liquid are of hydrogen,* as soon as the zinc begins to go into solution in the form of positively charged zinc ions which repel the positive hydrogen ions away from the zinc plate and toward the copper plate, these repelled hydrogen ions begin to accumulate about the copper plate and there to change into neutral hydrogen molecules, and thus to alter completely the character of the surface of the plate. In effect the plate of copper is replaced by a plate of

* $H_2SO_4 + H_2O \longrightarrow H_2O + 2\overset{+}{H} + \overset{--}{SO_4}$ (see Chap. XVI).

hydrogen. This alters completely the E.M.F. of contact at the surface of this plate, and it also alters the internal resistance of the cell. A cell of this sort is said to be a *polarizing* cell, because the current which it furnishes diminishes rapidly as the hydrogen accumulates.

A nonpolarizing cell is one in which the plate toward which the ions are urged is immersed in a solution of a salt of the same metal as the plate; for example, a copper plate in a copper sulphate solution. The character of the surface of this plate is then quite unchanged by the passage of the current, for the substance deposited is the same as the substance of which the plate is composed. The most familiar of the nonpolarizing cells is the Daniell cell (Fig. 88). It consists of a jar partly filled with a copper sulphate ($CuSO_4$) solution in which is placed a porous earthenware cup containing a solution of zinc sulphate ($ZnSO_4$). In the zinc sulphate is a zinc plate and in the copper sulphate a copper plate. The porous cup is simply to keep the liquids from mixing and yet to permit the passage of the ions, which travel with relative ease through its pores. Such a cell, if freshly set up, will have both an E.M.F. and an internal resistance which are very nearly constant. Since the copper is continually passing out of solution, the compartment S is usually kept filled with copper sulphate crystals so as to keep the solution saturated.

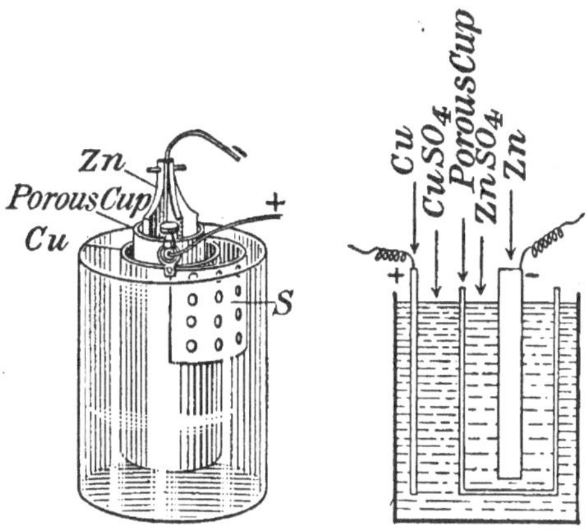

FIG. 88

79. Methods of measuring the E.M.F. and internal resistances of Daniell cells. One of the simplest and most satisfactory ways of measuring the E.M.F. and internal resistance of a Daniell cell is as follows. First connect with, say, No. 18 copper wire, the terminals of the cell directly to the terminals of a voltmeter and read the P.D. indicated. Then replace the voltmeter by an ammeter and read the current furnished. If R_V and R_A represent the resistances of the voltmeter and ammeter respectively (quantities which may be assumed to be known), PD the voltmeter reading, and I the ammeter reading, then, by Ohm's law, the current sent through the voltmeter in the first of the above observations is PD/R_V. But this same current is also equal to $\frac{EMF}{R_V + r}$. Hence our first equation connecting EMF and r is

$$\frac{EMF}{PD} = \frac{(R_V + r)}{R_V}. \tag{3}$$

The equation which corresponds to the second observation is

$$\frac{EMF}{(R_A + r)} = I. \tag{4}$$

We have only to solve (3) and (4) in order to obtain both EMF and r.

When r is small as compared to R_V, equation (3) gives $EMF = PD$ (the voltmeter reading) as a first approximation. Now for the determination of a quantity so little constant as r it will, in general, be sufficiently accurate to make the assumption that the E.M.F. of the cell is the voltmeter reading, and then to solve equation (4) for r. The value of the E.M.F. correct to a second approximation, which will generally be as accurate as the voltmeter can be read, may then be obtained by substituting the value of r as thus found in equation (3), and solving for EMF.

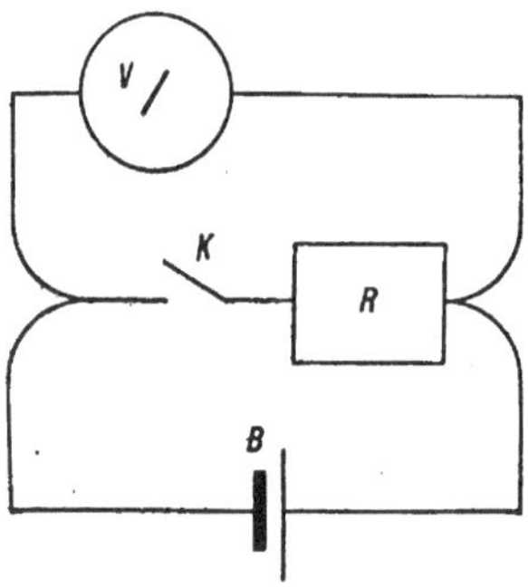

FIG. 89

A second method of measuring internal resistance, and one which has the advantage of making it possible to determine whether or not, with the cell employed, r varies largely with the current used, is the following. The cell B, a voltmeter V, a resistance box R, and a key K are connected as in Figure 89, the resistances of contacts and connections between K and R being made as small as possible. The voltmeter reading when the key is up is approximately the E.M.F. of the cell. The voltmeter reading when the key is down is the P.D. across the resistance R. If R is small in comparison with the resistance of the voltmeter, then the current furnished by the cell when the key is down is the reading of the voltmeter (which we shall call simply PD) divided by R. But this same current is also given by dividing the E.M.F. of the cell by $R + r$. Hence we have

$$\frac{EMF}{(R+r)} = \frac{PD}{R}, \tag{5}$$

or

$$r = \frac{R\,(EMF - PD)}{PD}. \tag{6}$$

In the use of this method, if results correct to even so much as 5 per cent, for example, are to be obtained, R should never be made more than a twentieth of the resistance of the voltmeter. Furthermore, instead of using the voltmeter reading when the key is open as the E.M.F. of the cell, it is, of course, more accurate to raise this reading to the true E.M.F. by substituting in equation (3) the approximate value of r, known by a rough preliminary observation either by this method or its predecessor.

EXPERIMENT 10

Object. To determine the E.M.F. and the internal resistance of a Daniell cell.

Directions. I. Clean the zinc of the Daniell cell carefully so as to remove any deposit of copper or copper oxide which may have accumulated because of the diffusion of the copper sulphate through the porous cup, and which is a fruitful source of polarization in the cell. Clean also the porous cup and put in a fresh solution of zinc sulphate.

II. Determine the resistance of the coarser register of the milliammeter either by the Wheatstone-bridge method or else by sending a current from the cell through it and the voltmeter arranged in parallel. In the latter case the reading of the voltmeter divided by the reading of the ammeter will be the resistance of the latter. If the current is too large for the ammeter, reduce it by inserting a few feet of German silver wire.

III. Determine the resistance of the voltmeter either by the bridge method, or else by putting the milliammeter (low register) and the voltmeter in series in the circuit of the cell. The reading of the voltmeter divided by the reading of the milliammeter will be the resistance of the voltmeter.

IV. Connect the high register of the milliammeter directly to the terminals of the cell, or insert, if necessary to keep the deflection on the scale, a small length of German silver wire of known resistance per meter and read the current; then replace the ammeter by the voltmeter and read the *P.D.* Compute r and EMF as indicated in section 79, equations (3) and (4), subtracting, of course, the resistance of the German silver wire if it is used.

V. Connect the cell, a resistance box, a key, and a voltmeter, as indicated in the second method of section 79, using a short, thick wire (e.g. No. 16) to connect the key to the resistance box, and seeing to it that the contacts at all binding posts and plugs are thoroughly good.

Compute r from equation (6) for at least three different values of R, e.g. 1, 2, and 5 ohms, and decide whether or not the internal resistance of a Daniell cell varies appreciably with the current taken from it.

EXAMPLE

I. The ammeter reading was .382 ampere when the reading of the voltmeter shunted across it was .04 volt, hence $R_A = .10$ ohm.

II. A current of .0027 ampere, as measured by the low register of the milliammeter, was passed through the voltmeter, which read 1.04 volts. Hence $R_V = 385$ ohms.

III. The voltmeter connected directly to the Daniell cell used read 1.05 volts. The ammeter when connected directly read .427 ampere. Hence $r = 2.35$ ohms, and $EMF = 1.065$ volts.

IV. A P.D. of .33 volt was observed across the terminals of a resistance of 1 ohm. Substitution of these values and the value of the E.M.F. found above gave

$$r = \frac{1.065 - .33}{.33} = 2.23 \text{ ohms.}$$

Similarly for $R = 2$ ohms, $PD = .85$ volt and $r = 2.39$ ohms. Also for $R = 3$ ohms, $PD = .60$ volt and $r = 2.32$ ohms.

CHAPTER XI

THE COMPARISON OF ELECTROMOTIVE FORCES

80. E.M.F.'s of polarizing cells. The methods described in the preceding chapter can obviously be employed only with cells, like the Daniell, which do not polarize. But having once found the E.M.F. of any nonpolarizing cell, the E.M.F.'s of any other cells, polarizing or nonpolarizing, can be very easily and very accurately determined by the use of a comparison method in which the first cell is taken as a standard.* We shall consider two such methods.

81. The ballistic galvanometer method. Since the E.M.F. of a cell is the P.D. which it maintains between its terminals on open circuit, it is only necessary to connect the terminals of any cell to the plates of a condenser in order to charge these plates to a P.D.

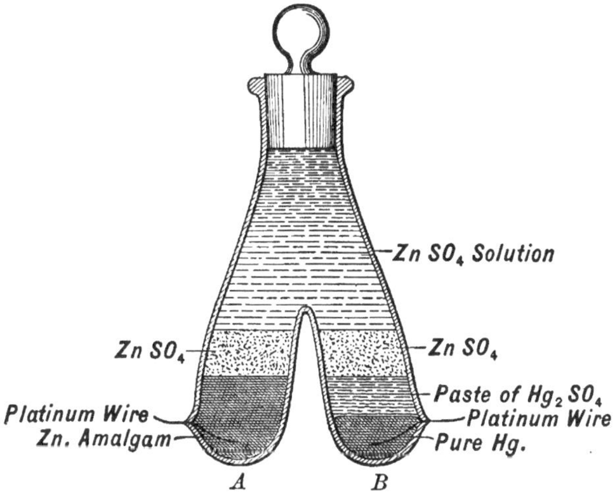

FIG. 90

* The cell most commonly used as a standard for the comparison of E.M.F.'s is some modification of the Clark cell shown in section in Figure 90. When made under certain standard conditions it has an E.M.F. expressed by the following relation where t is the temperature.

$$E = 1.434 \times [1 - .00077\,(t - 15°)].$$

The cell, however, is incapable of supplying an appreciable current without considerable polarization. It is therefore serviceable for accurate work only if kept on open circuit. It will be seen that it may be used according to either of the two methods of this chapter.

equal to the E.M.F. of the cell. Since, further, the quantity of charge on the plates of a given condenser is proportional to the P.D. between them ($Q = C \times PD$), we have only to charge the same condenser successively by means of different cells in order to obtain charges which are strictly proportional to the E.M.F.'s of the cells. Since, finally, the throws of a given ballistic galvanometer are proportional to the charges sent through it ($Q = \frac{Kt\theta_1\rho^{\frac{1}{2}}}{\pi}$, see p. 96), we have, if θ_1 and θ_2 are the respective throws produced by discharging a given condenser through a ballistic galvanometer when the condenser has been charged, first by a cell of E.M.F. E_1, and second by a cell of E.M.F. E_2,

$$\frac{E_1}{E_2} = \frac{\theta_1}{\theta_2}. \quad (1)$$

82. The potentiometer method. The only disadvantage of the preceding method is that it is a deflection rather than a zero method. The following method is just as faultless theoretically, and it can be made accurate to the fifth or sixth place of decimals if temperature and other conditions can be kept constant enough to make such a degree of accuracy desirable. A storage battery B, or any combination of batteries having an E.M.F. which is constant and somewhat higher than that of any of the cells to be compared, is connected as in Figure 91 to the ends of a wire ab of sufficiently high resistance to prevent a current from flowing which is large enough to heat the wire appreciably. The resistance of ab must also be so high in comparison with the internal resistance of B that the P.D. maintained by B between a and

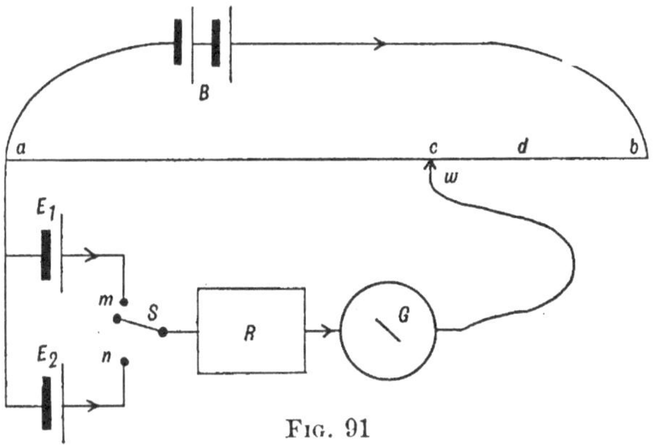

FIG. 91

b is greater than the E.M.F. of either E_1 or E_2, the two cells to be compared. These cells are connected as in the figure, so that their negative terminals are joined to the same point a to which the negative terminal of B is connected, their positive terminals being connected to the contact points m and n of a double switch S, through which either cell can be put into connection with a resistance box R, a galvanometer G, and a wire w which can be touched at any point along the wire ab. The comparison of the E.M.F.'s is made as follows. Suppose that the switch S is turned so as to touch the contact m and thus put the cell E_1 into the galvanometer circuit. If now the free terminal of the wire w were to be touched to any point on ab, for example c, a current would always flow through G and R from right to left provided there were no cell in the circuit $cGRma$. The cell E_1 which is in this circuit, however, tends to force current in the opposite direction, namely from left to right through R and G. If, then, the P.D. which already exists between c and a, when the wire is touched at c, is greater than the E.M.F. E_1, a current will actually flow through G and R from right to left; but if the E.M.F. of the cell E_1 is greater than the P.D. which is maintained between the points c and a by the battery B, then a current will flow through the galvanometer from left to right. If the P.D. between c and a is exactly equal to the E.M.F. E_1, then no current whatever will flow through the circuit of the cell, that is, through G. We have then only to find the point on ab which can be touched by the free end of the wire without producing any galvanometer deflection whatever, in order to obtain the point such that the P.D. between it and a is exactly equal to the E.M.F. of the cell.

Suppose now that the switch S is turned so as to make contact with the terminal n of the other cell. If it is now found that some other point d is the point for which the galvanometer shows no deflection, then, if the wire ab is uniform, we have

$$\frac{E_1}{E_2} = \frac{PD \text{ from } c \text{ to } a}{PD \text{ from } d \text{ to } a} = \frac{\text{length } ac}{\text{length } ad}.$$

While the point of no deflection is being found, the resistance R should be made very large (e.g. 20,000 ohms), for then no appreciable current will flow through the cell circuit, and hence this cell will not polarize, even if it be one of the polarizing kind.

After the point of zero deflection is found, the resistance R may be varied, or in fact entirely removed, without altering the point of balance, for obviously at this point the cell is in exact equilibrium with the P.D. between c and a; that is, it is virtually on open circuit. The only reason for introducing R at all was to prevent the cell from polarizing while the point of balance was being found, and to protect the galvanometer from too violent deflections. Varying the value of R will then alter nothing save the sharpness with which the point of zero deflection can be located.

EXPERIMENT 11

(A) Object. To find the E.M.F. of a Leclanché* cell by comparing it with a Daniell cell by the ballistic-galvanometer method.

Directions. First determine the E.M.F. of the Daniell cell E_1 by the method of Experiment 10. Then set up E_1, the Leclanché cell E_2, the double switch S, the condenser C, the discharge key K, the galvanometer G, and the damping key, in the manner shown in the diagram (Fig. 93). Be very careful in so doing not at any time to short circuit the Leclanché cell, for if it is once allowed to become polarized, it may take it hours to recover entirely its original E.M.F. Take in succession the throws of each cell. From the known E.M.F. of the Daniell cell and the mean of at least six throws, compute the E.M.F. of the Leclanché cell.

* The Leclanché cell, shown in Figure 92, is of wide commercial use for circuits closed only intermittently. It consists of zinc and carbon electrodes in a solution of ammonium chloride (NH_4Cl). The carbon electrode is commonly in a porous cup filled with manganese dioxide (MnO_2). The action of the cell is as follows. The zinc goes into solution in the ammonium chloride in the form of positively charged zinc ions. This gives a positive charge to the solution about the zinc plate, and hence the positive ions already in solution (NH_4) are driven toward the carbon plate. Instead of going out of solution, however, they decompose the water (H_2O) about this plate and form ammonium hydroxide (NH_4OH) and free H ions, which are driven out of solution at the plate in the form of hydrogen gas. The function of the MnO_2 is to dispose of this hydrogen by uniting with it to form Mn_2O_3 (manganic oxide) and water. This depolarizing action of the MnO_2 is quite slow, however, so that the cell polarizes rapidly on short circuit, but recovers completely when left to itself for a few hours.

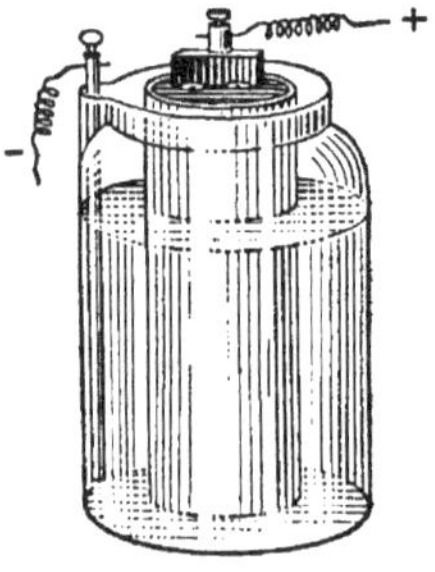

FIG. 92

(B) Object. To compare the E.M.F.'s of the Daniell and Leclanché cells by the potentiometer method.

Directions. Connect as in Figure 91, being very careful not at any time to short circuit the Leclanché cell. Let R be a resistance of from 10,000

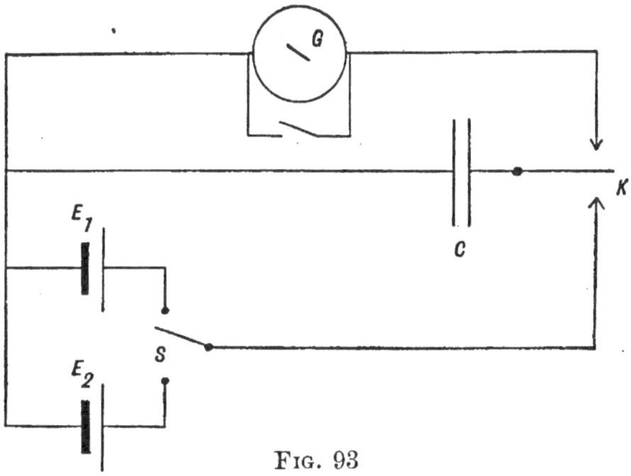

FIG. 93

to 100,000 ohms while the point of balance is being found. If the point of no deflection is not sharply marked, reduce R after the point of balance has been approximately located. Compare results with those obtained in (**A**).

EXAMPLE

(**A**) The voltmeter reading for a Daniell cell was 1.080 volts. The voltmeter resistance was 197.5 ohms and the internal resistance of the cell as found by the first method of section 79 was 2.48 ohms; hence the E.M.F. of the Daniell cell was 1.094 volts. A condenser of .1-microfarad capacity charged by this E.M.F. gave a throw of 3.00 cm. on the galvanometer used. The same condenser when charged by the Leclanché cell caused a throw of 4.28 cm. Hence the E.M.F. of the Leclanché cell $= \frac{4.28}{3.00} \times 1.094 = 1.563$ volts.

(**B**) For the Daniell cell a balance was obtained when ac was 50.06 cm., for the Leclanché cell when ad was 71.10 cm. Hence the E.M.F. of the Leclanché cell $= \frac{71.10}{50.06} \times 1.094 = 1.556$ volts. The values found by the two methods differed by .45 of one per cent.

CHAPTER XII

ELECTRO-MAGNETIC INDUCTION

83. The principle of electro-magnetic induction. Thus far our discussion of electrical phenomena has been centered about Oersted's discovery that an electric current is surrounded by a magnetic field. We now come to the discussion of the applications of an exceedingly important discovery, published for the first time by Faraday in 1831, although the discovery itself was doubtless made a year earlier by Joseph Henry, at that time a high-school teacher in Albany. It is to the discovery of the principle of electro-magnetic induction that we owe the modern dynamo, the induction coil, the transformer, the telephone, and most of the other practical applications of electricity to-day. This principle may be stated as follows: *Whenever a conductor moves in a magnetic field in such a way as to cut lines of magnetic force, an E.M.F. is induced in it.* Thus when the vertical wire *ac*, shown in Figure 94, is moved across the field, an E.M.F. is induced in it, and since, in this case, the circuit is completed by the wire *abc*, this E.M.F. causes a current to flow about the circuit. No current whatever is found to be induced when the conductor moves parallel to the lines of force. These statements can be easily verified by moving a single wire, to the ends of which a sensitive galvanometer is attached, first in a direction perpendicular to, then in a direction parallel with, the field of a strong magnet.

84. Direction of the induced E.M.F. Since energy is expended when an electrical current flows in a wire, it follows from the principle of the conservation of energy that work must be done in inducing this current. Now we have already seen in sections 50 and 51 (p. 78) that any current in *ac* (Fig. 94) interacts with the magnetic field in such a way as to tend to push the wire across the field. If, then, work must be done in inducing the current in *ac*, it follows that the direction of the induced current must be such that the mechanical force with which it is urged across the field constitutes the reaction against which the work is done; i.e. this mechanical force must be opposite to the direction of motion. For if the induced current could be in the opposite direction, the wire would be urged across the field in the direction in which it is going and electrical energy would be created indefinitely without any expenditure of mechanical energy whatever. *The induced*

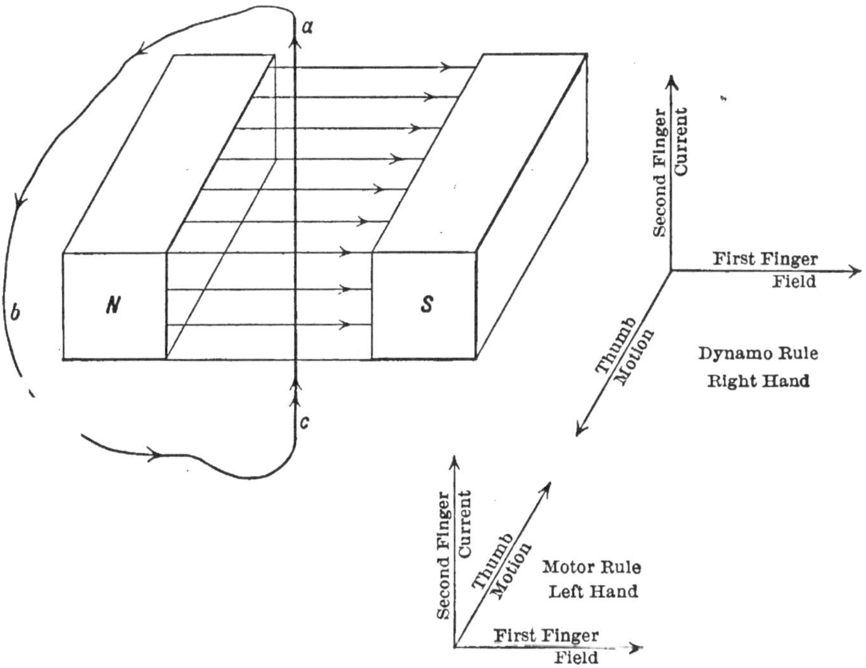

FIG. 94

current must then be in a direction such as to call into play a force which opposes the motion inducing it. This statement, which is seen to follow from the principle of the conservation of energy, is known as Lenz's law.

A current, of course, cannot flow unless the circuit is complete, but an E.M.F. which would cause a current to flow in a direction such as to oppose the motion inducing it, is nevertheless always induced when a conductor cuts lines of force, and this E.M.F. causes positive and negative charges to appear upon the ends of the open circuit just as the E.M.F. of a cell causes such charges to appear upon the terminals of the cell.

It is possible to formulate for induced currents a rule connecting the directions of the motion, of the magnetic field, and of the current, similar in form to the motor rule, but making use of the right hand instead of the left. *Thus if the thumb and the first two fingers of the right hand are extended in directions at right angles to one another, and if the forefinger is made to point in the direction of the magnetic lines and the thumb in the direction of the component of the motion of the conductor which is at right angles to these lines, then the second finger will point in the direction of the induced current.* It will be obvious from a consideration of Figure 94 that this rule is only another form of statement of Lenz's law. From its especial application to the dynamo it is known as the *dynamo rule.*

85. The value of an induced E.M.F. The quantitative expression for an induced E.M.F. may be found by a consideration of Figure 95. A and B represent heavy conducting bars of negligible resistance, between the terminals of which is connected the coil D. The line ab represents a sliding bar, also of negligible resistance. The dots represent the cross section of a uniform magnetic field of intensity $\mathcal{H}$ perpendicular to the plane of the paper, in which this metal framework is supposed to lie. A force F acts on the slider ab, moving it with uniform speed through a small distance dx to a new position $a'b'$. This motion induces an E.M.F. which causes to flow around the circuit $a'Db'$ a current of I absolute units. The reaction of the field $\mathcal{H}$ on the l cm. of length of the slider is a force equal to F and of value $Il\mathcal{H}$ dynes (sect. 50, p. 78). The mechanical work done in moving the coil dx cm. against this reaction is $Il\mathcal{H}dx$ ergs. If E

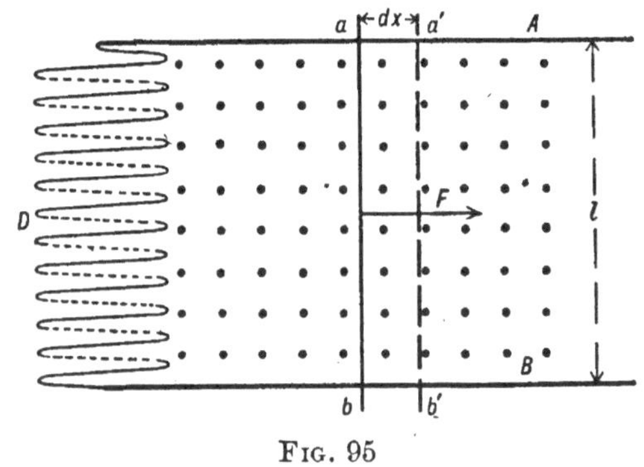

FIG. 95

represents in absolute units the E.M.F. causing this current to flow, and if dt is the time during which the motion is taking place (i.e. the time during which we are considering the flow of current), then the electrical energy expended by the current in this time is $EIdt$ ergs (see Chap. IV). Now since no part of the original mechanical energy is transformed into forms other than electrical (e.g. chemical), these two expressions, the one for the work done and the other for the electrical energy developed, must be equal. That is,

$$Il\mathcal{H}dx = EIdt, \tag{1}$$

or

$$E = \frac{l\mathcal{H}dx}{dt}. \tag{2}$$

Now since ldx represents the area across which the slider has moved, and since $\mathcal{H}$ represents the number of lines per square centimeter, it is clear that $l\mathcal{H}dx$ is the total number of magnetic lines cut in the time dt, and hence that *an induced E.M.F. is numerically equal to the rate at which magnetic lines of force are cut.* That is, *one absolute electro-magnetic unit of E.M.F. is induced in a conductor which cuts one magnetic line of force per second.* Since a volt is 10^8 absolute units, a volt is induced in a conductor when it is cutting lines at the rate of 100,000,000 per second.

The total number of magnetic lines which pass through any area is known as the *magnetic flux* through that area, or simply as the *flux*. Algebraically stated, if Φ represents flux, $\mathcal{H}$ field strength, and A area, then, by definition, in case the direction of the field is normal to the area,

$$\Phi = A\mathcal{H}. \tag{3}$$

In case the field makes an angle θ with the normal to the area, then

$$\Phi = A\mathcal{H}\cos\theta. \tag{4}$$

That is, the flux across any area is the product of the field strength by the projection of this area upon a plane normal to the direction of the field. In the expression for E just found, $l\mathcal{H}dx$ evidently represents the small increment of flux added (algebraically) during the element of time dt to the flux already inclosed by the circuit. Denoting the increment of flux by $d\Phi$, we have

$$E = \frac{d\Phi}{dt}. \qquad (5)$$

That is, *the E.M.F. in absolute units induced in any closed circuit is the rate of change of the included flux.* To be adaptable to this expression, Lenz's law may be restated as follows: *The E.M.F. induced in any closed circuit will cause a current to flow in a direction such as to oppose the change in flux inducing it.*

86. General statement of Lenz's law. As was seen above, the one and sufficient condition necessary for the electro-magnetic induction of an E.M.F. in a conductor is that the conductor shall cut magnetic lines of force. This cutting may take place, however, in a variety of ways, and it is of some importance to deduce

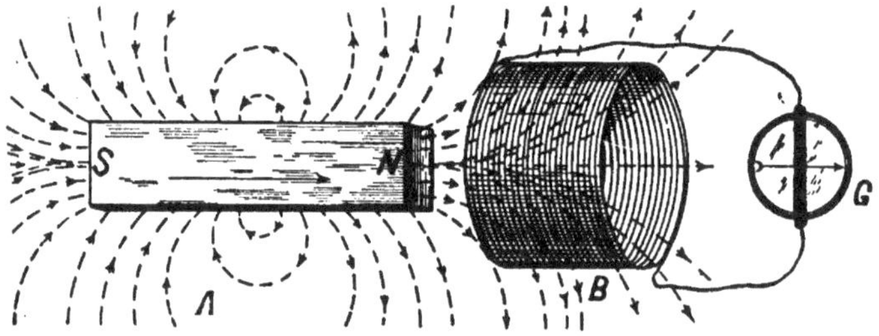

FIG. 96

a statement for Lenz's law which will be applicable to all cases. Thus if instead of moving a wire through a magnetic field, as in the above illustration, we push a magnet NS from left to right up to a coil B, as in Figure 96, some of the lines of force which exist in the space surrounding the magnet, and move with it, are cut by the coil, and consequently a current flows in it in such a direction that the field it sets up about the coil opposes the approach of the magnet; for this case is obviously merely the converse of that considered in the preceding paragraph. Furthermore, pulling the coil away from the magnet must cause a cutting of lines in the opposite direction, and hence a current in such a direction as to attract the receding magnet; that is, to oppose the separation.

Again, if instead of pushing up the magnet *NS* we set near *B* a second coil *A* (Fig. 97), which is connected to the circuit of a battery *D* through a key *K*, and then close *K*, we obviously cause *B* to be cut by the lines of force which spring into existence about *A*, precisely as though we had thrust *NS* up to *B* as above. The induced current in *B* must then be in such a direction as to cause repulsion between *A* and *B*; that is, it must thrust lines of force through *B* in a direction opposite to that in which the rising current in *A* is thrusting lines through *B*. Conversely, when *K* is opened, the current induced in *B* must be in such a direction as to cause attraction between *A* and *B*; that is, in such a direction

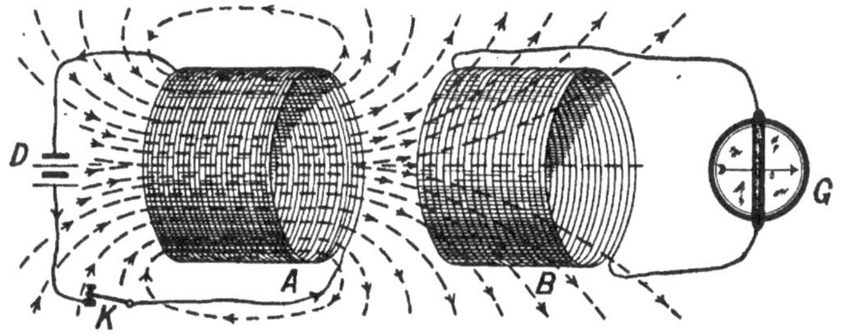

FIG. 97

as to thrust lines through *B* in the same direction as that of the lines which are disappearing from *B* because of the decay of the current in *A*. In its most general form Lenz's law may now be stated thus: *Whenever an E.M.F. is induced by the relative motion of a conductor and magnetic lines, the induced current will always flow in such a direction as to oppose the change which is inducing it.* If the change is a mechanical one, the induced current is in such a direction as to oppose the motion; if the change is a magnetic one, the induced current is in such a direction as to oppose the magnetic change which is taking place.

87. The ideal dynamo. The most practical application of this principle of electro-magnetic induction is to the dynamo, a machine for converting mechanical energy into electrical energy. In its simplest form it consists of a rectangular coil of wire, rotating about the axis of its length in a uniform magnetic field, the direction of which is perpendicular to its axis. Suppose the coil to consist of one turn of wire and to be represented in cross section by the points A and B of Figure 98. The field of strength $\mathcal{H}$, in which it is free to rotate, is represented by a few of its lines. Suppose the coil to be rotated from a position $A'B'$ to that of $A''B''$ through a small angle 2α. Let its mean angular displacement from its original position at right angles to the field be denoted

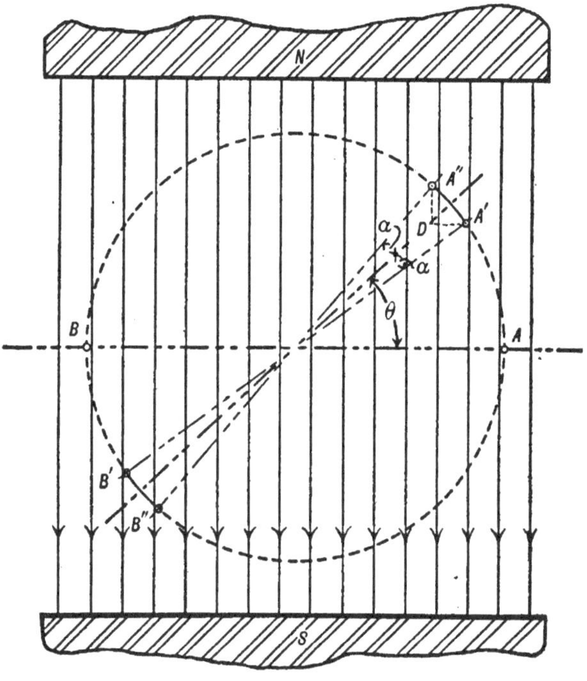

Fig. 98

by θ. The number of lines cut by the wire A is proportional to $A'D$, and if l represents the length of the coil, it is equal to $l\mathcal{H}\,(A'D)$. This expression divided by the time, dt seconds, during which the motion took place, gives the average value of the E.M.F. induced in the wire A when displaced by an angle θ from its original position. Now, since the wire B cuts an equal number of lines in the same time, it has induced in it an equal E.M.F. But this E.M.F. is on the opposite side of the loop and also in the opposite direction; hence it tends to cause a current about the loop in series with that in A; i.e. the total E.M.F. E_θ induced in the given position is twice the amount given above. That is,

$$E_\theta = \frac{2\, l\, \mathcal{H}\, (A'D)}{dt}. \tag{6}$$

This may be put in a different form by the following substitutions. Since the angle $2\,\alpha$ subtended by the chord $A'A''$ is small, the chord may be written for the corresponding arc. Therefore $A'D = A'A'' \times \sin A'A''D$. But angle $A'A''D = \theta$. Therefore, upon substitution,

$$E_\theta = \frac{2\, A'A''\, l\, \mathcal{H} \sin\theta}{dt}. \tag{7}$$

Now $A'A''$, the arc or chord, divided by dt, is the linear speed of the moving wire. Representing the radius of the coil by r and the uniform angular velocity by ω, since linear speed equals angular speed times the radius, we have

$$\frac{A'A''}{dt} = r\omega. \tag{8}$$

This gives

$$E_\theta = 2\, rl\, \mathcal{H}\omega \sin\theta. \tag{9}$$

Now $2\,rl$ is the area of the coil, and when multiplied by $\mathcal{H}$ it is the maximum flux through it, — that is, the flux through the coil when in its original position perpendicular to the field. Representing this flux by Φ, we get from equation (9)

$$E_\theta = \Phi\omega \sin\theta. \tag{10}$$

If the coil consists of N loops in series instead of a single loop, the same E.M.F. is induced in each, and these separate E.M.F.'s must be added to get the total value of E_θ. That is,

$$E_\theta = N\Phi\omega \sin\theta. \tag{11}$$

This is the E.M.F. in absolute units. Expressed in volts it is

$$E_\theta = N\Phi\omega \sin\theta\, 10^{-8}. \tag{12}$$

Since, for any given case, N and Φ are constants, if ω, the angular velocity, is kept constant, the induced E.M.F. varies directly as the sine of the angular displacement of the coil from a position at right angles to the field. A method for showing this relation will now be described.

88. Experimental analysis of dynamo induction. The ideal dynamo shown in Figure 99 is arranged so that at every revolution of the crank the ratchet is lifted, allowing a spring to rotate the coil through 10°. The terminals of the coil are connected to two separately insulated copper rings attached to the shaft. Copper strips or "brushes" bearing upon this ring are connected to a moving-coil galvanometer. As the coil moves through any ten-degree interval, an E.M.F. is induced which corresponds to the mean angular displacement of the coil from its original position. Since the total resistance through which this E.M.F. causes a current to flow remains the same throughout the experiment, the quantity of electricity passing through the galvanometer, and therefore the galvanometer deflection, is proportional to this E.M.F. If

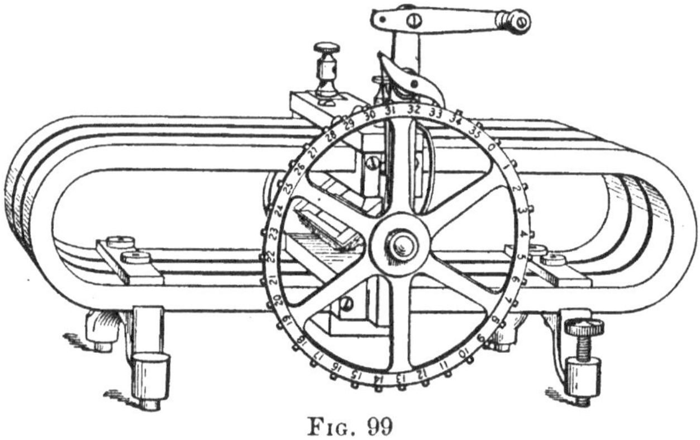

FIG. 99

the observed deflections are plotted as ordinates and the corresponding values of the angular displacements as abscissas, the smooth curve drawn through these points will in general have practically a sine form.

89. The molecular theory of magnetism. The phenomenon of an induced E.M.F. as a result of the cutting of magnetic lines may be made use of to determine the magnetic condition of iron, for example the distribution of magnetism in a bar magnet. Before considering the method of doing this, it is desirable to consider the general nature of magnetism.

It is a matter of experimental knowledge that into however small parts a magnet is divided each part still retains magnetic properties. The assumption which naturally follows is that the molecules of magnetic metals are themselves magnets. This hypothesis offers an easy explanation of the phenomenon of induced magnetism, — that is, the phenomenon of the acquisition of magnetic properties by a bar of iron, or of steel, as a result of its presence in a magnetic field. For we have only to assume that

Fig. 100

in an unmagnetized bar the molecular magnets have various orientations, so that their magnetic effects neutralize one another, and consequently the bar as a whole shows no magnetic properties. It is not improbable that the molecular magnets are arranged in small groups, as shown in Figure 100, the magnets of each group being in equilibrium. When the bar is under the influence of a magnetic field this condition of equilibrium is disturbed, and

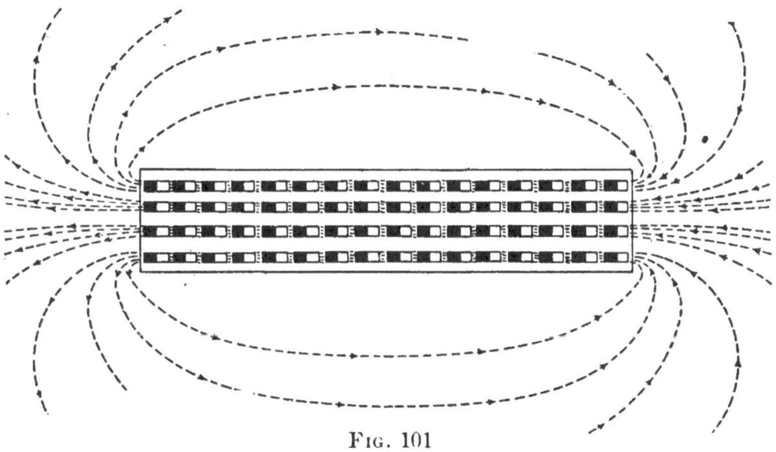

Fig. 101

equilibrium does not again exist until some, at least, of the molecules have formed new groupings, oriented in the general direction of the impressed magnetic field. If the field is very intense, the alignment might perhaps be well-nigh perfect, as in the ideal diagram of Figure 101, which corresponds to a condition of entire *saturation;* but in an ordinary magnet the alignment would be very imperfect, and doubtless, also, some of the closed molecular groups would still exist. Figure 102 is an attempt to picture something more or less similar to this condition. When the directive force of the field is withdrawn, some of the molecules probably form new closed groups, but a large number retain the direction into which they have been rotated by the field. The fact that causes which facilitate a rearrangement of the molecules, such as heating or jarring by blows, facilitate also either the induction of magnetism in a bar placed in a magnetic field, or its demagnetization when withdrawn from the field, may be taken as evidence in favor of this hypothesis as to the molecular nature of magnetism.

90. The distribution of magnetism. Since through the space surrounding a magnet the magnetic lines of force are known to pass from its *N* end to its *S* end, it is in accordance with the

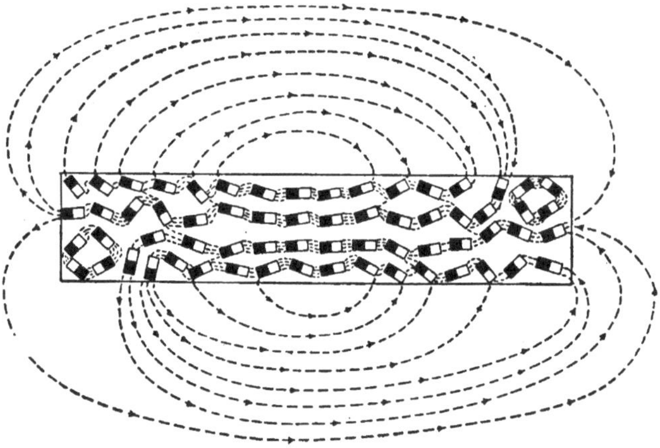

Fig. 102

molecular hypothesis to assume that these lines form closed curves passing through the iron from the *N* end of each molecule to the *S* end of the next molecule (see Fig. 101). Wherever, then, lines of force *leave* the large magnet there are free molecular *N* poles, i.e. *N* poles of molecules not in immediate conjunction with neighboring *S* poles (see Fig. 102). And similarly, wherever magnetic lines *enter* the large magnet there are free *S* poles. In the case of a bar magnet these lines of force leave the magnet along the entire *N* half and enter it along the entire *S* half after the manner shown in Figure 102. The distribution of these lines may then be taken as a measure of the distribution of magnetism along the bar. Within the bar the actual number of lines is of course greatest at the center where the largest number of molecules are aligned in the direction of the axis of the magnet. But the magnitude of any action at a point in the immediate neighborhood of the bar depends upon the number of magnetic lines entering or leaving the bar at that point, i.e. upon the distribution of magnetism. The distribution is not uniform along either half of the bar, and except for a bar which is carefully magnetized it is also asymmetrical with respect to the center of the bar.

91. Experimental determination of the distribution of magnetism. The distribution of magnetism in a bar magnet may be found by the use of a "test coil." This is a small coil *c* (Fig. 103)

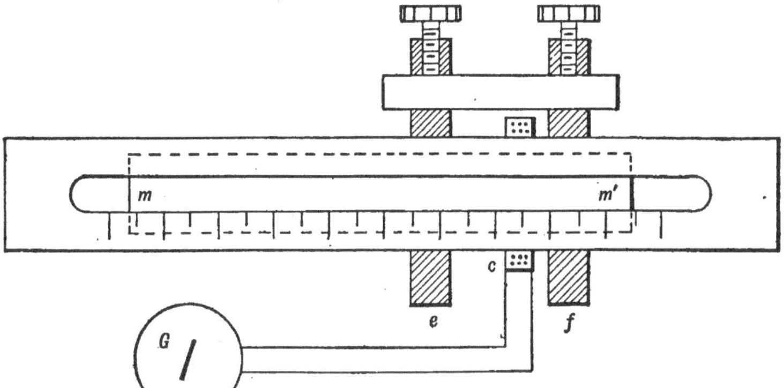

FIG. 103

which fits closely over the bar and which may be slipped along it, thus cutting the lines which enter or leave the magnet. The terminals of the coil are connected to a moving-coil galvanometer, the deflections of which are proportional to the quantity of induced electricity which flows through the galvanometer, and hence to the number of lines entering or leaving the magnet in the space over which the coil has moved. These deflections are plotted as ordinates with a direction up or down depending on the direction of the galvanometer throw (see Fig. 104), the abscissas corresponding to these ordinates being the distances from one end of the magnet to the middle of the space over which the test coil is moved. A smooth curve is then drawn through these points. The intersection of this curve with the axis of distances gives the distance from one end of the bar to the point of no free magnetism, — that is, to the magnetic center of the bar.

92. Location of the poles in a bar magnet. Now the moment of force which would act upon the magnet if placed at right angles to a uniform magnetic field is by definition proportional to the magnetic moment of the magnet (p. 18). It is the sum of the

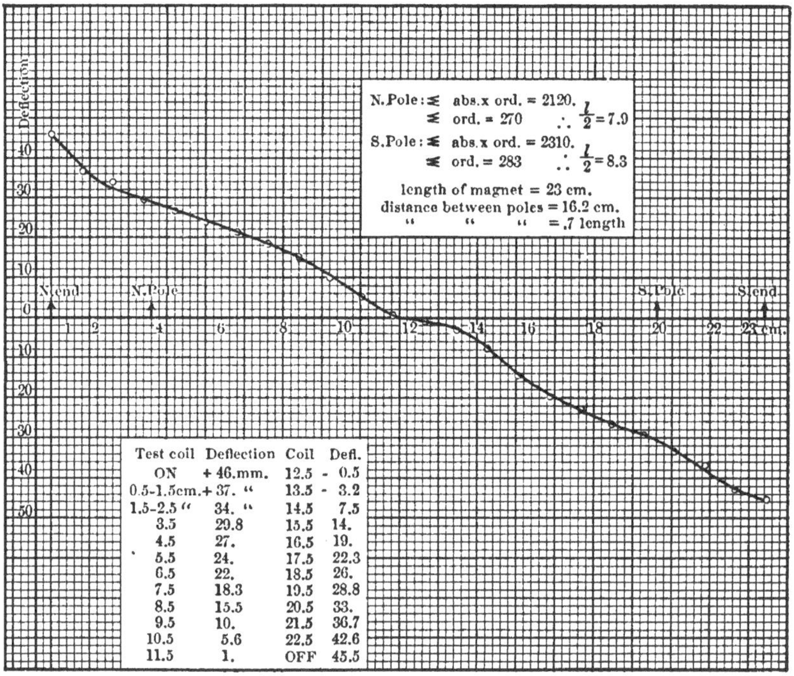

Test coil	Deflection	Coil	Defl.
ON	+ 46.mm.	12.5	- 0.5
0.5-1.5cm.	+ 37. "	13.5	- 3.2
1.5-2.5 "	34. "	14.5	7.5
3.5	29.8	15.5	14.
4.5	27.	16.5	19.
5.5	24.	17.5	22.3
6.5	22.	18.5	26.
7.5	18.3	19.5	28.8
8.5	15.5	20.5	33.
9.5	10.	21.5	36.7
10.5	5.6	22.5	42.6
11.5	1.	OFF	45.5

Fig. 104

moments due to all the separate quantities of magnetism in the bar; that is, it is proportional to the sum of the products of all these quantities of magnetism by their lever arms taken with respect to an axis at the magnetic center of the bar. Now the ordinates of the curve are proportional to the quantities of magnetism. Therefore a number proportional to the magnetic moment may be found by adding all the products of these ordinates by their corresponding lever arms as found from the curve. The distance from the point of intersection of the curve with the axis of abscissas to the foot of any ordinate is the lever arm corresponding to the quantity represented by that ordinate. But by definition the magnetic moment is the product of the pole strength of the magnet and the distance between the poles. Now the sum of all the ordinates on one side of the magnetic center is proportional to the pole strength. Therefore the distance between the poles may be found by dividing the number found as above to be proportional to the magnetic moment by the number proportional to the pole strength. This distance between the poles of a magnet may be any fraction whatever of the length of the bar. In a permanent magnet it never exceeds five sixths of the length and is usually much less than this.

EXPERIMENT 12

(A) Object. To plot the curve showing the variation of the E.M.F. induced in the coil of an ideal dynamo with the angular displacement of the coil.

Directions. By connecting to the galvanometer terminals a battery which is short circuited through a piece of copper wire, find the direction of the current which corresponds to a given direction of deflection. Connect the galvanometer to the dynamo and allow the coil of the latter to turn through a ten-degree interval. Applying to this motion the dynamo rule, note the verification of the rule. Then following the method outlined in section 88, note the deflections of the galvanometer for ten-degree intervals during one complete revolution of the dynamo coil. Choose as the axis of abscissas a line about the middle of the sheet of coördinate paper and in the direction of its length. Choose the scale for plotting the curve as large as the size of the sheet will allow. Tabulate the observed throws and the corresponding angular displacements in one corner of the sheet and let this sheet be the record of the experiment (see Fig. 105).

(B) Object. To plot the curve representing the distribution of the magnetic lines leaving a bar magnet and to find the actual distance between the poles of the magnet.

Directions. The magnet *mm′* to be used is contained in a slotted brass tube on which a scale is ruled or pasted (see Fig. 103). The test coil *c* may be moved suddenly between the stops *e* and *f* of the frame. These should be so set as to admit of a motion of the coil of about 1 cm. For convenience of manipulation this distance may be made an even submultiple of the length of the magnet (e.g. $\frac{1}{16}$).

For the first deflection set the frame so that *e* and *f* are equidistant from the right end of the magnet. Now remove the stop *f* and the test coil from the tube. Then note the deflection caused by suddenly slipping the test coil back over the end of the magnet to a position against the stop *e*. This motion cuts all the lines that leave the end of the magnet, and the corresponding deflection is therefore a measure of the magnetism at the end of the bar. Plot it from the point on the X axis of your diagram which corresponds to the end of the bar. Replace the stop *f* and shift the frame until when the coil is against *f* it is in exactly the same position as when it was against *e* before. Place *e* so that the coil moves, say, 1 cm. between stops, then move it quickly from *f* to *e* and plot the throw at the point on the X axis which corresponds to the mid-point of its motion.

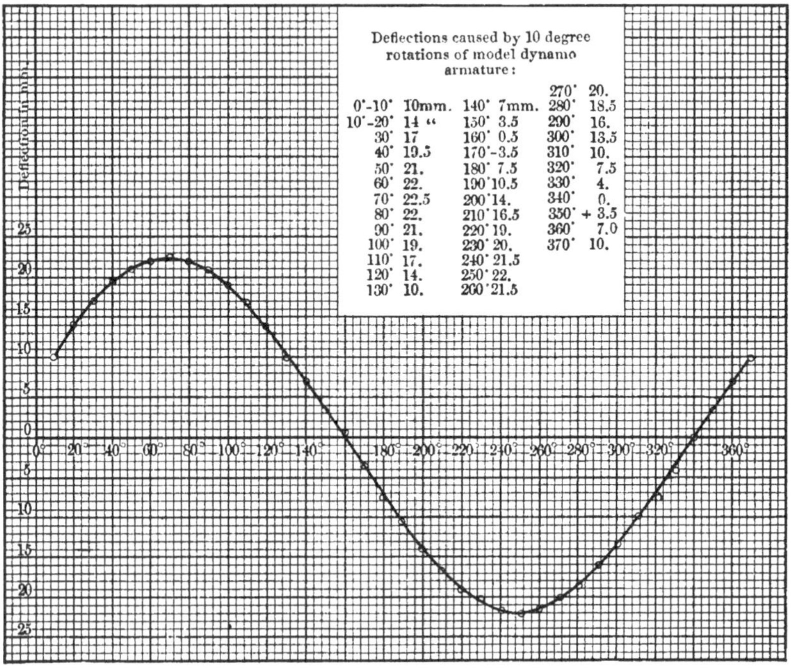

FIG. 105

Following the method outlined in section 91, continue until the opposite end of the magnet is reached. Then remove the stop *e* and take the deflection caused by slipping the test coil entirely off the tube.

Plot the curve showing the distribution of magnetism, utilizing in so doing the full size of the sheet of coördinate paper. Tabulate upon the same sheet the values used in plotting the curve. Also indicate on this sheet the sum of the products of the ordinates by their corresponding abscissas, the sum of the ordinates, and the value determined for the distance between the poles. Let the curve and these tabulated values be the record for the experiment (see Fig. 104).

CHAPTER XIII

THE CONSTANTS OF THE EARTH'S MAGNETIC FIELD

93. Direction of the earth's magnetic field. A magnetic needle placed in the earth's magnetic field, and free to rotate only in a horizontal plane, i.e. about a vertical axis, is not found, in general, to set itself exactly in coincidence with the geographical meridian. The angle by which the direction of the needle differs from the geographical north-and-south line is known as the *variation* at the point at which the needle is placed.

Again, if the needle is free to rotate about a *horizontal* axis, it is not found to assume a horizontal position, but in the northern latitudes the N end dips downward. The angle which the needle makes with the horizontal at any point is known as the *magnetic dip* at that point. For most localities within the United States the angle of dip is between 60° and 75°.

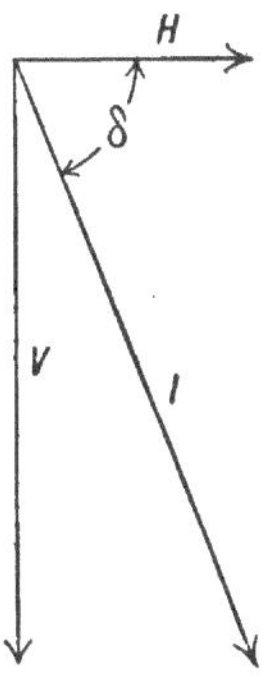

FIG. 106

Let the intensity of the earth's field be denoted by I. Since, in most of the instruments in which a magnetic needle is employed, such, for example, as the tangent galvanometer, the needle is suspended so as to be free to turn only in a horizontal plane, it is usually only the component of I parallel to the horizon which it is necessary to determine. In the experiment which follows, however, we shall measure both the horizontal component H and the vertical component V of the earth's field, and from these measurements we shall deduce the absolute intensity I of this field in the direction in which it acts, and the angle δ which the direction of this field makes with the horizontal. The relations which exist between these four quantities, and by means of which these deductions can be made, are obviously the following (see Fig. 106):

$$H = I \cos \delta. \tag{1}$$

$$V = I \sin \delta. \tag{2}$$

$$\frac{V}{H} = \tan \delta. \tag{3}$$

The quantities H, V, and δ, with the variation, are known as the *magnetic constants* of any given locality.

94. Absolute measurement of *H* and *V* by means of an earth inductor. The most rapid and convenient and, on the whole, one of the most satisfactory methods of measuring H or V is based upon the principle of electro-magnetic induction. It requires a ballistic galvanometer, a resistance box, and a "test coil" known as an "earth inductor" (see Fig. 107). Such an instrument consists simply of a circular coil of large area and large number of turns, so mounted that it may be rotated very suddenly through 180°. The principle involved is that stated in section 85 (p. 130), namely, that the E.M.F. induced in a conductor is equal to the rate at which the conductor cuts lines of magnetic force. Thus consider the coil of the earth inductor to have an average area of a square centimeters and to be composed of n loops of wire. Let it be placed so that at the beginning of the motion its plane is at right angles to the horizontal component of the earth's field. In Figure 108 the heavy line represents the coil in this position. As then it rotates through one fourth revolution about a vertical axis lying in the plane of the coil, each loop cuts Ha lines. Its position is then represented by the dotted lines of the figure. In the next one fourth revolution each loop

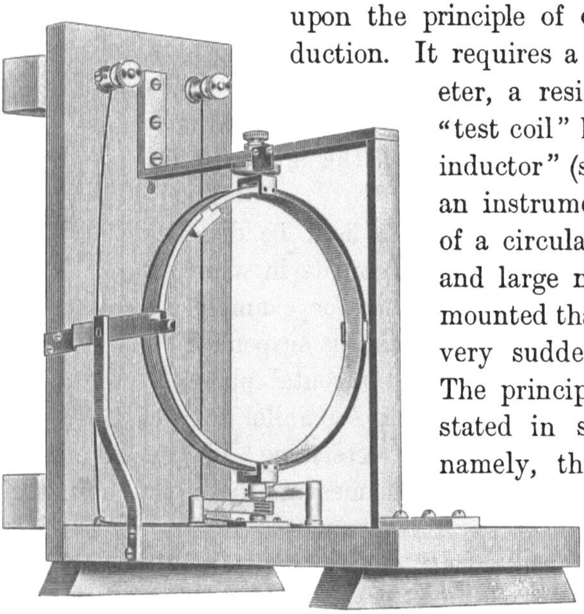

FIG. 107

also cuts Ha lines, but in such a manner that the E.M.F. induced is in the same direction as that induced by the first one fourth revolution. In one half revolution, then, each loop cuts $2\,Ha$ lines. If τ represents the time of this half revolution of the coil, the average E.M.F. induced in each loop is $\frac{2\,Ha}{\tau}$. Hence the average E.M.F. for the coil of n loops is $\frac{2\,Han}{\tau}$. The product an will be denoted by A and known as the total area of the coil. If the terminals of the coil are connected to a ballistic galvanometer circuit such that the total resistance of the circuit is R C.G.S. units, this E.M.F. causes to flow a current of average value $\frac{2\,HA}{R\tau}$. The quantity Q of electricity caused to pass through the galvanometer is τ times this current. Hence

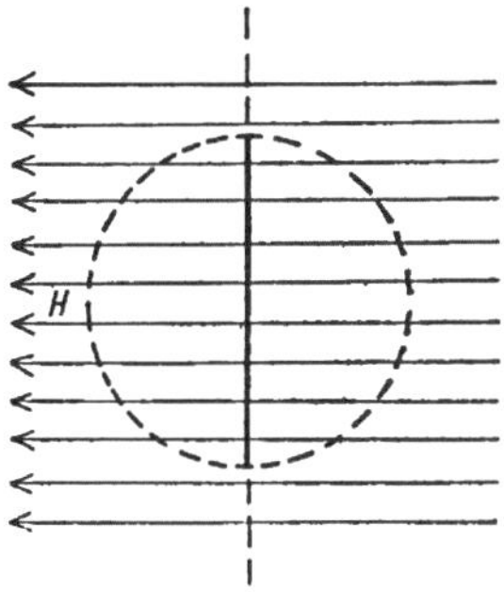

FIG. 108

$$Q = \frac{2\,HA}{R} \text{ C.G.S. units.} \tag{4}$$

If the time τ is so small, as compared with the period of the galvanometer, that this quantity Q may be assumed to have passed before the coil turns appreciably, then Q may be obtained from the throw θ_H and the galvanometer constants, as explained in section 61 (p. 93). Thus $Q = \frac{Kt\theta_H \rho^{\frac{1}{2}}}{\pi}$. From this and equation (4) the value of H is obtained in the form

$$H = \frac{RQ}{2\,A} = \frac{RKt\theta_H \rho^{\frac{1}{2}}}{2\,\pi A}. \tag{5}$$

Both ρ, the damping factor, and t, the half period of the galvanometer, must be determined under the conditions prevailing in the circuit at the time of the observation of θ_H; for the ratio $\rho\,[=\theta_1/\theta_2]$ will not have at all the value which it had in Experiment 8, where the galvanometer was swinging on open circuit. In the present case the circuit of the galvanometer coil is at all times closed through the resistance R. It is desirable to make R

so large, through the insertion of resistance boxes into the circuit, that the coil will make at least ten or fifteen swings before coming to rest. Otherwise neither ρ nor t can be determined with sufficient accuracy.

If the coil is turned over so that in its original position its plane is perpendicular to the vertical component of the earth's field, a similar set of observations will of course give the value of V. Thus

$$V = \frac{RKt\theta_V \rho^{\frac{1}{2}}}{2\pi A}. \tag{6}$$

If we do not care to determine either H or V absolutely, but wish to know simply the value of the angle of dip δ, we have only to take the throws θ_H and θ_V corresponding to the motions of the coil which cause it to cut the horizontal and vertical components respectively. We obtain, then, from equations (5) and (6),

$$\frac{V}{H} = \frac{\theta_V}{\theta_H}, \tag{7}$$

and from equation (3),

$$\delta = \tan^{-1}\frac{\theta_V}{\theta_H}. \tag{8}$$

EXPERIMENT 13

(A) Object. To find the angle of dip.

Directions. Connect in series the ballistic galvanometer, a resistance box, and the earth inductor, the last instrument being placed in the position occupied by the tangent galvanometer in Experiment 3. Adjust the resistances of the box until the throw produced by cutting the horizontal component of the earth's lines is from three to six centimeters. Take the mean of six throws, three to the right and three to the left, as the correct value of θ_H. Take similarly six observations upon θ_V. Then compute δ from equation (8) above. Change the value of the resistance in the box and see how well you can duplicate the ratio θ_V/θ_H.

(B) Object. To determine the absolute values of V and H.

Directions. Insert as much as 10,000 ohms into the circuit. Set the coil so as to cut the vertical component of the earth's field and take observations upon θ_V, t, and ρ.

Measure the resistance of the earth inductor and the galvanometer, either by the bridge method or by determining the E.M.F. of a Daniell or a storage cell and observing with a milliammeter the current which this cell sends through the inductor and the galvanometer in series (see Fig. 63,

p. 84). Either determine K or take its value from the results of Experiment 7. Measure the mean area a of the inductor and multiply by the number of turns to find A. Compute V from equation (6), remembering that R in C.G.S. units is R in ohms $\times 10^9$, and that K in C.G.S. units is K in amperes $\times 10^{-1}$.

Having found V, compute H by multiplying V by the value of θ_H/θ_V found in (A).

From V and δ compute I.

EXAMPLE

(A) The earth inductor was placed in the position occupied by the tangent galvanometer in Experiment 3. When there was no additional resistance in the galvanometer and earth-inductor circuit, the throws corresponding to V and H were 20.51 cm. and 6.53 cm. Hence $\tan\delta = 3.139$. When there was 4000 ohms additional resistance, making the total 4919 ohms, the throws were 6.35 and 2.00 cm. Hence $\tan\delta = 3.175$. The mean value of 3.157 corresponds to an angle of 72° 28′ 30″, the angle of dip.

(B) Making R now equal to 10,919 ohms, and placing the scale 146.7 cm. from the galvanometer mirror, an average throw of 3.30 cm. was observed when the vertical component of the earth's field was being cut. Hence $\theta_V = .01125$ radians. The period and damping factor of the galvanometer found under these conditions were $t = 3.20$ seconds and $\rho^{\frac{1}{2}} = 1.103$. The constant K was found to be, in amperes per radian, .00001055. The average area of the earth inductor was 208.7 sq. cm. and there were 600 turns. Hence

$$V = \frac{10919 \times 10^9 \times .00001055 \times 10^{-1} \times 3.20 \times .01125 \times 1.103}{2\pi \times 600 \times 208.7} = .5823.$$

Also $$H = \frac{V}{\tan\delta} = \frac{.5823}{3.157} = .1845 \quad \text{and} \quad I = \frac{V}{\sin\delta} = .6106.$$

The value of H as found by the earth inductor agreed, therefore, to within .6 per cent with the value .1856 determined in Experiment 2, for the same locality, by using a magnetometer.

CHAPTER XIV

SELF-INDUCTION

95. Nature of self-induction. In section 83 (p. 126) the necessary and sufficient condition for the electro-magnetic induction of an E.M.F. in a conductor was found to be the cutting of magnetic lines of force by the conductor. It follows, then, that in general any conductor which lies in a magnetic field will have an E.M.F. induced in it by any change in the intensity of this field, for such a change must cause lines of force either to appear or to disappear, and in so doing they will, in general, sweep across the conductor. Now, since a straight conductor carrying a current lies in its own magnetic field (see Fig. 109, which represents a section of the field about a conductor carrying a current into the plane of the paper), and since the strength of this field varies as the current varies, it is to be expected that an induced E.M.F. will be set up in every element of this conductor by any change in the current which it carries. This inference is completely confirmed by experiment. We may picture this so-called E.M.F of self-induction to arise as follows: when the field shown in the figure collapses, i.e. when the current in the wire dies out, the circular lines of force may be thought of as shrinking to points at the center of the wire, and in so doing sweeping across the conductor from outside to inside. Similarly, when the current rises we may imagine these lines as springing from the center outward, and in so doing sweeping through the conductor from inside to outside. In each case, in accordance with Lenz's law, the induced E.M.F. must be in such a direction as to oppose the change which is inducing it, i.e. in such a direction as

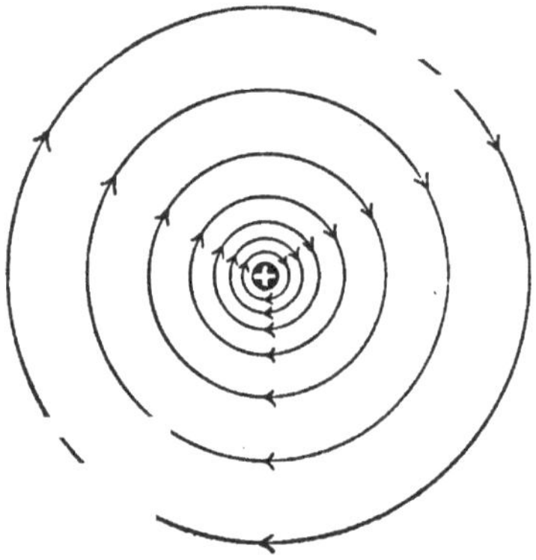

FIG. 109

to oppose the growth of the current when the latter is rising and to retard its decay when it is falling. We may reach this conclusion also by the direct application of the dynamo rule, remembering that the direction in which the conductor is cutting the lines is opposite to the direction in which the lines are moving across the conductor.

It will be seen that the self-induction of an electrical current is exactly analogous to the inertia of a mechanical system, for, just as inertia is the property of such a system by virtue of which it resists any attempt to change its condition of rest or of motion, so self-induction is the property of an electrical system by virtue of which it resists any attempt to change its existing state.

96. Self-induction in various forms of circuit. Since self-induction is due to the cutting of the circuit by its own magnetic lines, it will be seen that we have it in our power to increase or decrease it almost at will, for we can easily give the circuit such a form that its own magnetic lines will sweep across it many or few times.

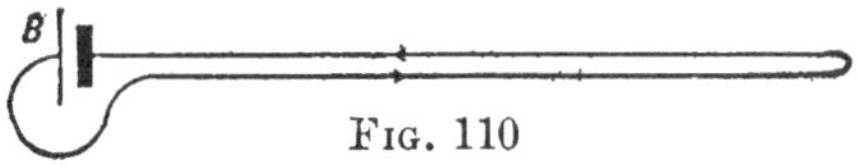

FIG. 110

Thus if we wish to remove self-induction entirely from a circuit, we have only to arrange it, as in Figure 110, so that the outgoing and incoming currents from the generator *B* travel over practically the same path. In this case there will be no magnetic field whatever about the conductor, since the fields due to the outgoing and incoming currents are everywhere equal and opposite. Otherwise stated: if the current changes, the E.M.F. induced in each element of the outgoing conductor by the cutting of its own lines is exactly neutralized by the opposite E.M.F. due to the cutting of the oppositely directed lines of the adjacent element in the other conductor. The coils used in resistance boxes are always made noninductive in this way. The wire is first doubled on itself and then wound on the spool as a double strand, in the manner shown in Figure 111.

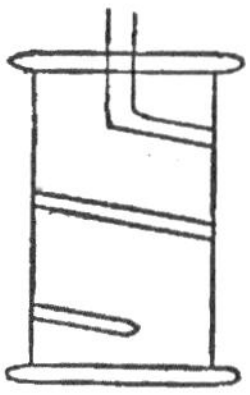
FIG. 111

On the other hand, if we wish to make a circuit of very large self-induction we have only to arrange it in the form of a closely packed coil of many turns, through all of which the current passes in the same direction (see Fig. 112), for in this case the lines of force which the rise of the current in each loop would thrust through that loop are also thrust through all the other loops, so that the E.M.F. of self-induction which opposes the rising current is enormously increased. If, for example, the coil has 100 turns, the lines of force due to any one turn sweep across the entire 100 turns. Across any turn, then, there sweeps 100 times the number of lines which it itself produces, and therefore, other things being equal, there is induced in each turn 100 times the E.M.F. which would be induced in a single turn. But since there are 100 turns in series, the E.M.F. of self-induction in the whole coil is 100^2 times as large as it would be for a coil of a single turn.

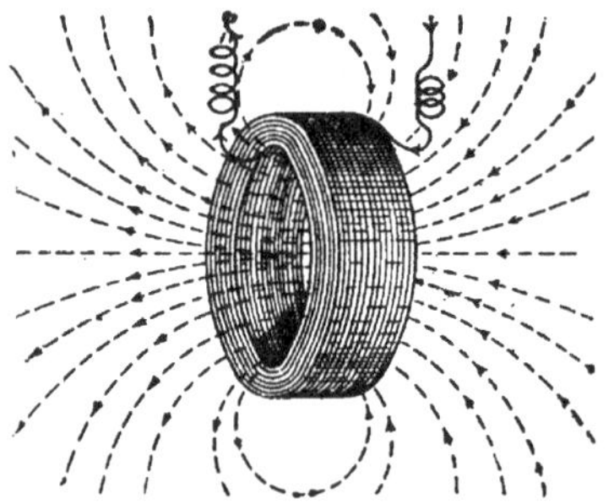

FIG. 112

97. Units of self-induction. It is evident from the last section that the total number of cuttings by lines of force which a given conductor undergoes because of the appearance or disappearance of a given current within it depends upon the form of the conductor. For a given conductor it is also directly proportional to the current, since the total number of lines, i.e. the magnetic field strength, about the conductor is proportional to the current within it. The factor by which the current must be multiplied to give the number of cuttings is then a constant of the conductor. It is called *the coefficient of self-induction* of the conductor. If L represents this coefficient for a given conductor, I the current flowing through it, and N the total number of cuttings by lines of force which the conductor experiences when this current is stopped, then the definition of the coefficient of self-induction is given by the equation

$$N = LI \qquad \text{or} \qquad L = \frac{N}{I}. \tag{1}$$

In words, *the coefficient of self-induction of a conductor may be defined as the total number of times that this conductor is cut by lines of force when the current which it carries undergoes an increase or a decrease of one unit.*

From equation (1) it is obvious that L is 1 when I is 1 and N is 1; i.e. the absolute unit of self-induction is the self-induction of a conductor which is cut once by a line of force when the current within it undergoes a change of one absolute electro-magnetic unit of current. The practical unit of self-induction, the *henry*, named in honor of the American physicist Joseph Henry (1799–1850), is taken as 10^9 times this unit. It is the self-induction of a conductor which is cut by lines of force 10^9 times when the current within it undergoes a change of one absolute unit, or by 10^8 lines of force when the current undergoes a change of one ampere (see also sect. 98). If this cutting takes place in just one second, the mean E.M.F. of self-induction developed is evidently one volt (see sect. 85, p. 129).

98. The measurement of self-induction. One of the simplest ways of determining the coefficient of self-induction of a coil of wire for which this coefficient is not too small is the following. The coil S (Fig. 113), the coefficient of which is sought, is joined in series with a variable noninductive resistance r_1, and S and r_1 are then made one arm of a Wheatstone's bridge, the other arms of which, M, O, and T, are also noninductive resistances. After suitable values have been given to M, O, and T, r_1 is adjusted until, when K_1 is first closed and then K_2, the galvanometer shows no deflection. This is of course merely the balancing of resistances for steady currents, such as has already been done in connection with the Wheatstone's bridge. If, however, while K_2 is closed, K_1 is suddenly opened, the condition of steady currents no longer holds. Since r_1, T, O, and M are noninductive resistances, the decaying current can cause no cutting of lines by these conductors. In the case of S, however, the interruption of the current causes the field about S to collapse, and there is thus induced in S

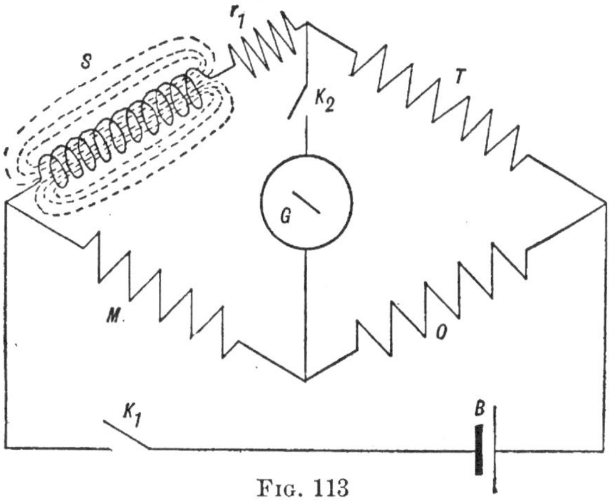

FIG. 113

an E.M.F. which causes a momentary current to flow through the circuit. This circuit is composed of the series resistances S, r_1, M, and the divided circuit, one branch of which is G and the other $T+O$. If we let R represent the total resistance of this divided circuit, then the total resistance of the circuit through which the self-induced current flows is $S + r_1 + R + M$. If τ represents the time required for the disappearance of these magnetic lines in S, then, since the induced E.M.F. is numerically equal to the rate of cutting of these lines of force, we have

$$\text{the mean E.M.F. of self-induction} = \frac{N}{\tau} = \frac{LI}{\tau}. \qquad (1)$$

$$\therefore \text{ the mean current of self-induction} = \frac{LI}{\tau(S+r_1+R+M)}. \qquad (2)$$

If then Q_1 represents the total quantity of electricity which flows through the circuit when these LI lines of force cut it because of the stopping of the current I, we have, since quantity is the product of current by time,

$$Q_1 = \frac{LI}{S+r_1+R+M}. \qquad (3)$$

The fraction of this quantity which passes through the galvanometer G is of course determined solely by the character of the conductors T, O, and G.

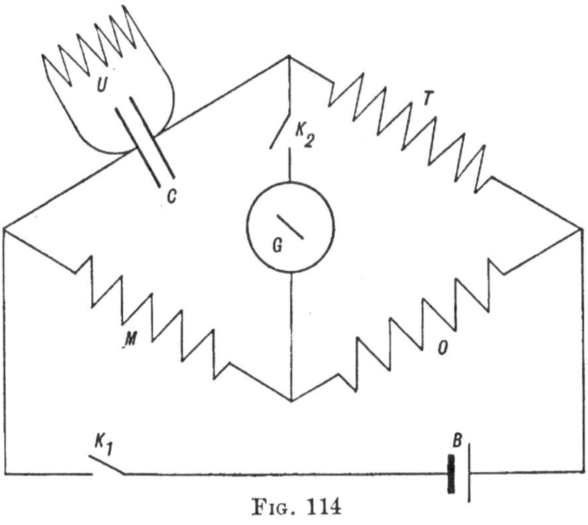

Fig. 114

The coil S is now replaced by a condenser C (Fig. 114), across which is shunted a noninductive resistance U. U is then given such a value that a balance is again obtained when K_1 is first closed, then K_2. This means, of course, that $U = S + r_1$. The quantity Q of electricity which is now upon the condenser plates is the capacity C of the condenser times the

P.D. to which it is charged. But this P.D. is by Ohm's law $U \times I$. When now K_1 is opened, K_2 being kept closed, this quantity UCI discharges partly through U and partly through R and M. By the law of parallel connections (see eq. 11., p. 83), the fraction Q_2 of Q which follows the latter path is given by

$$Q_2 = Q\frac{U}{U+R+M} = \frac{U^2CI}{U+R+M}. \tag{4}$$

Since the fraction of Q_2 which passes through the galvanometer is the same as the fraction of Q_1 which passed through it in the discharge of the coil, and since these fractions are proportional to the two throws θ_1 and θ_2 of the galvanometer, we have

$$\frac{\theta_1}{\theta_2} = \frac{Q_1}{Q_2} = \frac{LI}{S+r_1+R+M} \times \frac{U+R+M}{U^2CI}; \tag{5}$$

or, since $S + r_1$ has been made equal to U, we have

$$L = U^2C\frac{\theta_1}{\theta_2}. \tag{6}$$

We have then but to know the shunt resistance U, the capacity C, and the two throws θ_1 and θ_2, in order to determine the coefficient of self-induction L.

If U and C are expressed in absolute electro-magnetic units, then L will also be expressed in absolute units of self-induction. If U and C are expressed in practical units, i.e. U in ohms and C in farads, then L will be expressed in practical units of self-induction, i.e. in henrys. Since an ohm is 10^9 and a farad 10^{-9} absolute units, it is clear from equation (6) that a henry is 10^9 absolute units; i.e. that a henry is the self-induction of a conductor which is cut by 10^8 lines of force when there is a change of 1 ampere in the current carried by it (see also sect. 97).

99. Practical illustrations of self-induction. One of the most striking experimental proofs of the existence of an E.M.F. of self-induction is the following. If a 100-volt lamp L is shunted as in Figure 115 across an electro-magnet U, which is energized by a 25-volt or even a 50-volt battery, the lamp will not glow appreciably so long as the key K is closed, but when K is opened the

E.M.F. of self-induction will force a large enough momentary current through the lamp to make it glow brightly. Furthermore, if K is opened very suddenly, as by a blow struck with a mallet, the brilliancy of the momentary glow will be very greatly increased. The experiment furnishes very satisfactory demonstration of the fact that an induced E.M.F. is proportional to the *rate* of cutting of lines of force. For when K is opened slowly the E.M.F. of self-induction causes a spark to jump between the opening contact points, and as these points draw farther and farther apart, this spark is drawn out into an arc of gradually increasing resistance. Since, on account of the formation of this arc, the current dies out slowly rather than suddenly as K is opened, the induced E.M.F. is small. Increasing the suddenness of the break decreases the time required for the magnetic field to collapse, and hence increases the *rate of cutting* of the circuit by lines of force; i.e. it increases the E.M.F. of self-induction.

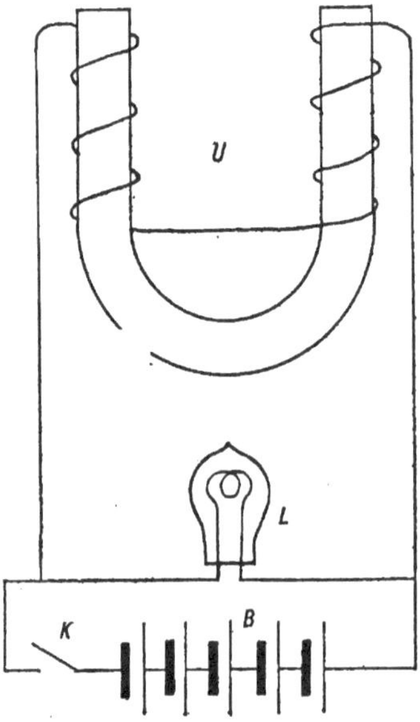

FIG. 115

100. The induction coil. One of the finest illustrations of self-induction is found in the operation of an induction coil. A primary coil P (Fig. 116) is wound upon a soft iron core and connected into the circuit of a battery B through the contact point o. A secondary coil S is wound about the same core as the primary. A condenser C is connected across the spark gap o. When the current starts in the primary it magnetizes the core and draws the iron

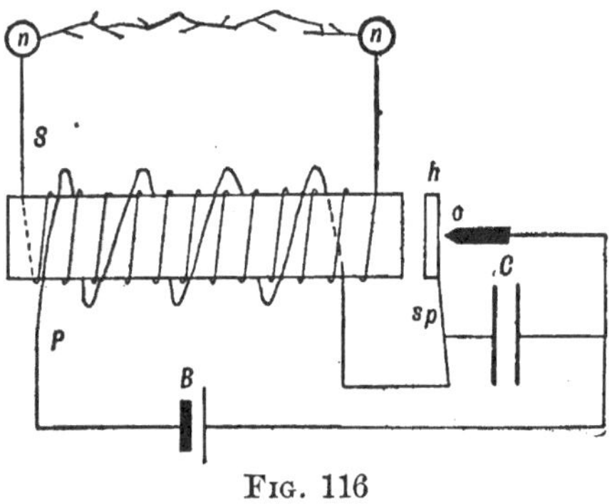

FIG. 116

hammer h away from the contact point o. This breaks the circuit and the spring sp then restores contact at o. The operation then begins over again. At every *break* of the primary at o a spark passes between the terminals of the secondary, but no spark passes at *make*. The cause of this unidirectional character of the discharge between n and n is found in the fact that at *make* the effect of the self-induction of the primary is simply to retard the growth of the primary current, and hence to render the growth of the field within the secondary slow. This means, of course, a small induced E.M.F. in the secondary. At *break*, however, the presence of the condenser causes the current to fall to zero in an exceedingly short time. For as soon as the contact points begin to separate, the self-induction of the primary, which would normally create an arc between the points, now drives its induced current into the condenser for an instant, and thus gives the contact points time to get so far apart that, by the time the condenser is charged, the spark can no longer leap across the gap and set up an arc. The function of the condenser is then to prevent sparking at o and thus to make a sudden, rather than a slow, collapse of the field. Thus, while the same number of lines of force cut the secondary at *make* as at *break*, they take, at *make*, perhaps a thousand times as long to do it, and hence the E.M.F. at *make* is much too weak to force a spark across the terminals of the secondary, unless these terminals are extremely close together, for example .01 of an inch.

101. The transformer. The transformer is a modified form of the induction coil. In it the core to be magnetized by the primary current is given some shape such that the magnetic lines of force have a continuous iron path instead of being obliged to push out into the air, as in the induction coil. Further, it is an alternating rather than an intermittent current which is sent through the primary of a transformer. The effect of such a current is first to magnetize the core in one direction, then to demagnetize it, then to magnetize it again in the opposite direction, and so on. These changes, of course, induce in the secondary an alternating current similar to that which is being sent through the primary. The E.M.F. induced in the secondary is obviously proportional to the number of turns of wire upon it, for each turn is threaded by all of the lines which pass through the core.

In practice the transformer is largely used as a "step-down transformer"; that is, the mean potential at which the primary is supplied is many times higher than the mean potential which is induced in the secondary. In this case the primary coil has a large number of turns and the secondary coil has a smaller number. For example, if the main conductors are kept at a mean P.D. of 1100 volts, while the lamps of the secondary require 110 volts, there must be 10 times as many turns upon the primary as upon the secondary.

Transformers are usually kept connected in parallel to the high potential mains which connect with the alternating-current dynamo,

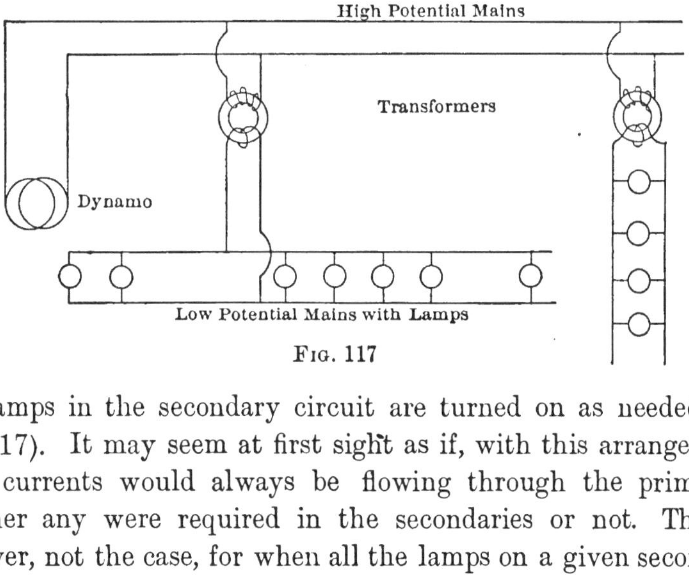

Fig. 117

and lamps in the secondary circuit are turned on as needed (see Fig. 117). It may seem at first sight as if, with this arrangement, large currents would always be flowing through the primaries, whether any were required in the secondaries or not. This is, however, not the case, for when all the lamps on a given secondary are turned out, so that no current is delivered by it, the self-induction of the primary "chokes off" practically all current from this primary itself. For, since the effect of self-induction in any circuit is simply to retard the growth of a rising current, or the decay of a dying current, when an alternating E.M.F. is applied to a circuit of sufficiently large self-induction, the current due to the impressed E.M.F. in one direction scarcely gets started before the reversal comes; and this reversed E.M.F. in its turn only begins to change this current when another change in direction occurs. So that, in general, a circuit of sufficiently large self-induction behaves toward an alternating E.M.F. just as a body of sufficiently large mass behaves toward a rapidly alternating mechanical force; i.e. in neither case is any appreciable effect produced upon the system. But suppose that lamps are turned on in the secondary of such a transformer circuit. The currents which are induced in the secondary are, by Lenz's law, in such direction as at all times to thrust lines of force through the core in a direction opposite to the direction in which the primary E.M.F. is attempting to thrust lines through the core. The actual number of lines sent through the core by the rise of a given current in the primary will then be less if the secondary is closed than it will be if the secondary is open. Closing the secondary circuit is therefore equivalent to withdrawing a certain amount of self-induction from the primary. Hence a larger current flows through the primary when a lamp is turned on in the secondary than when the latter is open. Turning on two lamps in the secondary withdraws twice as much self-induction from the primary and thus increases in like amount the current which the impressed E.M.F. sends through it, and so on. Thus the current taken from the mains by the primary of a transformer automatically adjusts itself to the demands of the secondary. If no energy is taken from the secondary, the primary takes practically no current from the dynamo; but if a large amount of energy is demanded by the secondary, a large current is taken from the dynamo by the primary. The transformer is therefore merely a device for transferring energy from one circuit to another without the intervention of any mechanical motions of any sort. Its efficiency is commonly as high as 97 per cent, the remaining 3 per cent being transformed into heat in the core and coils of the transformer.

EXPERIMENT 14

Object. To find the coefficient of self-induction of a coil.

Directions. In order to avoid the necessity of obtaining a perfect balance for steady currents, — a tedious operation at best, — modify as follows the ideal arrangement described above. Insert a double key K_1 as in the diagram (Fig. 118),* so that when K_1 is depressed and the switch K_2 closed, the battery circuit is closed through the bridge, and at the same time the galvanometer is short-circuited; and also so that when K_2 is closed, lifting the finger from K_1 opens the short circuit on the galvanometer an instant before it opens the battery circuit. Give the resistances M, O, and T values which are of about the same order of magnitude as the galvanometer resistance. Observe the zero of the galvanometer, then close in succession K_2, the lower contact of K_1, and then the upper contact of K_1. Adjust r_1, all these contacts being kept closed, until the galvanometer stands approximately at its zero. If now opening K_2 produces a large deflection, adjust

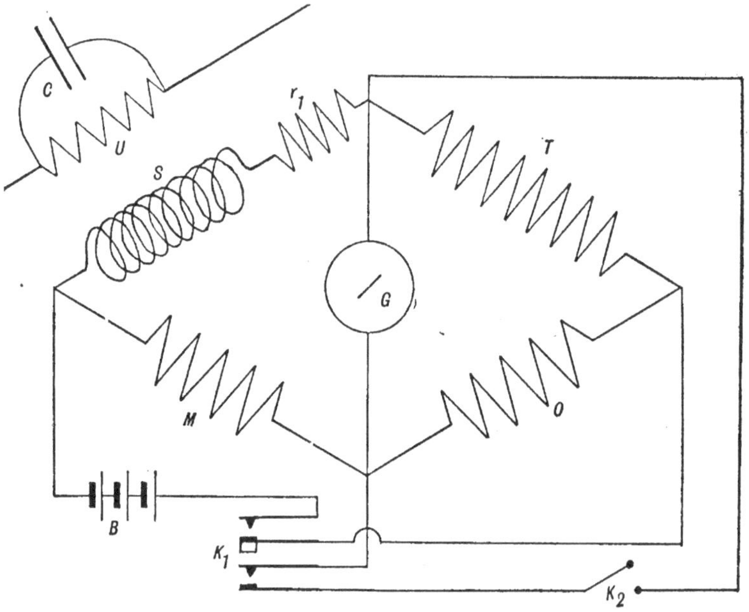

FIG. 118

r_1 still further until this deflection is reduced to not more than a centimeter or two. In order to effect this adjustment it is usually necessary to make the wire which connects r_1 to T a few feet of bare German silver wire (say No. 22), and to slip this along through the binding post of r_1 in such a way as to include a greater or smaller length of it in the branch which contains r_1. When the balance has been made correct to within a centimeter close K_2 permanently, take the reading, then open K_1 and read the throw. Repeat several times; then, in order to eliminate thermal effects, reverse the battery terminals with a commutator or otherwise, and read the throw in the opposite direction. Take the means of several

* This scheme of connections is due to Fleming.

throws in opposite directions as θ_1. This throw will be practically all due to the self-inductive current from S, since the interval between the opening of the galvanometer short circuit in K_1 and the battery circuit is so brief that the galvanometer coil has no appreciable impulse given to it because of the want of exact balance in the bridge arms. This statement may be tested by seeing whether the same throw is not obtained when the balance is exact as when it is inexact. If θ_1 is not a throw of several centimeters at least, the number of dry cells in B should be increased.

Now replace S and r_1 by C and U. Make U the same as M, O, and T, and adjust still further with the German silver wire, if necessary, until as good a balance is obtained as before.

Take the throw θ_2 precisely as θ_1 was taken, and substitute in equation (6) to obtain the coefficient of self-induction of S. This will be expressed in henrys if C is in farads and U in ohms. It is, of course, desirable to make θ_1 and θ_2 of the same order of magnitude. If θ_2 is too small, use a condenser of larger capacity. If it is too large, place C in shunt with a fraction only of U and use the number of ohms across which C is shunted as the resistance to substitute in equation (6).

EXAMPLE

The coil, the self-induction of which was determined, was 3.7 cm. high, 9.9 cm. in external diameter, 4.1 cm. in internal diameter, and was wound with about 4000 turns of No. 25 copper wire. The resistance of the coil was about 29 ohms, that of the galvanometer about 600 ohms. The resistance in each of the four branches of the bridge was made 1000 ohms. The mean value of θ_1 was 27.9 mm., the current being furnished by three dry cells. The condenser used was of $\frac{1}{3}$-microfarad capacity and produced a deflection, when U was 1000 ohms, of 45.3 mm. Therefore $L = .205$ henrys. This differed by less than half of one per cent from the value marked upon the coil, which had been previously obtained by a careful comparison with a standard of self-induction by a method similar to that described in section 67 for comparing condensers.

CHAPTER XV

MAGNETIC INDUCTION IN IRON

102. The test coil. In section 94 a method was described for measuring the intensity of a magnetic field by the use of a ballistic galvanometer and a test coil, known in that experiment, from its special application to the measurement of the earth's magnetic field, as the "earth inductor." The principles involved may be applied to a general study of magnetic induction and are here restated with particular reference to the experiment which follows this section.

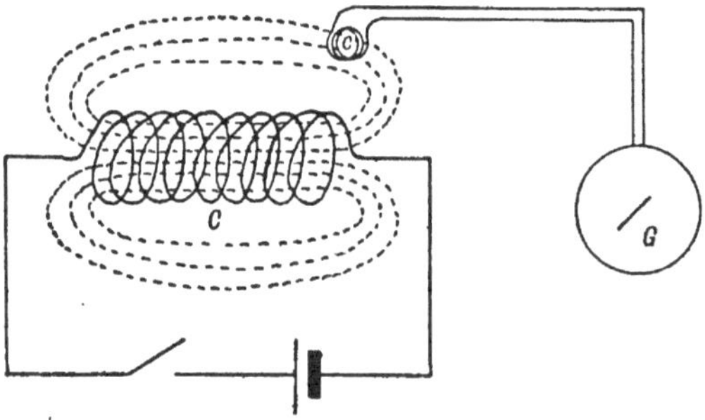

FIG. 119

Consider a magnetic field which may be established or destroyed at will by making or breaking the circuit of a conductor *C*. In the neighborhood of this conductor (see Fig. 119) there is placed a small coil of wire *c*, the test coil. Let the average area of this coil be such that when multiplied by the number of turns of the coil the total area is *A* square centimeters. The coil is connected to a ballistic galvanometer, the constants of which are known. The combined series resistance of the galvanometer and coil, in C.G.S. units, will be represented by *R*. If in the time τ the number of magnetic lines passing through the test coil is changed from zero to $\mathcal{H}$ lines per square centimeter by the establishment of the magnetic field, then the mean E.M.F. induced in this coil during the time τ is $\mathcal{H}A/\tau$ C.G.S units. The quantity of electricity Q which this E.M.F. causes to flow through the galvanometer is given by the relation $Q=\mathcal{H}A/R$. Q can be found as shown in section 94 from the throw of the galvanometer. Transforming the expression gives

$$\mathcal{H} = \frac{QR}{A}; \tag{1}$$

that is, QR/A is a measure of the number of magnetic lines per square centimeter passing through the test coil.

103. Magnetic field-intensity. Now the number of lines passing through the test coil as determined from the galvanometer throw is found to depend upon the medium within the coil. If the medium is air, the number of magnetic lines per square centimeter is known as the magnetic field-intensity. And even if the medium is not air, the number which *would* exist were the medium replaced by air is still known as the magnetic field-intensity and is commonly denoted by the symbol $\mathcal{H}$.

104. Diamagnetic and paramagnetic substances. The number of magnetic lines per square centimeter determined as above is for most substances very slightly less than it is for air. Substances for which this is true are called *diamagnetic.* That the difference is very slight may be seen from the fact that for bismuth, the most highly diamagnetic substance known, the ratio of the number of lines established to the number which would be established in air by the same cause is approximately .9998.

Those substances for which the number of lines, other things being equal, is greater than it is in air, are called *paramagnetic.* Most of these substances also differ but very slightly from air. The exceptions, iron, nickel, and cobalt, with certain of their compounds or alloys, admit of such exceptionally large values for the number of magnetic lines that they really constitute a separate group. In comparison with members of this group, which will be known as *magnetic,* all other substances may be considered as magnetically indifferent, and for practical purposes identical in their behavior with air.

105. Magnetic induction. The name *magnetic induction* is given to this phenomenon of the establishment in a magnetic substance of a number of magnetic lines different from the number which would exist in air from the same cause. To study this phenomenon consider two identical coils of wire, the first wound about an annular ring composed of some magnetic material (e.g. soft iron), and the second about an identical ring of some magnetically indifferent material (e.g. wood) (see Fig. 120). These two coils are connected in series so that identical conditions of current, and consequently of magnetic field, exist for both coils. The magnetic conditions within these rings are observed by similar test coils S and S'. Upon completing the circuit through the primaries the number of magnetic lines within the wood core, as measured by the expression QR/A, gives the magnetic field-intensity $\mathcal{H}$ in lines per square centimeter. The test coil wound over the iron core has induced in

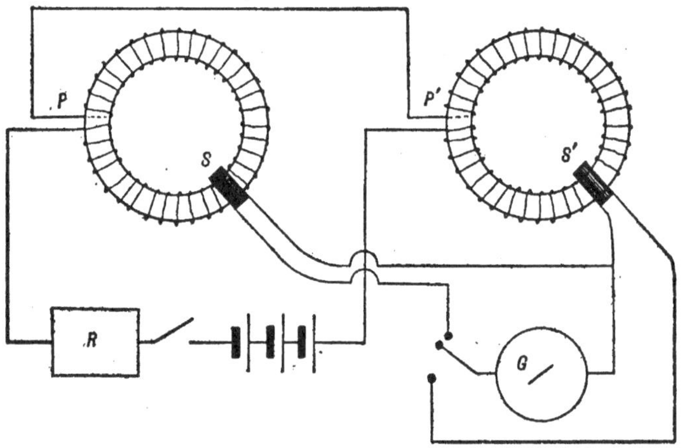

FIG. 120

it a quantity of electricity Q'; which is larger than Q, and which, when substituted in the expression $Q'R/A$, gives a measure of the number of lines per square centimeter through the iron. This number will be represented by the symbol $\mathcal{B}$ and will be known as the *induction* in the iron due to the magnetic field of intensity $\mathcal{H}$. That is, *induction is the total number of lines per square centimeter passing through a substance as a result of its presence in a magnetic field.* Induction is also frequently known as *flux density,* since it follows from its definition that it is the flux (or total number of lines through the iron) divided by the area of cross section normal to their direction.

106. Magnetization. If a larger current is used, a larger value of $\mathcal{H}$ is obtained, and a larger value of $\mathcal{B}$. In this way it is possible to study the variation of the induction in iron with the magnetic intensity of field which causes it. The curve representing the general nature of the variation is shown in Figure 121. Now, since there are $\mathcal{H}$ lines per square centimeter in the air core, there must be $\mathcal{H}$ of the lines in the iron core which are due to the current. The difference, $\mathcal{B} - \mathcal{H}$, represents, then, the number of lines per square centimeter which are due solely to the presence of the iron in the magnetic field inside the circular coil. It is taken as a measure of the *magnetization* of the iron. The technical definition of magnetization, commonly denoted by $\mathcal{I}$, is this number of lines divided by 4π. Thus, by definition,

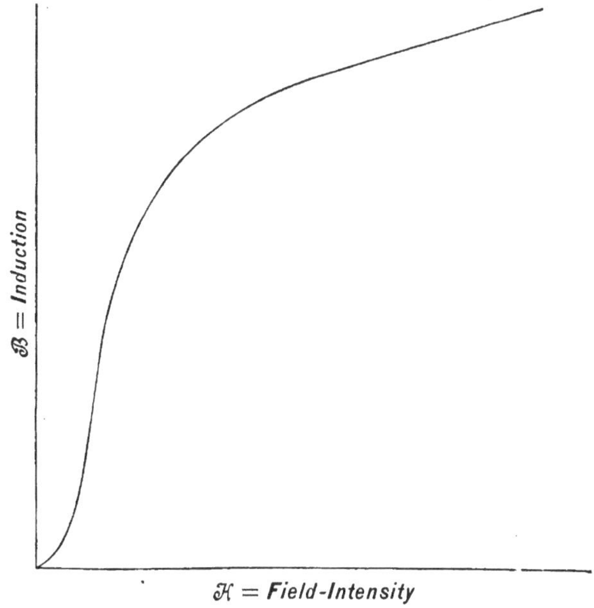

FIG. 121

$$\mathcal{I} = \frac{(\mathcal{B} - \mathcal{H})}{4\pi}. \qquad (2)$$

The reason for the introduction of the factor 4π will be discussed in a later paragraph.

107. Three stages of magnetization. When successive values of $\mathcal{I}$ are computed from the last equation and successive values of $\mathcal{B}$ and $\mathcal{H}$ as given in Figure 121, and when then the successive values of $\mathcal{I}$ are plotted as ordinates with the corresponding values of $\mathcal{H}$ as abscissas, a curve of the type shown in Figure 122 is obtained. This latter curve shows that the process of magnetizing iron may be divided into three stages which are represented respectively by the portions of the curve from *b* to *c*, from *c* to *d*, and from *d* to *e*. During the first stage, which corresponds to very small values of the magnetizing force $\mathcal{H}$, the magnetization increases in nearly direct proportion to $\mathcal{H}$; then during the second stage, namely from *c* to *d*, the magnetization of the iron increases exceedingly rapidly, and during the final stage, from *d* to *e*, it becomes practically constant, showing that the magnetization of iron cannot be pushed higher than a certain limit, however strong fields may be used. The explanation of this behavior in terms of the molecular theory (see sect. 89, p. 134) is that while the field is still too weak to break up any of the existing molecular groups, the molecular magnets turn slightly against the mutual magnetic actions of the surrounding molecules, much as a solid body which is not distorted beyond its elastic limits yields under the influence of an external force. But as the magnetizing force $\mathcal{H}$ increases, a point is reached at which the molecular groups begin to break up, and the molecules to set themselves parallel to the direction of the field $\mathcal{H}$. The part of the curve between *c* and *d* represents the corresponding values of $\mathcal{J}$ and $\mathcal{H}$ when this breaking down of the molecular groups is taking place most rapidly. It will be seen that within this region a very small increase in $\mathcal{H}$ produces a very large increase in the magnetization. As $\mathcal{H}$ is still further increased the

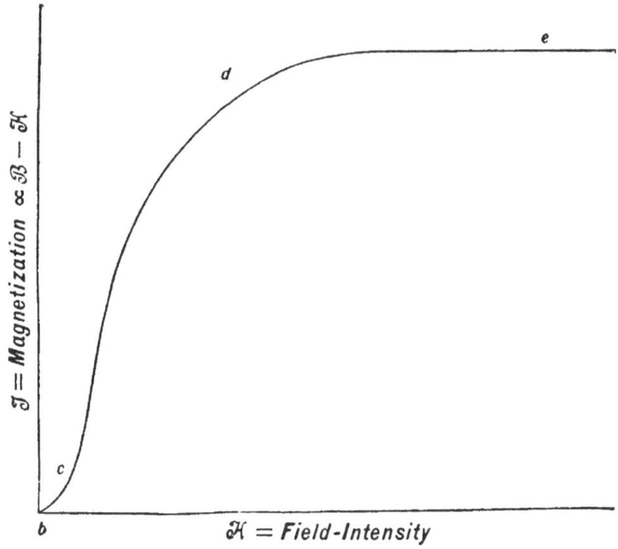

FIG. 122

magnetization enters upon its third stage represented by the portion *de* of the curve. In this stage the molecular groups are all broken up, but the molecules have not yet all assumed exact parallelism with the field. As $\mathcal{H}$ still further increases, the molecules gradually assume the exact direction of the field, and the iron is then said to be magnetically saturated, since any further increase in $\mathcal{H}$ produces no increase in $\mathcal{I}$.

108. Origin of the factor 4π. The reason that the magnetization $\mathcal{I}$ is not taken as numerically equal to the number of lines due to the iron, namely $\mathcal{B} - \mathcal{H}$, but rather to $(\mathcal{B} - \mathcal{H})/4\pi$, is as follows: Since, by definition, unit magnetic pole, when placed at the center of a sphere of unit radius, produces a magnetic field of unit strength at every point on the surface of the sphere, and since a field of unit strength is represented by one line of force per square centimeter, it is clear that one line of force from the pole must be thought of as piercing each of the 4π sq. cm. on this unit sphere. Hence we must imagine 4π lines of force as emanating from every unit *N* pole, and $4\pi m$ lines of force as emanating from any *N* pole which contains *m* units of magnetism. Now we imagine that in an ideal magnet which is fully saturated there are no free poles except at the very ends of the magnet, so that all of the lines which are associated with the north pole emerge from the face itself, pass around in closed curves to the *S* face, and then return through the magnet to the *N* face (see Fig. 101, p. 135). If the strength of the magnet's poles is *m*, then, as has just been shown, the number of these lines is $4\pi m$. The number of unit poles in one square centimeter of the face is then not equal to the number of lines which pass out of this face, but is rather equal to this number divided by 4π. Now it has been decided to regard *magnetization* as the number of unit poles per square centimeter, rather than as the number of lines per square centimeter. Hence it was that we divided the number of lines per square centimeter due to the iron, namely $\mathcal{B} - \mathcal{H}$, by 4π in order to obtain the numerical value of $\mathcal{I}$ inside the ring. Even in an ordinary unsaturated magnet in which lines emerge all along the sides, the intensity of magnetization at any point within the iron is defined as the number of lines per square centimeter there present divided by 4π. That is, it is the number of units of molecular magnets which we must imagine to be present in this particular square centimeter in order to account for the number of lines found in this square centimeter because of the magnetization of the iron.

109. Hysteresis. The curve *bcde* (Fig. 122), representing the values of the magnetization corresponding to a magnetic field of constantly increasing magnetization, is shown again in Figure 123. As now the magnetic field is allowed to decrease to its original value of zero, the magnetization assumes successive values represented by the curved portion *ef*. The explanation, in terms of the

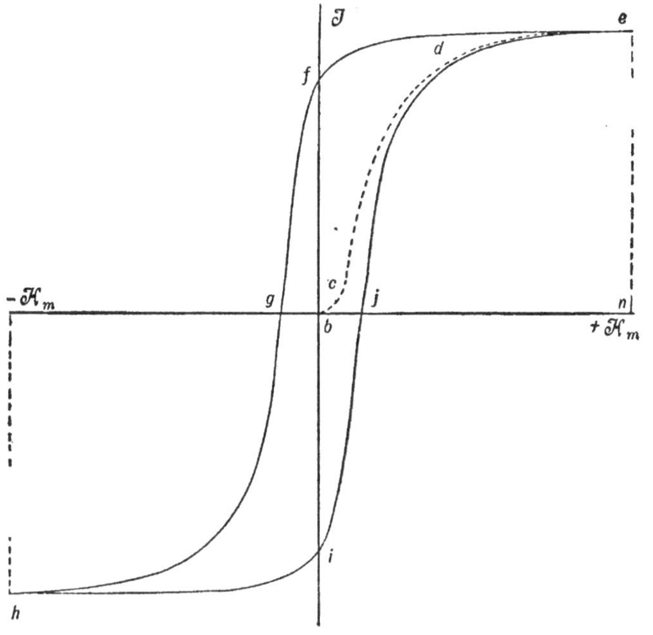

Fig. 123

molecular theory, of the fact shown by the curve, that the successive values of the decreasing magnetization do not lie on the original curve of increasing values, is as follows: When the magnetizing field $\mathcal{H}$ is withdrawn or diminished, some of the molecules return under their mutual actions to their original groupings, but many of them retain their alignment until this is broken up by jars or other outside forces. In hard steel the difficulty which the molecules experience in moving out of any positions which they have once assumed is very large, while in soft iron it is small. In general, then, the magnetization is said to "lag behind" the field-intensity which induces it. This phenomenon, known as *hysteresis*, is an important characteristic of all magnetic substances.

When the field has been decreased to zero the value of the magnetization retained by the iron is known as the *residual magnetization*. In practice, however, the knowledge of this quantity is of little value. It is in general more desirable to know the ratio which the residual magnetization, represented by the ordinate *bf*, bears to the maximum value of the induced magnetization, represented by the ordinate *en*. This ratio is defined as the *retentivity* and is usually stated in per cent.

In order that the magnetization shall be again zero, the direction of the field must be reversed, and the value of its intensity caused to increase in this opposite direction. Under the action of this reversed field, the values of which will for convenience be called negative and plotted to the left of the vertical axis, the value of the magnetization rapidly decreases in a manner represented by the curve *fg*. The value of the negative intensity which must be applied to reduce the residual magnetization to zero is known as the *coercive force*. Numerically, it is represented by the abscissa *bg*.

It is evident that if this oppositely directed field is still further increased in intensity, there must result a magnetization opposite in direction to the original magnetization. Now let the maximum value of the original intensity, shown in the figure by the abscissa *bn*, be represented by the symbol $+\mathcal{H}_m$. It is of interest to consider the effect of hysteresis upon the successive values of the magnetization corresponding to a cyclic change of $\mathcal{H}$, in which $\mathcal{H}$ passes continuously through all the values from $+\mathcal{H}_m$ to $-\mathcal{H}_m$ and back again to $+\mathcal{H}_m$. After a sufficient number of repetitions of this cycle, the values of $\mathcal{I}$ are found to lie on a closed curve of the general form represented by *efghije* and known as the *hysteresis loop*.*

* In the commercial transformers described in section 101, page 153, the iron core undergoes such cyclic changes in its magnetic state. Work is done in producing these changes, and this loss of energy due to hysteresis is of practical importance in the design of transformers.

110. Formula for field strength within a long solenoid. It is generally desirable in experiments on $\mathcal{B}$ and $\mathcal{H}$ to determine $\mathcal{B}$ by a test-coil method, applying the principles of section 105 in a manner which will be explained in the next section; but instead of using a wooden ring for the determination of $\mathcal{H}$ it is customary to calculate this quantity from the number of turns z per centimeter of length of the solenoid and the current I flowing about it. The relation between these quantities is as follows:

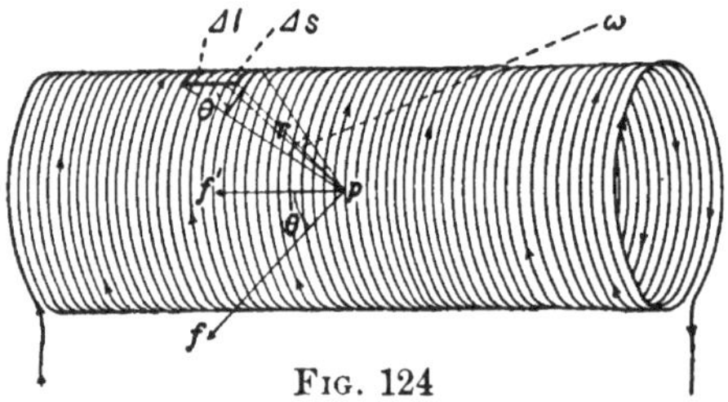

FIG. 124

$$\mathcal{H} = \frac{4\pi z I}{10}.$$

The origin of this formula may be seen from the following considerations.

The force which a unit magnet pole placed at a point p (Fig. 124) on the axis of a long solenoid experiences because of a current I flowing around it is, of course, simply the resultant of all the magnetic forces exerted upon it because of the magnetic fields surrounding all elements of the wire constituting the solenoid. Consider first the value of the force exerted upon the pole at p by the elements of current which lie on a small element of area $\Delta l \Delta s$ of the surface of the solenoid. If there are z turns per centimeter, there will be $z\Delta l$ elements of current in the distance Δl. Each of these has a length Δs. The force which each element exerts on a unit magnet pole at the point p a distance r from the element is, by definition of unit current (see sect. 30, p. 36), $\frac{I\Delta s}{r^2}$, and the total force f due to all the elements $z\Delta l$ is given by

$$f = \frac{Iz\Delta l \Delta s}{r^2}. \tag{3}$$

Since, from considerations of symmetry, it is obvious that the resultant force exerted on the pole at p by the whole solenoid must be parallel to the axis of the solenoid, we are here concerned only with the component of f in the direction of the axis. Calling this f', we have

$$f' = \frac{Iz\Delta l \Delta s \cos\theta}{r^2}. \tag{4}$$

But precisely as in section 73, page 108, $\frac{\Delta l \Delta s \cos\theta}{r^2}$ is the solid angle subtended at p by the elementary area $\Delta l \Delta s$. Hence, representing this solid angle by u, we have

$$f' = Izu. \tag{5}$$

*That is, the magnetic force exerted parallel to the axis by any element of the solenoid is the solid angle subtended at the point by the element, multiplied by the current and by the number of turns per centimeter of length of the solenoid.** The total force $\mathcal{H}$ acting upon the unit pole at p is the sum of the forces due to all the elements of the surface, i.e. $\mathcal{H} = \Sigma Izu$. Since I and z are the same for all elements, we may write this in the form $\mathcal{H} = Iz\Sigma u$, and if the solenoid is so long that we may neglect the solid angle subtended by its open ends in comparison with the solid angle subtended by its surface, we have, since the solid angle about a point is 4π,

$$\mathcal{H} = 4\pi z I. \tag{6}$$

If I is expressed in absolute units of current, the field strength $\mathcal{H}$ will be expressed in gausses. If I is in amperes, in order to obtain $\mathcal{H}$ in gausses we must write

$$\mathcal{H} = \frac{4\pi z I}{10}. \tag{7}$$

Since this analysis holds for any point within the solenoid which is not too close to its ends, it is obvious that the field strength within a long solenoid is uniform. If the solenoid is a ring, the deduction is rigorously correct for all points within the ring. If it is a cylinder, the general rule is that the length must be as much as twelve times the diameter in order that equation (7) may be applied without appreciable error to find $\mathcal{H}$ at the center. In order to see the reason for this rule it is only necessary to consider for what ratio of length to diameter the solid angle subtended at the center by the ends becomes negligible in comparison with the solid angle subtended by the surface of the solenoid.

* Although this conclusion has been arrived at by considering a point *on the axis*, it holds for all points within the solenoid, as can be shown by resolving each element of current into two components, one parallel and one perpendicular to the line connecting the point and the element.

111. Ballistic method of determining the magnetization curve. The method to be described for plotting the curve of magnetization discussed in section 107 consists, first, in finding a set of corresponding values of $\mathcal{B}$ and $\mathcal{H}$, and then, from the relation $\mathcal{I} = (\mathcal{B} - \mathcal{H})/4\pi$, finding the values of $\mathcal{I}$ corresponding to each value of $\mathcal{H}$. The material to be examined is given the form of a ring about which is wound uniformly a coil of wire represented by P in Figure 125. In series with the coil P there is an ammeter A and a resistance R' which may be varied without causing an interruption of the current passing through it. A commutator c admits of a reversal of the current in the coil. At some point of the coil P is wound a small test coil S which is connected through a resistance R to a ballistic galvanometer G. Starting with zero

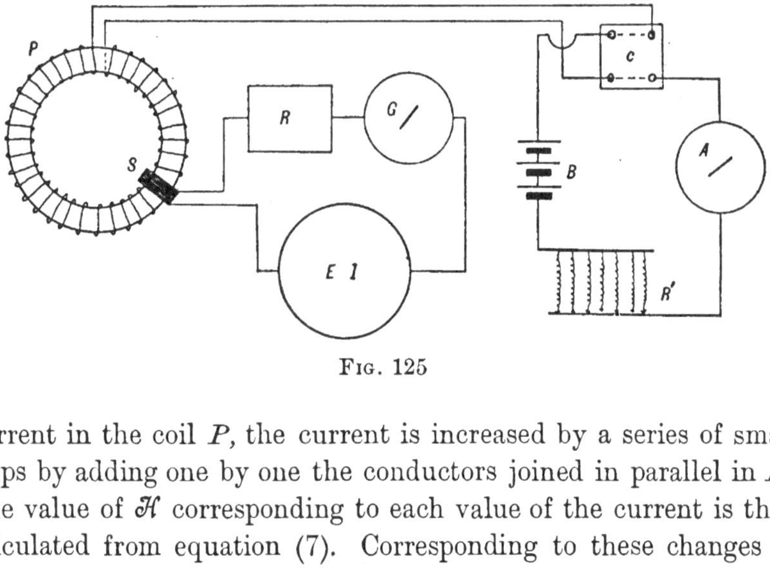

FIG. 125

current in the coil P, the current is increased by a series of small steps by adding one by one the conductors joined in parallel in R'. The value of $\mathcal{H}$ corresponding to each value of the current is then calculated from equation (7). Corresponding to these changes in the value of $\mathcal{H}$ there are changes in the induction $\mathcal{B}$. Each change in induction, which will be represented by $\Delta\mathcal{B}$, is proportional to the galvanometer throw which it produces, and might be measured by the relations given in equation (1), section 102, but is most easily found by inserting an earth inductor EI into the galvanometer circuit and comparing the throw produced by it as it rotates in the known field of the earth with the throws produced by $\Delta\mathcal{B}$. Obviously the total induction $\mathcal{B}$ in the iron at any time is the algebraic sum of all the preceding values of $\Delta\mathcal{B}$. Corresponding values of $\mathcal{B}$ and $\mathcal{H}$ are then plotted. The magnetization curve, i.e. the curve connecting $\mathcal{I}$ and $\mathcal{H}$, may be found from these values by the method of the next section.

112. Graphical transformation of $\mathcal{B}$ and $\mathcal{H}$ curve to an $\mathcal{I}$ and $\mathcal{H}$ curve. The number of lines representing the induction $\mathcal{B}$ is in general so many times greater than the number representing the

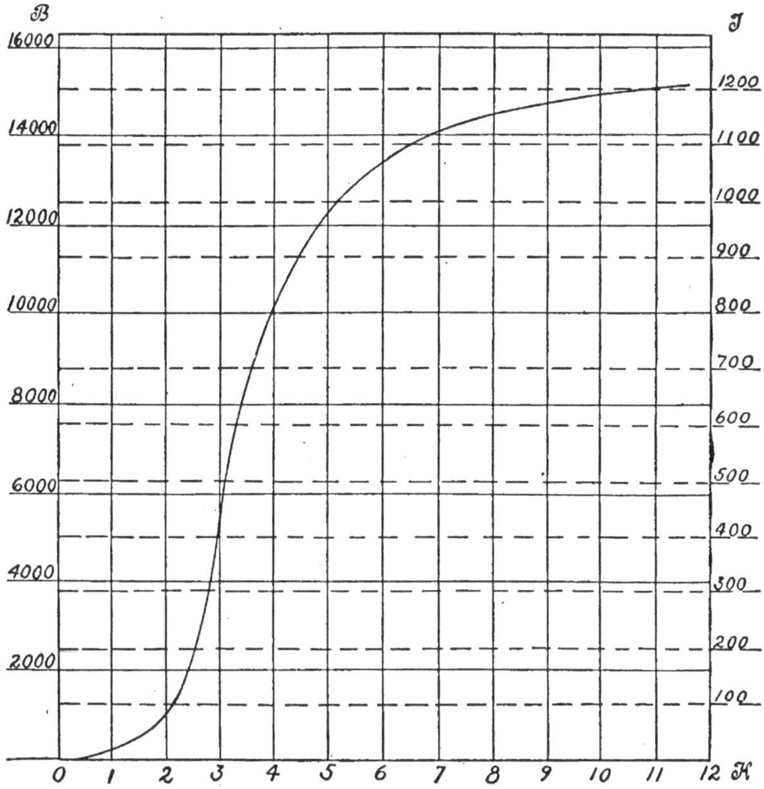

FIG. 126

corresponding value of $\mathcal{H}$ that they are plotted on very different scales. Thus in Figure 126 is shown the relation of $\mathcal{B}$ and $\mathcal{H}$ for increasing values of $\mathcal{H}$. The scale of $\mathcal{B}$ is about one thousandth of that for $\mathcal{H}$. To find the value of $\mathcal{I}$ corresponding to any value of $\mathcal{H}$ it is necessary to subtract $\mathcal{H}$ from $\mathcal{B}$ and divide by 4π; that is, $\mathcal{I} = (\mathcal{B} - \mathcal{H})/4\pi$. For all practical purposes, however, in dealing with values of $\mathcal{I}$ below saturation, it is sufficient to neglect $\mathcal{H}$ in comparison with $\mathcal{B}$ and to write $\mathcal{I} = \mathcal{B}/4\pi$. Now, if the scale of the axis of $\mathcal{B}$ in Figure 126 is changed by letting each division represent the $1/4\pi$ part of its original value, the division of $\mathcal{B}$ by 4π is at once performed for all values of $\mathcal{B}$. In other words, within these limits, one single curve represents both the relations of $\mathcal{B}$ and $\mathcal{H}$ and of $\mathcal{I}$ and $\mathcal{H}$. The new scale for the axis of ordinates is shown at the right of the scale for $\mathcal{B}$. The value of $\mathcal{I}$ corresponding to any value of $\mathcal{H}$ is then read at once from the curve on this new scale. This method of transforming a $\mathcal{B}$ and $\mathcal{H}$ curve to an $\mathcal{I}$ and $\mathcal{H}$ curve may of course be applied to the entire hysteresis loop.

113. Demagnetization of the specimen to be tested. As is evident from the preceding discussion of hysteresis, the fact that the magnetizing force $\mathcal{H}$ is zero is no evidence that the magnetization $\mathcal{I}$ is also zero. And further, even though both be zero, the previous magnetic history of the specimen under examination may be such that it has a distinct set or tendency toward a magnetization in some particular direction. Now it is of practical importance to compare the magnetic qualities of various substances by their magnetization curves. In order that such a comparison may be made, it is evident that the original state of the specimens must be the same at the time the observations are begun. This is made possible by demagnetizing as follows: starting with a value of the current in the coil P (Fig. 125) such that the value of $\mathcal{H}$ is higher than any value to which the specimen has been recently subjected, the current is gradually decreased to zero. During this decrease the direction of the current is made to undergo a series of rapid alternations produced at the commutator. As a result of this operation the substance shows no predisposition toward a magnetization in any one direction.

114. The magnetometric method of testing magnetic substances. If the specimen to be tested is in the form of a bar, it may be magnetized by placing it inside a long solenoid. The magnetization, which is the pole strength per unit area of cross section, may then be found by the use of the magnetometer as shown in Chapter II. The chief objection to this method lies in the fact that the free magnetic poles formed near the ends of the bar set up within the solenoid a field in the opposite direction to the magnetizing field $\mathcal{H}$, and thus tend to weaken this field. The actual amount of the weakening at any instant depends upon the pole strength of the bar at that instant and upon its form. It may be calculated by an analysis which is, however, beyond the scope of this text.

The ballistic method is free from this objection, for in a closed ring there are no free poles. But it is open to the objection that it takes into account only *sudden* changes in the induction. The growth of the magnetization is, however, subject to a small gradual increase, which continues for some time after a sudden change in the magnetizing force. This phenomenon is known as "magnetic creeping."

115. Permeability and susceptibility. In the study of the magnetic properties of various materials, two quantities, known as permeability and susceptibility, are of particular interest. The ratio of the induction $\mathcal{B}$ to the corresponding value of the magnetizing force $\mathcal{H}$, as taken from the ascending curve of magnetization for a specimen previously demagnetized, is defined as the *permeability* of the specimen for that value of $\mathcal{B}$ or $\mathcal{H}$. That is, representing permeability by μ, we have

$$\mu = \frac{\mathcal{B}}{\mathcal{H}}. \tag{8}$$

In the same manner the ratio of the magnetization $\mathcal{I}$, taken from the ascending curve of magnetization for a specimen previously demagnetized, to the corresponding value of the magnetizing force $\mathcal{H}$, is known as the *susceptibility*. Hence, representing susceptibility by κ, we have

$$\kappa = \frac{\mathcal{I}}{\mathcal{H}}. \tag{9}$$

And since $\mathcal{B} = \mathcal{H} + 4\,\pi\mathcal{I}$, we have

$$\mu\mathcal{H} = \mathcal{H} + 4\,\pi\kappa\mathcal{H}, \tag{10}$$

or

$$\mu = 1 + 4\,\pi\kappa. \tag{11}$$

The permeability of a specimen of soft iron attains a maximum of about 2500 at some point of the second stage of magnetization referred to in section 107, page 161. As the values of the induction are increased still further the permeability decreases. As saturation is approached ($\mathcal{B} = 16{,}000$) the value of μ is very low, only about 500. For very high values of the induction, for example, $\mathcal{B} = 45{,}000$ (which corresponds to $\mathcal{H} = 24{,}000$), μ is reduced nearly to one. This is what we should expect; for beyond saturation $\mathcal{H}$ becomes large as compared to $\mathcal{J}$, which remains practically constant, and $4\pi\kappa$ in equation (11) approaches zero. A curve showing the relation between μ and $\mathcal{B}$ for a specimen of soft iron is shown in Figure 127.

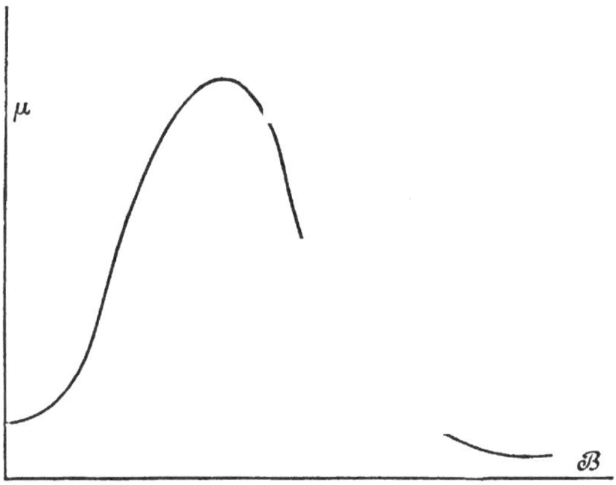

FIG. 127

EXPERIMENT 15

Object. To plot the curve of magnetization for a sample of iron; also to plot the hysteresis loop for the same sample.

Directions. The sample to be tested is in the form of a ring (Fig. 128), the mean peripheral length of which is some 25 or 30 times the diameter of its cross-sectional area. The ring is wound uniformly in two layers with about 20 turns of wire per centimeter of mean peripheral length. The layers have separate terminals and are to be connected in series. Connect as in Figure 125. The milliammeter A is one of range 0 to 500. The resistance R' is a specially constructed set of resistances all in parallel and controlled by individual knife switches (see Fig. 129). The battery B should contain a number of cells such that with all these resistances in the circuit the current is about .490 ampere. The commutator c is a specially constructed rotary commutator shown in Figure 130.

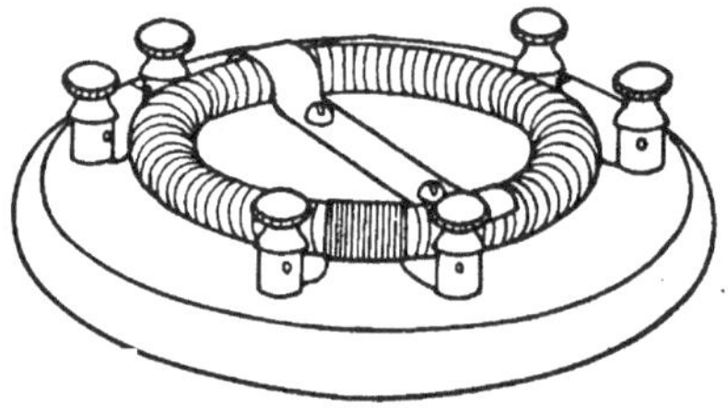
FIG. 128

The terminals of the battery circuit are to be connected to the binding posts *e* and *f*, those of the coil to *g* and *h*.

The test coil, consisting of a number of turns of fine wire, is wound over the primary, covering a space of 5 to 10 mm. of length. In series with this there is an earth inductor *EI* (Fig. 125), a resistance *R* (about 20,000 ohms), and the ballistic galvanometer *G*.

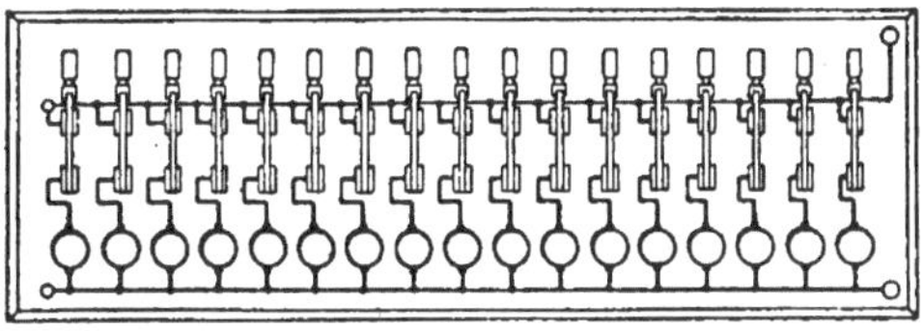

FIG. 129

Open the galvanometer circuit and demagnetize the iron ring as follows. Starting with all the resistances in, rotate the commutator rapidly and one by one open the resistance switches, opening the highest resistance last.

Now connect the galvanometer circuit. Close the switches one by one in the opposite order to that in which they were opened, observing as each switch is closed the galvanometer deflection caused and the value of the current. This group of observations constitutes the data for the ascending $\mathcal{B}$ and $\mathcal{H}$ curve.

Without altering any connections, obtain the data for the hysteresis loop as follows. Open the switches, one at a time, and observe the corresponding deflections and currents. When they are all open turn the commutator carefully so as to reverse the direction of the current through the primary, then increase it to its maximum value by successively closing, and afterward decrease it by successively opening, the switches, observing the corresponding deflections and current values as before. Call the deflections of the galvanometer in the original direction plus, and in the opposite direction minus. When the current has been brought to zero reverse again the direction and obtain a set of readings for the current increasing in the original direction.

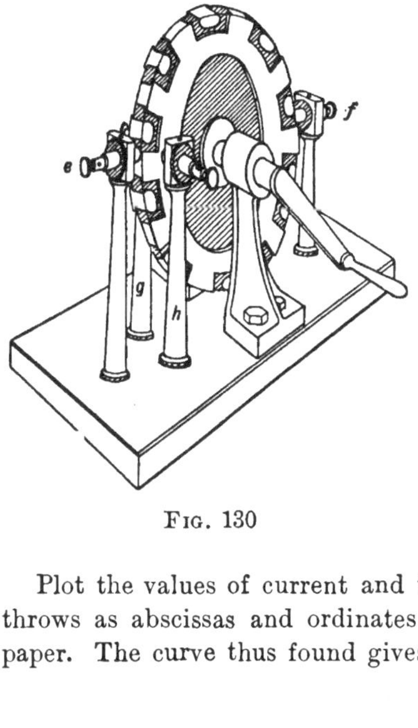

FIG. 130

Without altering any connections, take a set of three readings for the throw d' produced by the earth inductor in cutting the horizontal component of the earth's magnetic field.

Plot the values of current and the algebraic sum of the galvanometer throws as abscissas and ordinates respectively on a sheet of coördinate paper. The curve thus found gives the relation between the magnetizing current I in amperes and the galvanometer throw due to the magnetic induction thus caused.* Since $\mathcal{H}$ is proportional to this current and $\mathcal{B}$ proportional to the throw, this curve may at once be transformed into a $\mathcal{B}$ and $\mathcal{H}$ curve by changing the scale. Thus from the relation $\mathcal{H} = \frac{4\,\pi z I}{10}$ find the value of I which corresponds to a value of $\mathcal{H} = 10$. Through points on the axis of current values that correspond to the plus and minus of this value of the current draw vertical lines in red ink. With a scale, or a pair of hairspring dividers, divide the space between each line for which $\mathcal{H} = 10$ and the axis, or line for which $\mathcal{H} = 0$, into ten equal parts. Lay off on either side of these two vertical lines other points corresponding to $\mathcal{H} = 11$, $\mathcal{H} = 12$, etc. Through all these points draw vertical lines in red ink and number accordingly.

Similarly, the transformation of the vertical axis into an axis giving values of $\mathcal{B}$ may be made. Thus if H represents the value of the horizontal component of the earth's field at the point where the earth inductor is placed, A' the total area of the inductor, and d' the average throw of the galvanometer when the inductor is rotated, and if n is the number of turns on the test coil, a the average area of the iron ring (for all the lines pass through it), and d the total throw which the galvanometer would make because of the insertion of $\mathcal{B}$ lines per square centimeter through the iron ring, then, since the throws are proportional to the change in the number of lines, we have

$$\frac{d}{d'} = \frac{\mathcal{B} na}{2 HA'}, \qquad \text{or} \qquad \mathcal{B} = \frac{2\,A'H}{nad'}\,d. \tag{12}$$

Solve this relation to find the value of the galvanometer throw d which corresponds to a change of induction $\mathcal{B}$ of 10,000. Call the point on the vertical axis which corresponds to this value of the deflection 10,000, and divide the space between this point and the horizontal axis into ten equal parts as before. Continue this division to either side of the points representing $\mathcal{B} = +10{,}000$ and $\mathcal{B} = -10{,}000$ in order to obtain $\mathcal{B} = 11{,}000$, etc., and rule through these divisions horizontal red ink lines. Reading on the scale represented by the red lines, we now have the $\mathcal{B}$ and $\mathcal{H}$ curve for the specimen.

Plot on the same sheet and draw the curve in red ink, showing the relation $\mu = \mathcal{B}/\mathcal{H}$, as found from the ascending curve of magnetization. Use the axis of $\mathcal{B}$ already drawn as one of the axes of this new curve. Tabulate the values used in plotting the $\mathcal{B}$ and $\mathcal{H}$ curve in one corner of the sheet. Below these place the value of the retentivity and of the coercive force as found from the curve after the manner of section 109.

* If the hysteresis loop is not found to close completely at the upper end, it is because the iron has not been carried through this particular cycle a sufficient number of times to lose previous tendencies which were not entirely eliminated by demagnetizing.

EXAMPLE

The coil surrounding the ring had a total of 579 turns. The average radius of the ring was 4.75 cm. Hence

$$z = \frac{579}{2\,\pi\,4.75} \qquad \text{and} \qquad \mathcal{H} = 4\,\pi\,\frac{579}{2\,\pi\,4.75}\cdot\frac{I}{10}.$$

Hence a current of .4102 amperes produced a field of $\mathcal{H} = 10$ gausses. The test coil consisted of 80 turns. The area a through which the flux passed was that of the iron core, since all the lines were in the iron ring. The average diameter of the ring was 1.434 cm. Hence the area was $\pi \times .717^2$ sq. cm. The earth inductor had a total area found by multiplying 600, the number of turns, by 208.7, the average area. The mean throw produced by the earth inductor was 1.40 cm. The value of H as found in Experiment 13 was .1845 gausses. Hence a deflection of $d = 39.14$ cm. on the scale corresponded to a value of $\mathcal{B} = 10{,}000$ (see eq. 12).

The curves plotted were of the form of those in Figures 123 and 126. The maximum value of I used was .495 ampere. This corresponded to a value of $\mathcal{H}$ of 12.05 gausses, and a value of $\mathcal{B}$ of 16,300 lines. The coercive force was found to be 1.5 gausses and the retentivity 61 per cent. The permeability reached a maximum value of 2470 when the magnetizing force $\mathcal{H}$ was 2 gausses.

CHAPTER XVI

ELECTROLYTIC CONDUCTION

116. Early views of electrolytic conduction. Faraday and his immediate successors regarded the decomposition which attends the passage of a current through an electrolyte * as an immediate *result* of the electric current. Thus, when hydrochloric acid (HCl) was dissolved in water, they imagined the acid molecules to remain intact in the solution, each molecule consisting of a positively charged atom of hydrogen (H) and a negatively charged atom of chlorine (Cl) which clung together under the influence of their mutual attractions. When the positively and negatively charged terminals of a battery were immersed in the solution the molecules were supposed to be split up, by the influence of the charges on the battery terminals, into positive hydrogen ions and negative chlorine ions. These ions migrated through the solution in opposite directions to the two opposite electrodes. As the H ions collected about the negative electrode some of them touched it and gave up their positive charges. They received in return negative charges and were then in condition to unite with the free H ions which had not yet lost their positive charges, and thus to form neutral molecules of hydrogen gas (H_2), which collected as bubbles on the plate or rose to the air above. †

The mechanism by which the ions passed through the solution was supposed to be as follows. Under the influence of the electric

* The student may well reread in this connection sections 24 and 31.

† In terms of the electron theory this would mean that the positive H ion, which is positive because it lacks one electron, receives one from the electrode in giving up its charge, and then one in excess by virtue of which it is repelled from the plate, and is put into condition to unite with another positively charged H ion to form a neutral molecule of hydrogen. The result would of course be precisely the same if all of the H ions gave up their positive charges to the plate and then in some not understood way united to form hydrogen molecules.

field due to the electrodes the HCl molecules first oriented themselves as in the figure (Fig. 131), the + half turning toward the − electrode and the − half toward the + electrode. When the field became strong enough, decomposition took place; the + ion of molecule 5 went to the − electrode, while the − ion of 5 at once united with the + ion of 4, the − of 4 with the + of 3, and so on throughout the entire chain. Thus the current was transferred by virtue of the continual exchange of partners taking place in the HCl molecules in the solution. This picture is due to Grotthus (1805), and is known as "the Grotthus chain theory" of electrolytic conduction.

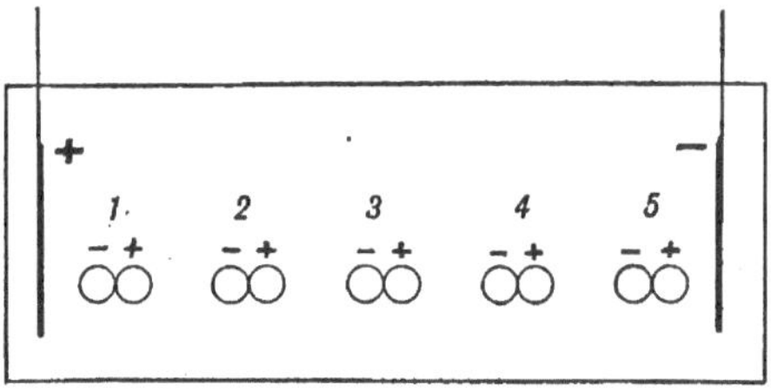

FIG. 131

117. Clausius's dissociation hypothesis. The first serious objections which were raised to this Grotthus chain theory were brought forward by the German physicist Clausius in 1857. The chief argument which he advanced against the point of view of Grotthus was as follows. The view demanded that a certain definite P.D. exist between the electrodes before any decomposition could occur, and therefore before any current whatever could flow. As soon as this critical P.D. has been exceeded the current should rise instantly to a considerable value. As a matter of fact, the current through an electrolyte does not behave in this manner. It is true that it does require a certain critical P.D. to produce a *continuous* decomposition of an electrolyte if the electrodes are of different material from that deposited on the cathode,* for then a back E.M.F. of polarization is introduced because of the formation of essentially new plates by the deposit; and this back E.M.F. must be exceeded before any continuous separation of the elements in the solution can take place. If, however, the electrodes are of the same nature as the metallic ion in solution, as when copper electrodes are dipped into a copper sulphate solution, then the current which

* See section 75, page 113.

flows is directly proportional to the P.D. applied. In other words, Ohm's law holds for such a solution as well as for metallic conductors, and there is no critical P.D. below which the current is zero.

In order to account for this fact, Clausius suggested the following modification of the theory of electrolytic conduction. He held that in every electrolyte the molecules of the substance in solution are *already dissociated,* in part at least, into their electrically charged constituents. The mechanism by which this dissociation was effected might be left unsettled. Clausius's own view was that in the impacts which the molecules make with the water molecules an occasional collision is so violent as to split up the molecule into its constituents. Each of these electrically charged particles then moves about among the water molecules until it meets another particle produced by dissociation, but of opposite kind, when it unites with it to form a new molecule of the compound. A condition of so-called active equilibrium is then set up when the number of molecules which become dissociated per second is equal to the number of new molecules formed per second by recombination.

According to this view at any instant in any conducting solution a fraction of the molecules is dissociated into its ions, and these free ions begin to move toward the electrodes the moment a P.D. is applied. *The current does not produce the dissociation, but rather the dissociation is the necessary condition for the passage of a current.* Substances which do not in part dissociate spontaneously upon going into solution cannot conduct. This view accounts for the fact that a critical value of the electrical field established between the electrodes is not required to start a current, for according to it the field, instead of having to pull the ions apart, has merely to set those which are already separated into motion toward the appropriate electrodes. Furthermore, under these circumstances Ohm's law would be expected to hold, for doubling the applied P.D. would merely double the *speed* with which the free ions move toward the electrodes, and hence the number which give up their charges per second to the electrodes.

118. Osmotic phenomena. This dissociation theory did not receive much attention until 1887 when it received strong support from an unexpected source, namely, the epoch-making work of Van 't Hoff upon *osmotic* phenomena.

The *osmotic pressure* of a solution is most directly measured by measuring the force with which the pure solvent (for example, water) tends to enter the solution through a membrane which is permeable to the solvent but not to the dissolved substance. It is most easily and most accurately measured, however, by observing the number of degrees the freezing point of the given solution is depressed below the freezing point of the pure solvent. Now it was discovered that while the osmotic pressure exerted by substances in solution is in general simply proportional to the number of molecules dissolved in a liter, and is not at all dependent upon the nature of the dissolved substance, this is not true for electrolytes, for these, in general, show abnormally high values of this pressure. Indeed, the osmotic pressure produced by introducing a given number of molecules of NaCl or KCl (for example, one gram-molecule*) into a given number of liters of water is, in dilute solutions, exactly twice as great as the osmotic pressure produced by introducing the same number of molecules of sugar into the same number of liters of water. The most natural interpretation of this phenomenon is that each of the molecules of NaCl or KCl breaks up into two parts upon going into solution.

When it is found that practically all electrolytes exhibit this abnormal osmotic pressure as measured by the depression of the freezing point per gram-molecule of dissolved substance; that no substances except electrolytes exhibit abnormally high osmotic pressures; that, in dilute solution, substances like NaCl and KCl, which can break up into two ions only, never show more than double the normal depression as observed for nonelectrolytes,

* A "gram-molecule" of any substance is a mass of the substance in grams equal to its molecular weight. The advantage in comparing gram-molecules instead of grams of different substances lies in the fact that the gram-molecule represents for all substances exactly the same number of molecules. A solution containing *one gram-molecule per liter* of the solution is called a *normal solution.*

while substances like $SrCl_2$ which might give three ions ($\overset{++}{Sr}$, $\overset{-}{Cl}$, $\overset{-}{Cl}$) actually do show between two and three times the normal depression, but never more than three times,* it is evident that Clausius's theory of dissociation has received powerful support. Furthermore, these phenomena seem to indicate that in dilute solutions the dissociation of some substances is practically complete; that is, that *all* of the molecules of the solute are split up into their constituent ions. Further light has been thrown upon this subject by a study of so-called "molecular conductivity."

119. Molecular conductivity. The significance of the term *molecular conductivity* may be seen from the following illustration. Imagine a vessel of indefinite height, of rectangular horizontal cross section, and provided with two platinum sides 1 cm. apart. Let a normal solution of some electrolyte be placed in this cell and the resistance between the platinum plates measured. The reciprocal of this resistance is called the molecular conductivity at a dilution of one. Let a liter of pure water be added to the contents of the vessel and the resistance measured again. The reciprocal of this is the molecular conductivity at a dilution of two. Similarly the molecular conductivity at any dilution may be found. It is evident that under these circumstances a comparison is made of the conductivities at different dilutions of the same number of molecules placed between plates which are the same distance apart; hence the name *molecular conductivity.*†

Now the molecular conductivities of nearly all electrolytes are found to increase with increasing dilution and to approach definite

* The lowering of the freezing point produced by .01 gram-molecule to one liter of water (that is, a 1/100 normal solution) is .01857° C. for sugar. (See Whetham, "Theory of Solution," p. 320.) Twice this value is .03714 and three times is .05571. The observed depressions produced by some of the substances which could split up into only two ions are as follows: KOH, .0371; HCl, .0361; KCl, .0360; NaCl, .0367. The depressions produced by substances which could split up into three ions are illustrated by the following:

$$H_2SO_4(\overset{+}{H}, \overset{+}{H}, \overset{--}{SO_4}).0449, \quad Na_2SO_4(\overset{+}{Na}, \overset{+}{Na}, \overset{--}{SO_4}).0509,$$

$$CaCl_2(\overset{++}{Ca}, \overset{-}{Cl}, \overset{-}{Cl}).0504, \quad MgCl_2(\overset{++}{Mg}, \overset{-}{Cl}, \overset{-}{Cl}).0508.$$

† This illustration is taken from Walker, "Introduction to Physical Chemistry," p. 227.

values as the dilution becomes very great. This is evident from the following table of observations on NaCl made by Kohlrausch.

Dilution	Molecular Conductivity
1	69.5
2	79.7
10	86.5
20	89.7
100	96.2
500	99.8
1,000	100.8
5,000	101.8
10,000	102.9
50,000	102.8
100,000	102.4

From Clausius's view as to the cause of dissociation the significance of the fact that the molecular conductivity approaches a limiting maximum value at infinite dilution is as follows. The chance of a molecule being split up by an unusually violent impact against a water molecule is the same at all dilutions, but the chance of unlike ions reuniting into molecules becomes infinitely small at infinite dilution. Hence at infinite dilution the dissociation ought to be complete, as the experiments on osmotic pressure given in the note on page 180 indicate is the case for some at least of these substances, even when the dilution is no greater than 100. Hence it was suggested by Arrhenius in 1887 that the per cent of dissociation of all electrolytes at infinite dilution might be considered 100, and the dissociations at any dilution determined solely from molecular conductivity measurements. The manner in which this can be done may be seen from the following considerations due to Arrhenius and Kohlrausch.

Molecular conductivity can vary with dilution only because of the variation of one of two factors. (1) The per cent of the molecules introduced which are dissociated into ions may change with the dilution; (2) the speed with which the ions move through the solution may change with dilution. Now if a change in dilution caused no change at all in the resistance which the individual ions experience in moving through the solution, that is, in the ionic speeds, then the molecular conductivities at two dilutions would obviously be proportional to the number of ions present in each case, that is, to the percentage of the total number of molecules introduced which have become dissociated into ions. If m represents the molecular conductivity of a substance at any given dilution, and m_∞ the molecular conductivity at infinite dilution, and if n and n_∞ correspond to the respective number of ions present in the two cases, then, if the above supposition were correct, we should have

$$\frac{m}{m_\infty} = \frac{n}{n_\infty}. \tag{1}$$

On the other hand, if the dissociation were uninfluenced by the dilution, but if the resistance which the ions meet in moving through the solution were proportional to the viscosity of the solution, then obviously the ionic speeds would be inversely proportional to the viscosities at the dilutions considered. That is, if η and η_∞ represent the viscosity coefficients at the given dilution and at infinite dilution respectively, then we should have

$$\frac{m}{m_\infty} = \frac{\eta_\infty}{\eta}. \tag{2}$$

If both the dissociation and the viscosity change with dilution, we have, by combining the two variations expressed in equations (1) and (2),

$$\frac{m}{m_\infty} = \frac{n}{n_\infty}\,\frac{\eta_\infty}{\eta}, \tag{3}$$

or

$$\frac{n}{n_\infty} = \frac{m}{m_\infty}\,\frac{\eta}{\eta_\infty}.\text{*} \tag{4}$$

If at infinite dilution the dissociation is complete, then obviously n/n_∞ represents the fraction of all the molecules introduced which are split into ions at the dilution corresponding to m. This is usually represented by the letter α. We have, then,

$$\alpha = \frac{m}{m_\infty}\,\frac{\eta}{\eta_\infty}.\dagger \tag{5}$$

As a matter of fact the dissociation of electrolytes as computed from equation (4) does not agree perfectly with the dissociation

* The molecular conductivities m and m_∞, as well as the viscosity coefficients η and η_∞, all refer to the same temperature.

† Arrhenius omitted the term η/η_∞, and wrote merely $\alpha = m/m_\infty$ for any given temperature. Indeed, since for dilutions greater than 8 the ratio η/η_∞ is found to differ but little from unity, it is not improper to write the dissociation equation in this form, provided its restriction to dilutions larger than about 8 is understood.

estimated from freezing-point determinations, although in general there is a fair agreement, and in some cases an exceedingly close one. In view of these uncertainties it must be said that the dissociation theory as here developed is not yet fully established.

120. Methods of measuring molecular conductivity. In practice molecular conductivities cannot be measured in the simple manner described in section 119, in which the immersed area of the plates increased proportionally to the dilution, and hence the conductivity changed only because of a change in the molecular conditions due to the dilution. The same result is obtained by measuring the resistance of the solution between plates 1000 sq. cm. in area and 1 cm. apart, and multiplying the reciprocal of this resistance by the dilution. If the molecular conductivity is to be obtained in absolute units, plates of 1 sq. cm. area are taken 1 cm. apart; that is, the absolute values of molecular conductivity are $\frac{1}{1000}$ of those obtained as described above. Since the resistance of a centimeter cube is called *specific resistance* (see p. 71) and since *specific conductivity* is the reciprocal of specific resistance, it will be seen that *molecular conductivity in absolute units is the dilution divided by the specific resistance of the solution under consideration.*

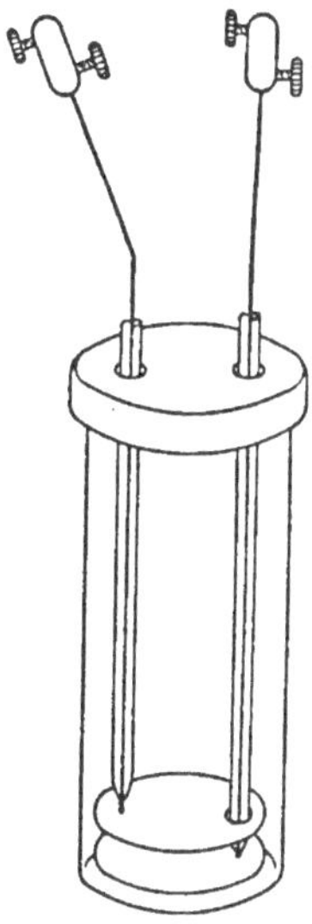

FIG. 132

The vessel most commonly used for molecular-conductivity experiments is shown in Figure 132. It consists of two platinum electrodes supported, as shown, in a vessel made of a kind of glass which is especially insoluble in water. If only *comparisons* of molecular conductivities are desired, they are made by introducing the different solutions to be compared successively into the vessel and dividing the given dilutions by the observed resistances. On account, however, of the fact that in this vessel the current is conducted to some extent by the portion of the solution which is outside the edges of the plates, an absolute determination is usually made on a standard solution in a vessel of another type, and then comparisons are made in the vessel shown above. However, absolute specific resistance determinations of approximate correctness may be made with the cell of Figure 132 by multiplying the observed resistance by the area of the plates and dividing by the distance between them.

The measurement of the resistance is made by making the vessel one branch of a Wheatstone's bridge, as in Figure 133. Although the electrodes are of a different substance from that of the metallic ion of the electrolyte, difficulties due to the back E.M.F. of polarization are eliminated by using a source of alternating current in place of the battery of the ordinary bridge scheme. This obviously

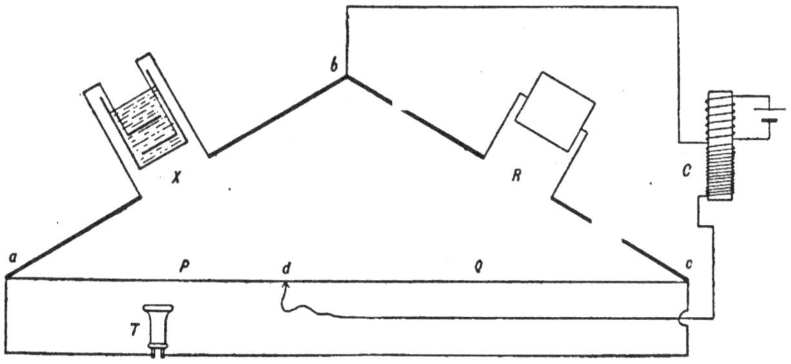

FIG. 133

allows no permanent separation of the elements of the electrolyte at the electrodes. The source of alternating current is the secondary of a small induction coil *C*. A telephone receiver replaces the galvanometer of the ordinary bridge arrangement. The connections of Figure 133 differ from those of Figure 45, page 63, in the fact that the position of the generator and of the detector are interchanged. But, as we have seen, this does not alter the equation of balance.*

It is sometimes difficult in an arrangement of this sort to get a sharp minimum. This is probably because a thin film of some insulating material, presumably gas, collects on the platinum

* That this is so is evident from the discussion in section 46. The reason for this interchange is given in the footnote on page 71. In this case the generator resistance is that of the secondary of the small induction coil and is higher than the resistance of an ordinary telephone receiver.

electrodes, and makes each of them act as a condenser, having the gas film as a dielectric, the platinum and the solution as the two opposite condenser plates. Such condensers in series with the electrolytic resistance would have no effect upon the current, provided their capacities were so large that the opposite sides of the condensers did not charge to any appreciable P.D. before the applied P.D. of the generator reversed its direction. If this were not the case, the introduction of such condensers would cut down the alternating current flowing through the branch containing them and hence destroy the balance. Hence these difficulties may be eliminated by using platinum plates of very great surface, for then the capacities are very large. The method usually employed for increasing the size of the plates is to cover the surfaces of the electrodes with spongy deposits of platinum called "platinum black." *

EXPERIMENT 16

Object. To compare the molecular conductivities of some salt, for example sodium chloride (NaCl), at different dilutions.

Directions. Set up a cell, similar to that of Figure 132, in a large bath of water at the room temperature. Connect as in the diagram of Figure 133, using a meter bridge of which the wire is No. 36 German silver. Prepare a normal solution of NaCl either by weight or by hydrometric measurements with a Mohr's balance,† using the values of the density given in table 5 of the Appendix.

Introduce two pipettes full of this solution into the cell. Measure the resistance and take its reciprocal. Carefully remove one pipette full of solution and replace by a pipette of distilled water at room temperature. The dilution is then 2. Stir by moving the electrodes up and down, but be very careful not to alter in any way their distance apart. Then measure the resistance again. Similarly introduce another pipette full of water, and measure the resistance at the dilution 4. Continue in this way until a dilution of 2048 is reached.

* To platinize the electrodes place them in a 3 per cent solution, by weight, of platinum chloride to which is added a drop or two of lead acetate solution. Pass a current through this cell of such a value that the bubbles rise freely from the solution, then reverse its direction occasionally. Wash in distilled water and allow the electrodes to soak in the water for some time before using them.

† See "Mechanics, Molecular Physics, and Heat," p. 175.

Empty and rinse the cell and determine the resistance of the distilled water used. Take its reciprocal and subtract from the reciprocals of the resistances obtained above, then multiply by the corresponding dilutions to obtain quantities proportional to the molecular conductivities.

Assume the number thus corrected which corresponds to a dilution of 2048 as m_∞, and find α for the other dilutions used from equation (5) on page 182. The values of η and η_∞ may be taken from table 6 of the Appendix, and the ratio assumed correct for the temperature at which the experiment was performed.

EXAMPLE

The distance between the plates of the cell used was about 1 cm., and the diameter of the plates about 2.5 cm. The resistance of the distilled water used was found to be 11,143 ohms. Hence its conductivity was proportional to .0000897. This correction was neglected for dilutions less than 64. The salt used was $ZnSO_4$. The values observed and calculated follow.

Dilution	Resistance	Corrected Conductivity	$\frac{\eta}{\eta_\infty}$	α or the % Dissociation
1	2.27 ohms	.480	1.362	.844
2	3.88	.515	1.166	.848
4	7.16	.559	1.080	.852
8	13.64	.586	1.040	.859
16	26.17	.611	1.020	.879
32	50.41	.635	1.010	.904
64	97.57	.655	1.005	.928
128	189.6	.671	1.	.946
256	368.1	.673	1.	.949
512	706.7	.679	1.	.958
1024	1293.	.701	1.	.989
2048	2294.	.709	1.	
4096	3747.	.709	1.	

$m_\infty \propto .709$ temperature 22°

CHAPTER XVII

VELOCITY OF SOUND IN AIR

121. Sound. Sound is the name applied to the sensations of the auditory nerves. It is the result of the propagation through the air to the ear of a disturbance produced by the sounding body. Thus when a pistol is fired the layer of air surrounding it is suddenly pushed outward and compressed by the expanding gases which emerge from the pistol. This compression is transmitted from layer to layer and constitutes a compressional pulse traveling outward from the pistol. It is this pulse which, giving up a portion of its energy to the eardrum, results in a stimulation of the nerves of the ear and the sensation of sound. For the purposes of physics the subject of Sound is commonly limited to the study of the causes and nature of these pulses and the mechanism of their transmission.

122. Velocity of a compressional pulse in any elastic medium. Consider, then, the velocity of propagation of a compressional pulse in an elastic medium. For convenience the medium will be assumed to be contained in a tube of one square centimeter cross section and of infinite length. Let the density of the medium be

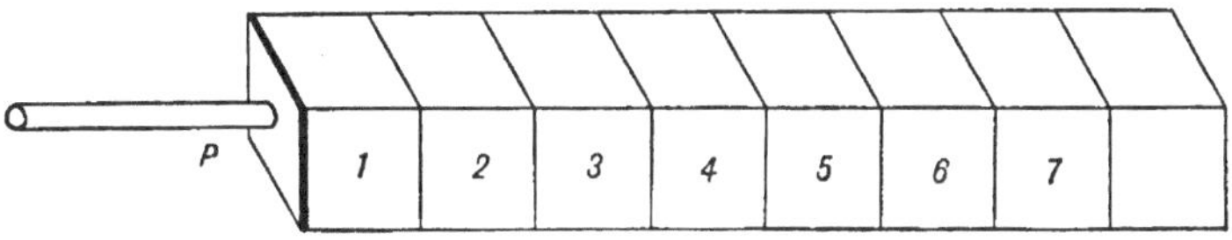

FIG. 134

represented by ρ and the pressure under which it stands before any compression is produced by p. For convenience of analysis let the medium be conceived to be divided into centimeter cubes, 1, 2, 3, 4, 5, etc. (Fig. 134). Let one end of the tube be closed by a frictionless, weightless piston. Now let the piston be suddenly started forward by the application to it of a constant pressure slightly greater than p, viz. $p + dp$. There will thus be started down the tube a compressional pulse which will travel through the medium with a velocity which will be represented by S. It is this velocity which it is desired to determine.

As soon as the piston begins to move forward, cube 1 begins to be compressed. The pressure inside of this cube rises. When this pressure has reached the value $p + dp$, the cube will cease to be compressed and will thenceforth merely transmit pressure to cube 2. Call dv the amount of compression, measured in fractions of a centimeter, which cube 1 thus experiences. Under the action of the pressure $p + dp$, which is thus transmitted by cube 1 to cube 2, the latter will also be compressed an amount dv and will thereafter merely transmit the pressure $p + dp$ to cube 3. Similar reasoning may be applied to the remaining cubes in their numerical order. Since the cross section of the tube is 1 sq. cm., while any particular cube, such as 6, is being compressed the amount dv cc., the piston will move forward dv cm., and each of the cubes previously compressed, namely 1, 2, 3, 4, and 5, will move forward dv cm. Thus, as the pulse moves down the tube, the piston, and with it all the compressed cubes, will move uniformly forward. Since S represents the velocity of the pulse, and since the cubes were chosen one centimeter long, it follows that in one second each of S cubes will experience the compression dv. During that second the piston will therefore have moved forward Sdv cm. That is, the velocity of the piston is Sdv cm. per second. Evidently this is also the expression for the velocity with which all of the cubes which have been compressed are moving forward at the end of a second.

Now the velocity with which the pulse moves forward may be found by an application of Newton's principle of work * as follows. In the above operation the acting force is the pressure applied to the piston, viz. $p + dp$. The work done by this force in one second is $(p + dp)Sdv$. The effects of the action of the force are of two kinds: (1) the compression of all the gas contained in S cubes

* This method of analysis is found in Edser's "Heat for Advanced Students."

from a condition in which it exerts a pressure p to one in which it exerts a pressure $p + dp$; and (2) the communication of a velocity Sdv to all the mass contained in S cubes. The mean force overcome in the first operation is half the sum of the initial and final pressure, namely $p + dp/2$, and the work done against this force is $(p + dp/2)\,Sdv$. In the second operation, since the mass of each cube is ρ, and since in one second S cubes are set in motion with a velocity Sdv, the kinetic energy imparted per second is $\frac{1}{2} \cdot \rho S\,(Sdv)^2$. Now by the scholium to Newton's third law of motion, the work done by the acting force is equal to the work done against the resisting forces; i.e. it is equal to the sum of the potential and kinetic energies imparted to the gas. Hence we have

$$(p + dp)\,Sdv = \left(p + \frac{dp}{2}\right) Sdv + \tfrac{1}{2} \cdot \rho S(Sdv)^2. \tag{1}$$

$$\therefore \ \frac{dp}{2} = \frac{\rho S^2 dv}{2}.$$

$$\therefore \ S^2 = \frac{dp}{\rho dv}, \quad \text{or} \quad S = \sqrt{\frac{\frac{dp}{dv}}{\rho}}. \tag{2}$$

Now since we are dealing with unit cubes the quantity dp/dv is the force applied per unit area divided by the change in volume per unit volume, and this is by definition the volume modulus of elasticity E of the medium. Hence, finally, we have for the velocity of a compressional pulse in any medium,

$$S = \sqrt{\frac{E}{\rho}}.$$

Since this expression involves only the constants E and ρ of the medium, it is evident that a pulse once started will travel on and on down the tube at a rate which has nothing whatever to do either with the size or shape of the tube, or with whether the piston continues to move forward or not. In other words, we have deduced a general expression for the velocity of a compressional pulse in any medium. The expression must hold for the velocity of propagation of a sound pulse which originates at a point within any medium and spreads radially from the center of disturbance;

for, in this case, as in the case just discussed, the pulse is one of pure compression, since the particles are free to move only in one direction, namely along radii emanating from the point of disturbance.

It will be interesting to see how well the results obtained by the use of this formula, which has been deduced from purely theoretical considerations, agree with the results of direct experiment. In 1893 Amagat in Paris made some experiments on the compressibility of water, and found that at 10°C. a change in pressure from 1 atmosphere to 50 atmospheres produced a change in volume from 1 cc. to .99757 cc.* From this data we get as the theoretical value of the velocity of sound in water, all quantities being expressed in absolute units,

$$S = \sqrt{\frac{49 \times 76 \times 13.6 \times 981}{1 \times .00243}}^{\dagger} = 143{,}000 = 1430 \text{ meters per second.}$$

The result of direct measurement made at 8°C. in Lake Geneva in 1827 by Colladon and Sturm gave 1435 meters per second.‡ The difference between the two values is well within the limits of observational error.

123. A train of waves. If in the case of the tube and piston of Figure 134 the applied pressure had been $p - dp$ instead of $p + dp$, the piston would have started back instead of forward, and cube 1 would then have expanded until its pressure reached the value $p - dp$, which expansion would have been followed by a similar expansion of cube 2, etc. Thus a pulse of rarefaction instead of one of condensation would have traveled down the tube, and reasoning in every respect identical with the above shows that the velocity of this pulse also would have been $\sqrt{E/\rho}$.

It is important to observe that *in a pulse of condensation the particles are always moving in the same direction as the pulse; whereas in a pulse of rarefaction the direction of motion of the particles is always opposite to the direction of propagation of the pulse.*

* Wüllner, "Experimental Physik," Vol. II, ed. 5, p. 100.

† 981 is the value of the acceleration of gravity in Paris.

‡ Wüllner, "Experimental Physik," Vol. I, ed. 5, p. 955.

If the piston is moved alternately forward and back at regular intervals, a succession of compressions and rarefactions will follow one another down the tube. Such a succession of compressions and rarefactions is called a train of waves. In this case it is evident that the motions of all of the particles of the medium follow, in succession, exactly the motions of the piston; that is, each particle moves forward during just the interval of time during which the piston is moving forward, and back during just the interval during which the piston is moving back. If then the piston is replaced by the vibrating prong of a tuning fork or by any body which vibrates under the influences of its own elasticity, the backward motion will begin at exactly the instant at which the forward motion ends, and hence at the end of one complete vibration of

Fig. 135

the prong, i.e. at the end of the time required for the prong to go from A to C and back again to A (Fig. 135), the whole of the medium between B and some point a to which a sound pulse travels during the period of one vibration, may be divided into two equal parts, ac and cB, such that all the layers between c and a are moving forward and are in a state of compression, while all the layers between c and B are moving back and are in a state of rarefaction. The relative velocities with which the layers are moving forward or back at the various points between B and a are represented by the lengths of the arrows in the figure. As the fork continues to vibrate the whole region about it becomes filled with a series of such waves, each consisting, as above, of a condensation and a rarefaction (see Fig. 135). The distance between the beginnings of two successive condensations or two successive rarefactions, or in general the distance between any two particles which are in the same condition or *phase* of vibration, is called a wave length. It is obvious that if S represent the velocity of sound, n the number of vibrations of the fork per second, and λ the wave length, then the following relation holds:

$$S = n\lambda. \tag{3}$$

The only distinction between a musical note and a mere noise is that the former consists of a train of waves, while the latter consists either of single pulses, or of irregularly timed pulses.

124. Newton's deduction of the velocity of sound in air. If dp represent the change in pressure which is applied to any body of volume V, and if dV be the change in volume produced by dp, then, by definition, the volume modulus of elasticity E is given by

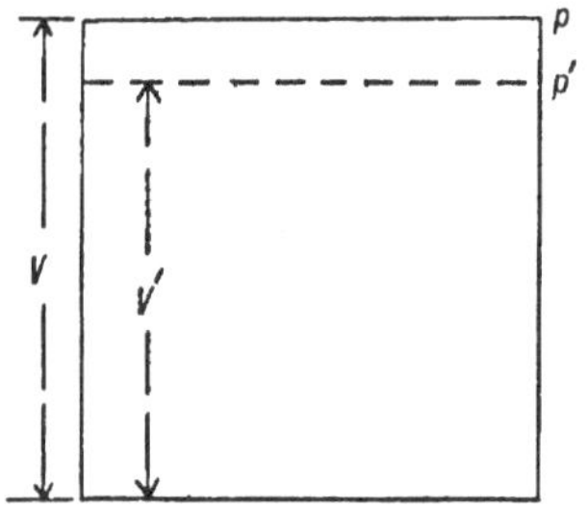

FIG. 136

$$E = \frac{dp}{\frac{dV}{V}}. \tag{4}$$

Let the body be a gas of volume V and pressure p, and let the applied force change the volume of the gas to V' and its pressure to p' (see Fig. 136). Then

$$E = \frac{dp}{\frac{dV}{V}} = \frac{p' - p}{\frac{V - V'}{V}}. \tag{5}$$

But if the heat which is developed by the compression is allowed to pass off, so that the final temperature is the same as the initial (i.e. if the change is isothermal), then Boyle's law may be applied. This gives $p/p' = V'/V$, or $V' = pV/p'$. Substitution of V' in (5) gives

$$E = \frac{p' - p}{\frac{p' - p}{p'}} = p'. \tag{6}$$

Now if, as is the case with sound waves, p' differs from p by an immeasurably small quantity,* then it is possible to replace p' by p.

* Lord Rayleigh estimates that for certain feeble, though distinctly audible, sounds dp is not more than 6×10^{-9} atmospheres.

In other words, *the volume coefficient of elasticity of a gas for isothermal changes is the pressure under which the gas stands.* Hence if a sound wave produces no temperature changes in a gas we have

$$\text{velocity} = \sqrt{\frac{\text{pressure}}{\text{density}}}. \qquad (7)$$

If the gas is air under ordinary conditions, then p is simply the barometric pressure, which must, of course, be reduced to dynes if S is to be determined in centimeters. This theoretical value for the velocity of sound in air, first deduced by Newton (1687), does not agree with experimental determinations, for the simple reason that the compression produced by a sound wave does not take place *isothermally,* as Newton assumed it to do.

125. Correct formula for velocity of sound in air. It was a hundred years after Newton's time when La Place first pointed out that a gas is heated by a compressional wave, and on account of the great velocity of sound the heat thus developed has no time to diffuse before the wave is past. When all the heat developed by compression is retained in the compressed body the change in pressure is said to take place *adiabatically.* Since bodies expand with heat, it is evident that the change in volume produced in a gas by a given change in pressure will be greater if the heat of compression is allowed to pass off than if it is retained. That is, dV/V is less if the change is adiabatic than if it is isothermal. Hence the quantity $E = (dp/dV)V$ is greater for an adiabatic change. Analysis which is beyond the scope of this text shows that the bulk modulus of a gas for an adiabatic change is equal to its bulk modulus for an isothermal change multiplied by a factor γ. This factor γ is the ratio of the specific heat of the gas at constant pressure and its specific heat at constant volume. It is always greater than unity; for air it has a value 1.403; for carbon dioxide (CO_2), a value 1.30. Hence *the theoretical value for the velocity of sound in a gas* is

$$S = \sqrt{\frac{\gamma p}{\rho}}. \qquad (8)$$

126. Effect of temperature upon the velocity of sound. If the pressure is changed without a change in temperature, then by Boyle's law the density change is directly proportional to the pressure change; that is, $p/\rho = \text{constant}$. Hence *the velocity of sound in gases is wholly independent of pressure.*

The dependence of the velocity upon the temperature of the gas may be found by the following reasoning. Suppose the pressure of the gas to remain constant. Let S_t represent the velocity at a temperature of $t°$ C., and ρ_t the corresponding density. Similarly, let S_0 and ρ_0 represent the velocity and density at 0°. The velocity at the temperature $t°$ is given by the equation $S_t = \sqrt{\gamma p/\rho_t}$, the velocity at 0° by the equation $S_0 = \sqrt{\gamma p/\rho_0}$. Therefore

$$\frac{S_t}{S_0} = \sqrt{\frac{\rho_0}{\rho_t}}. \tag{9}$$

Now by Gay-Lussac's law the density of a given gas is inversely proportional to its absolute temperature. Hence

$$\frac{S_t}{S_0} = \sqrt{\frac{\rho_0}{\rho_t}} = \sqrt{\frac{T}{T_0}},$$

or

$$S_t = S_0\sqrt{\frac{T}{T_0}}, \tag{10}$$

in which T_0 is 273 and T is the temperature of the gas on the Centigrade scale plus 273. By substituting in this formula 274 for T and for S_0 the best determination of the velocity of sound at 0°, namely 33,127 cm., it will be seen that the velocity of sound increases about 60 cm. for each degree Centigrade of rise in temperature.

127. Reflection of wave trains produced by a change in the density of the medium. Returning to the consideration of the tube and piston, suppose the latter to move forward suddenly a short distance and then stop. A compressional pulse will start down the tube. The progression of this pulse is due to the communication from layer to layer of the motion which is imparted by the piston to the layer next it. The case is precisely analogous to that of the collision of perfectly elastic equal balls. *So long as* *the pulse travels in a medium of uniform density each layer of particles gives up all its motion to the next layer,* which is precisely like it, just as a moving elastic ball striking a stationary elastic ball of the same size gives up all of its motion to the stationary ball and itself comes to rest. But if the pulse at some point *A* (Fig. 137) strikes a denser medium *D*, the case becomes analogous to the impact of a lighter ball upon a heavier one. The lighter layer adjoining *A* on the left, instead of coming to rest in the impact, reverses its motion and starts back. This backward motion is communicated from layer to layer so that a reflected compressional pulse travels from *A* back toward *P*. Let the long arrow represent the direction of the pulse and the short one the direction of motion of the particles in the pulse. It is evident that the direction of motion of the particles with respect to the direction of motion of the pulse is the same in the returning as in the advancing pulse. In other words, *a pulse of condensation is reflected from a denser medium as a pulse of condensation.*

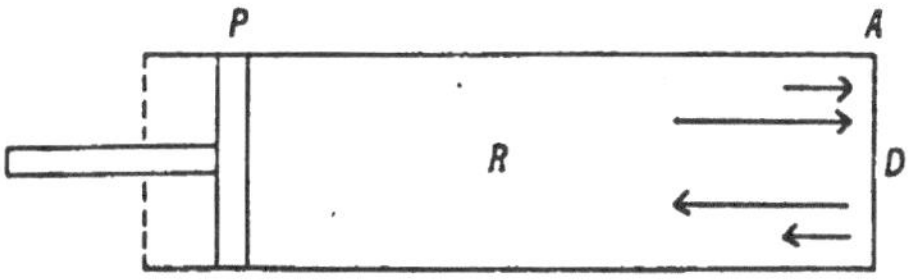

Fig. 137

But suppose the medium to the right of *A* is less dense than that in the tube (see *R*, Fig. 138), then the case is analogous to the impact of a heavy ball upon a light one. Until the pulse reached *A* each layer in the tube gave up its motion to the next and itself came to rest; but the layer at *A*, instead of coming to rest after the impact, continues to move forward and thus produces a rarefaction, i.e. a diminution of pressure at *A*. The excess of pressure in the layer to the left of *A* then drives particles toward the right. Thus a pulse of rarefaction moves back from *A* toward *P*. In other words, *a pulse of condensation is reflected from a rarer medium as a pulse of rarefaction.* In the pulse approaching *A* the particles move in the direction of propagation of the wave. In the pulse receding

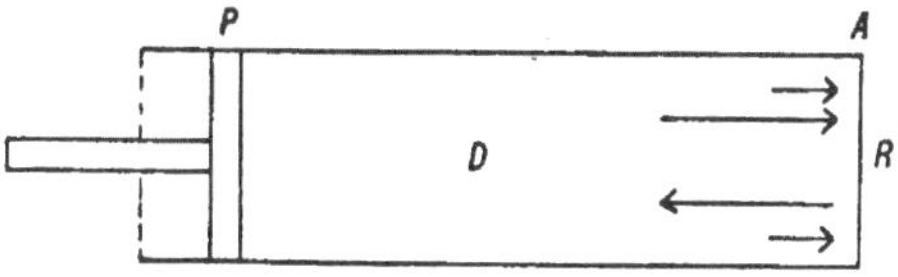

Fig. 138

from A the particles move in a direction opposite to that of propagation of the wave (see arrows, Fig. 138). Since the pulse changes instantly in the reflection from a condensation to a rarefaction, and since in a train of waves a rarefaction always follows just one half wave length behind a condensation, a wave is said to experience a loss of one half wave length in reflection from a rarer medium.

The same process of reasoning shows that a wave of rarefaction is reflected from a denser medium as a wave of rarefaction, but from a rarer medium as a wave of condensation.

128. Resonance of vibrating air columns. When an air wave traveling along a pipe reaches the open end it experiences the same sort of reflection as though it passed from a denser to a rarer medium. This statement can be easily proved experimentally (see sect. 130). It is also evident from the theoretical consideration that as soon as the wave reaches a point at which *lateral expansion* is possible, the forward movement of the particles is greater than inside the tube, where lateral expansion is not possible. This increased forward movement at the end of the tube means a wave of rarefaction starting back in the tube.

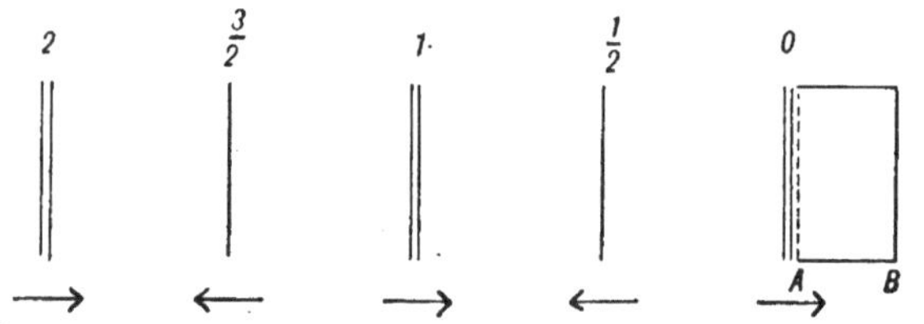

FIG. 139

Consider now a train of waves approaching the open end of a pipe the other end of which is closed (Fig. 139). In the condensations of the advancing train of waves the motions of the particles are all in the direction of motion of the wave, i.e. from left to right. In the rarefactions they are in the opposite direction. Suppose the pipe to have a length of exactly one fourth wave length. Then the condensation which is marked 0 will move down the pipe and be reflected at the closed end as a condensation, i.e. as a motion of the particles now from right to left. It will obviously return to the open end at the exact instant at which the rarefaction which is marked $\frac{1}{2}$, and which also consists of a motion of the particles from right to left, reaches the open end of the pipe. Since the reflected condensation which is returning from the closed end of the pipe now undergoes reflection at the open end as a rarefaction, i.e. as a motion of the particles from right to left, the newly reflected wave of rarefaction which starts back down the pipe unites with the rarefaction marked $\frac{1}{2}$ which is just entering the pipe, and a wave of rarefaction of increased amplitude is the result. This wave is reflected at the closed end as a rarefaction (motion from left to right), and again at the open end as a condensation (motion from left to right), exactly in time to unite with the condensation marked 1 as it enters the pipe. Thus by this process of continuous union of direct and reflected waves the motion in the pipe becomes larger and larger until it may be hundreds of times as large as the motion of the particles in the original wave. In fact, there would be no limit to the amplitude of the waves traveling up and down the pipe if at each end the motion were not partially given up to the outside air. The pipe thus becomes in a way the source of sound. The phenomenon is called *resonance.*

If the pipe had been only a trifle longer or shorter than one fourth wave length, the error in the coincidence of the first reflected wave with $\frac{1}{2}$ would have been but slight. Since, however, the reflected waves are now obliged to travel, each time they go up and down the pipe, a distance a little too great or too small, it takes them but a short time to get completely *out of step* with the advancing waves. In this condition the direct and reflected waves interfere with rather than assist each other, and no resonance is possible. This explains why, when the length is but a trifle more or less than the right amount, very little resonance is obtained.

If, however, the pipe is continually lengthened, other resonant lengths will be reached. The length above chosen was one which permitted 0 to return to the mouth of the pipe exactly in time to unite with $\frac{1}{2}$. This length is manifestly the shortest possible resonant length. It is clear that the next possible resonant length is one which permits 0 to return exactly in time to unite with $\frac{3}{2}$. Since $\frac{3}{2}$ is one wave length behind $\frac{1}{2}$, the second resonant pipe length must be one half wave length greater than the first; i. it must be three fourths wave length. Similarly, it is possibl to obtain resonance when the pipe length is five fourths wav length, seven fourths, nine fourths, and so on.

Experiment shows that the theoretical values of the pipe lengtl required for the fundamental resonance, namely one fourth wav length, is slightly too large. The discrep ancy is explained as follows. If the reflec tion of a wave traveling down the ope pipe *ac* (Fig. 140) took place exactly ir the plane *cd*, the resonant length woul be exactly one fourth wave length; bu in reality the point of free lateral expan sion is not reached at the instant at which the wave reaches *cd* hence the reflection does not occur until the wave has pushe out a short distance beyond *cd*. Thus the *true* pipe length i slightly greater than the *apparent* length. The amount of thi correction which must be applied at the open end is estimate by Rayleigh as the *radius of the pipe*, but in glass pipes it is somewhat less than this.

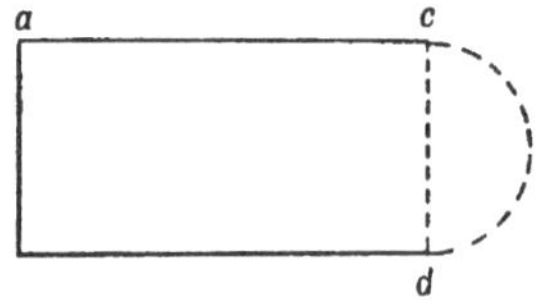

FIG. 140

EXPERIMENT 17

Object. To find the velocity of sound in air from resonance experiments upon a closed pipe, and to find the correction which must be applied to the open end of a pipe.

Directions. The pipe to be used consists of a long glass tube *A* (Fig. 141), one end of which is closed by water admitted by a tube from the vessel *B*. The length of the column of air contained in this pipe may be varied by changing the height of *B*. A rubber band around *A* may be slipped into any position along its length. A train of waves from a tuning fork of known rate is caused to enter the pipe by holding the fork over the mouth of the tube.

Set the fork *T* in vibration by striking it with a rubber mallet, and hold it over the mouth of the tube. It is not important in this part of the experiment that the fork be held farther than three or four millimeters from the end of the tube, provided it is always held in *precisely the same position.* Taking especial care about this point, raise or lower the vessel B and thus change the level of the water in A until the note of the fork is strongly reënforced. By causing the water to rise and fall rapidly several times in the vicinity of the position of reënforcement, the length of maximum resonance can be fairly accurately obtained. Set the rubber band at the level of the water in the tube and measure to it from the top of the tube. In this way determine two or three successive reënforcement points. If then l_1, l_2, and l_3 represent the first, second, and third measured resonant lengths, x the unknown correction which must be applied because the reflection at the open end does not take place exactly in the plane no, the preceding theory gives

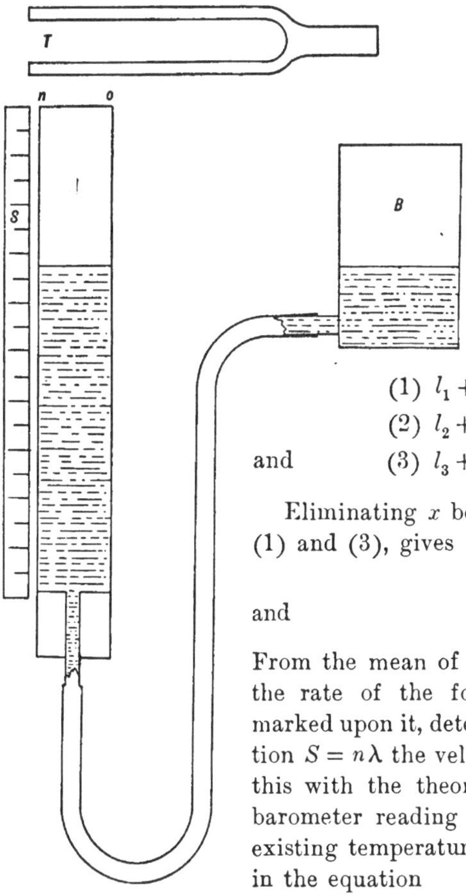

FIG. 141

(1) $l_1 + x = \frac{1}{4}$ wave length,

(2) $l_2 + x = \frac{3}{4}$ wave length,

and (3) $l_3 + x = \frac{5}{4}$ wave length.

Eliminating x both from (1) and (2), and from (1) and (3), gives

$$l_2 - l_1 = \tfrac{1}{2}\lambda,$$

and

$$l_3 - l_1 = \lambda.$$

From the mean of these two values of λ and from the rate of the fork n, which should be found marked upon it, determine with the aid of the equation $S = n\lambda$ the velocity of sound in air. Compare this with the theoretical value, deduced from the barometer reading and the density of air for the existing temperature and pressure, by substitution in the equation

$$S = \sqrt{\frac{\gamma p}{\rho}}.$$

Next repeat very carefully the determination of the first resonant length, the fork being now held at least as far away from the end as the radius of the tube. By subtracting this length from a true fourth wave length, as determined above, find the correction which must be applied at the open end of a pipe to make the first resonant length $\frac{1}{4}\lambda$. This might have been found from the preceding measurements had not the fork been held closer to the end than the radius of the tube. This was done for the sake of making it easier to locate accurately the second and third resonance points which otherwise would have shown but feeble reënforcement. Express the correction x as a fractional portion of the radius of the pipe.

EXAMPLE

The experiment was performed at a temperature of 24.0° C. and a corrected barometric pressure of 74.63 cm. The density of air at this temperature and pressure was found from Table 2 in the Appendix to be .001167. The pressure p, expressed in dynes per square centimeter, was found by multiplying 74.63 by 13.596, the density of mercury at 0°, and by g. Hence S, computed from the theoretical relation $S = \sqrt{\gamma \frac{p}{\rho}}$, was 34,570 cm. per second.

The resonant lengths observed were 15.9, 49.8, and 84.1 cm. Hence the average $\frac{1}{2}$ wave length was 34.0 cm. The fork was marked $n = 500$; but when it was sounded simultaneously with a standard 512 fork, and the two held close to the ear, it was found that there were four beats a second. Furthermore, loading the standard fork slightly with wax was found to decrease the number of beats. Hence the correct rate of the fork was 508. Therefore, since $S = n\lambda$, the observed value of S was $1016 \times 34.0 = 34{,}544$ cm., a value which differs from the theoretical result by but .07 per cent.

When the fork was held 3 cm. above the end of the pipe the first resonant length was found to be 15.7 cm. Hence $x = 17 - 15.7 = 1.3$ cm. The radius of the pipe was 1.8 cm. Hence this correction was .71 of the radius.

129. Notes to which a closed pipe will respond. It has been shown in section 128 that resonance is possible in a pipe closed at one end when the corrected length of the pipe (which will be denoted by P) is an odd multiple of one fourth of the wave length λ of the train of waves entering the pipe; i.e. when $P = \frac{1}{4}\lambda$, or when $P = \frac{3}{4}\lambda$, or when $P = \frac{5}{4}\lambda$, and so on. If the pipe length is kept constant while the wave length is changed, it follows at once that a succession of wave lengths, λ_1, λ_2, λ_3, λ_4, and so on, will be found to which the pipe will respond, and that $\lambda_1 = 4P$, $\lambda_2 = \frac{4}{3}P$, $\lambda_3 = \frac{4}{5}P$, and $\lambda_4 = \frac{4}{7}P$, etc. These wave lengths bear the ratios 1, $\frac{1}{3}$, $\frac{1}{5}$, $\frac{1}{7}$, etc. Hence, since vibration numbers, n, vary inversely as wave lengths (for $S = n\lambda$), the vibration numbers of the notes which are able to produce resonance in a closed pipe must bear the ratios 1, 3, 5, 7, 9, and so on. The note of longest wave length to which a given pipe can respond is called the *fundamental* of the pipe; the notes of higher frequency which give resonance are called its *overtones.*

130. Notes to which an open pipe will respond. If it is true, as stated in section 128, that a wave which is traveling down a pipe is reflected, upon reaching an open end, just as it would be if it had come to a new medium of smaller density than that within the pipe, then it ought to be possible to obtain the phenomenon of resonance with open as well as with closed pipes; and, furthermore, the shortest length of an open pipe which should produce the reënforcement of a given train of waves should be twice as great as the shortest resonant length of a closed pipe. For, if a pulse of condensation 0 (Fig. 142), in which the particles are moving from left to right, is reflected upon reaching B as a pulse of rarefaction, i.e. as a motion of the particles of the returning wave from left to right, it is at once evident that the shortest length of AB for which there can be a union of the direct and reflected waves in the same phase is that which permits 0 to return to A just in time to unite with 1. This means obviously that *the shortest resonant length must be one half the wave length of the train,* instead of one fourth, as in the case of a closed pipe. The fact that a length of open pipe can indeed always be found which will respond just as loudly to a given note as any closed pipe, and that this length is twice as great as that of the shortest resonant closed pipe, may be taken as complete experimental demonstration of the statement made in section 128 as to the nature of the reflection occurring when a wave reaches the open end of a pipe. If the open pipe is gradually lengthened, there should obviously again be resonance when 0 returns to A just in time to unite with

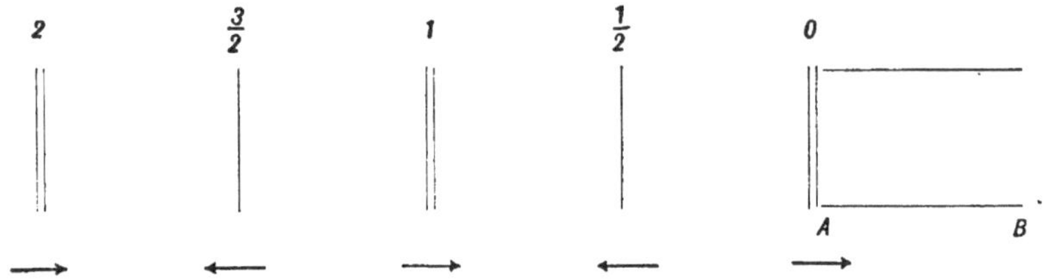

FIG. 142

2; i.e. when the pipe length has been increased by one half wave length, and again when 0 returns in time to unite with 3, etc. In a word, an open pipe should produce resonance when its length is any multiple whatever of $\frac{1}{2}\lambda$; i.e. when the pipe length bears to the wave length any of the ratios $\frac{1}{2}, \frac{2}{2}, \frac{3}{2}, \frac{4}{2}, \frac{5}{2}, \frac{6}{2}$, etc. This is equivalent to saying that the notes which will produce resonance in a given open pipe of fixed length must bear the frequency ratios 1, 2, 3, 4, 5, etc. In other words, while the fundamental and overtones of a pipe closed at one end must bear the frequency ratios represented by the odd numbers only, the fundamental and overtones of an open pipe should bear the frequency ratios represented by *all* the numbers, even and odd. This is often stated thus: *In closed pipes only the odd overtones are possible; in open pipes all the overtones, even and odd, are possible.*

131. Natural periods of pipes. Not only will a given pipe, open or closed, intensify, as explained above, trains of waves of certain definite wave length which present themselves at its mouth, but a single pulse entering such a pipe must be returned, by virtue of successive reflections at the ends, as a succession of pulses following one another at equal intervals. In other words, a single pulse must be given back by the pipe as a musical note, of very rapidly diminishing intensity, it is true, but of perfectly definite wave length. Furthermore, this wave length must be the wave length of the train which is capable of producing the fundamental resonance of the pipe. For, if the pipe is closed, for example, at the lower end, then the first time the pulse returns to the mouth after reflection at the closed end it will produce an outward motion of the particles near the mouth, the next time an inward motion, the next time an outward motion, and so on; i.e. the pulse must travel four times the length of the pipe in the interval between the appearance of two successive condensations at the mouth. The length of the pipe is thus one fourth of the wave length of the note given off by it, and this is the relation which exists in the case of a train of waves producing the fundamental resonance. The pipe is therefore said to have a *natural period*, or to be capable of producing a note of wave length four times as great as its own length.

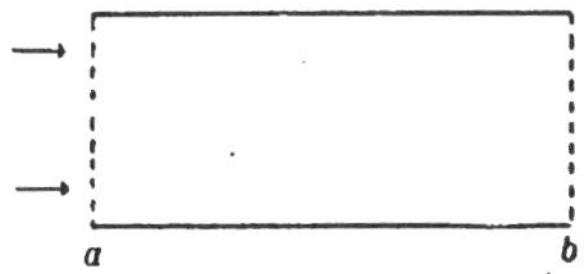

FIG. 143

If the pipe is open instead of closed at the farther end, a single pulse (a condensation) entering at *a* (Fig. 143) will emerge at *b* first as a motion of the particles from left to right. The reflected portion will then travel back through the tube as a motion of the particles from left to right (a rarefaction), which will in turn be reflected at *a*, still as a motion from left to right; and thus, after traveling the length of the tube twice, the pulse will again emerge at *b* in its original direction (a condensation). Thus, in this case, the wave length of the train of waves into which the pipe has transformed the single pulse is twice the length of the pipe; i.e. the note given off by the pipe has, as before, the same wave length as that which will produce the fundamental resonance in the pipe. This note is, of course, an octave higher than the note given off by a closed pipe of the same length. It is this ability of a pipe, open or closed, to pick up irregular pulses and transmute them by successive reflections into notes of definite pitch which explains the continuous humming in definite pitch which is heard when a tube, a sea-shell, or any sort of cavity of sufficient size is held close to the ear.

132. Production of the fundamentals of pipes by air jets. In order, however, that a pipe may be made to give forth its fundamental note distinctly, it is necessary to do more than to start a single pulse in at one end; for the energy of this pulse is dissipated so rapidly in the successive reflections and transmissions that only when the pipe is placed very close to the ear can anything which resembles a musical note be recognized at all. If, however, a gentle current of air is directed continuously against one edge of the pipe, as in Figure 144, the fundamental note can be made, with suitable blowing, to come out very strongly. In order to understand this action, consider first a pipe closed at the lower end, and suppose that the original current of air is so directed as to strike at a just inside the edge (see Fig. 144). A condensation starts down the pipe and is reflected, when it reaches the bottom, as a condensation, i.e. as an upward motion of the particles. When this condensation reaches the mouth it pushes the current of air outside of the edge. This starts a rarefaction down the pipe which, upon its return to the mouth as a rarefaction, draws the current of air inside the edge again. Thus the current is made to vibrate back and forth over the edge, the period of its vibration being controlled entirely by the natural period of the pipe; for between two instants of emergence of the jet from the pipe a rarefaction must travel twice the length of the pipe and then a condensation must do the same; i.e. a sound pulse must travel four times the length of the pipe. Hence the wave length of the emitted note is the same as that which corresponds to the natural period; i.e. it is four times the length of the pipe. The

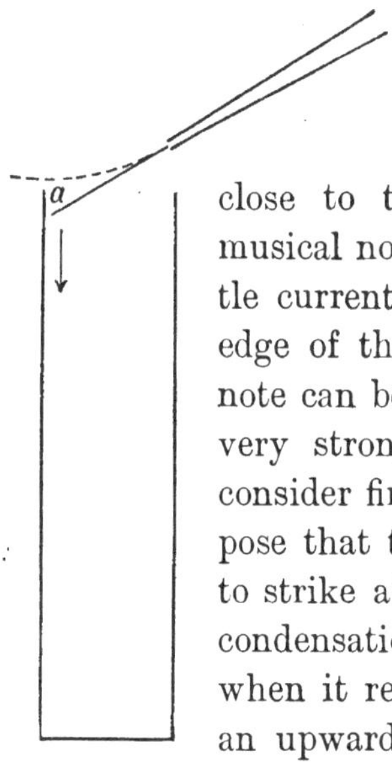

FIG. 144

source of the musical note is to be found, then, in the vibration of the air jet into and out of the end of the pipe. The pipe itself may be looked upon merely as a device for enabling the jet to send pulses to the ear with perfect regularity.

The theory of the open pipe differs only slightly from that of the closed. If the jet is directed just inside the edge, a condensation starts down the pipe, and at the same time, as is indeed also the case with the closed pipe, the pressure within the upper end of the pipe begins to rise because of the influx of air. If the blowing is of just the right intensity, this pressure may force the jet outside the edge at just the instant at which the original condensation reaches the lower end and starts back, in this case as a rarefaction. When this returning rarefaction reaches the mouth it draws the jet inside again. At this instant the rarefaction which started down the pipe when the jet first swung outside has just reached the lower end of the tube. Upon its return to the mouth as a condensation it drives the jet again outside, and thus the jet is alternately forced back and forth over the edge, its period being controlled entirely by the natural period of the pipe, for it will be seen that between two successive emergences of the jet from the mouth of the tube a sound pulse travels down the tube and back. If the blowing is not of just the right intensity so that the pressure reaction near the mouth throws the jet out for the first time at just the instant at which the first condensation reaches the lower end, then the pulses reflected from the lower end do not reach the mouth at the right instants to set up regular vibration of the jet over the edge, and consequently no note is produced.

133. Production of the overtones of pipes by air jets. If, in the case of the open pipe, the violence of the blowing is increased to just the right amount, the pressure within the top of the pipe may be increased so rapidly that the jet is thrown out in just one half its former period. In this case the reflected pulses will get back to the mouth in just the right time to keep the vibration going, but the note given forth will be the first overtone of the open pipe, namely the octave of the fundamental. Similarly, still harder blowing of just the right intensity will cause the jet to swing out in just one third its former period, and the returning pulses will then get back to the mouth in just the time to keep the jet vibrating in the period of the second overtone, the frequency of which is three times that of the fundamental, etc. Blowing of intermediate intensities will produce no notes at all, since the times of return of the reflected pulses are then such as to interfere with the period of vibration which is starting, instead of to keep it going.

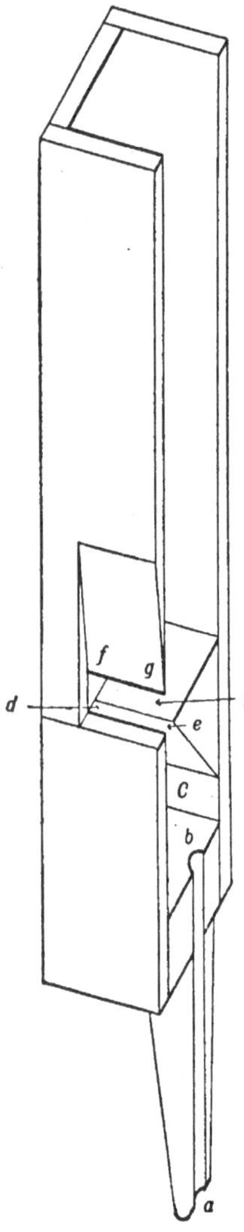

Fig. 145

The production of overtones in closed pipes is precisely similar, save that in order to produce the first overtone the blowing must be so hard as to cause the jet to swing out of the pipe in *one third* of the time required for the first condensation to travel to the bottom and back, for the first overtone of a closed pipe has a frequency three times that of the fundamental, the second five times, etc. (see sect. 129). By blowing with varying degrees of violence across either open or closed tubes, it is generally easy to produce three or four notes of different pitch which are found to have precisely the frequencies demanded by the above theory. If the pipe is long and narrow, it may be quite impossible to produce the fundamental for the reason that the jet is forced out by the increased pressure long before the first pulse returns from the remote end.

134. Types of wind instruments. The above theory explains the action of nearly all wind instruments. In organ pipes (Fig. 145) the current of air is forced through the tube *ab* into the air chest *C*, thence through the narrow slit *de*, into the *embrochure E* or mouth of the pipe, where it passes as a narrow jet toward the thin edge or lip *fg*. As a result of small differences in pressure inside and outside of the embrochure, the jet is caused to deviate to one side or the other of the lip. When the pipe is sounding, it vibrates back and forth across the lip precisely as the air jet vibrated back and forth across the edge of the pipe in the discussion of section 133.

Flutes and whistles of all sorts are precisely similar in their action to organ pipes. In any of them the air chamber may be either open or closed. In flutes it is open; in whistles it is usually closed; in organ pipes it is sometimes open and sometimes closed. In pipe organs there is a different pipe for every note, but in flutes, fifes, etc., a single tube is made to produce a whole series of notes either by blowing overtones or by opening holes in the side, — an operation which is equivalent to cutting off the tube at the hole, since a reflected wave starts back as soon as a point is reached at which there is greater freedom of expansion than has been met with before.

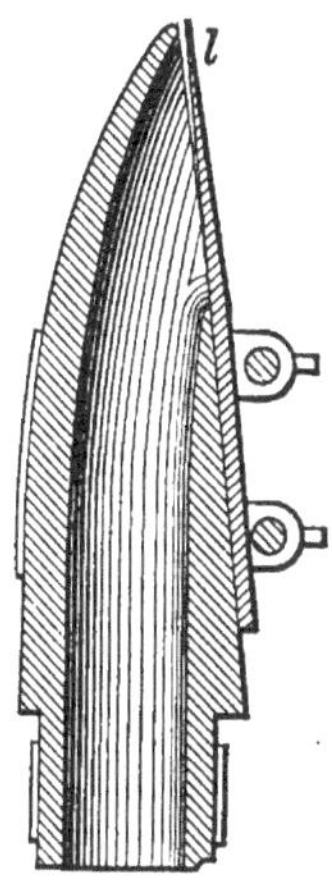

FIG. 146

In the case of some instruments, like the clarinet (Fig. 146), the end of the pipe against which the performer blows is almost closed by a reed l which is loosely pivoted at the base and free to swing, under the influence of an outside pressure, so as to close the opening entirely.

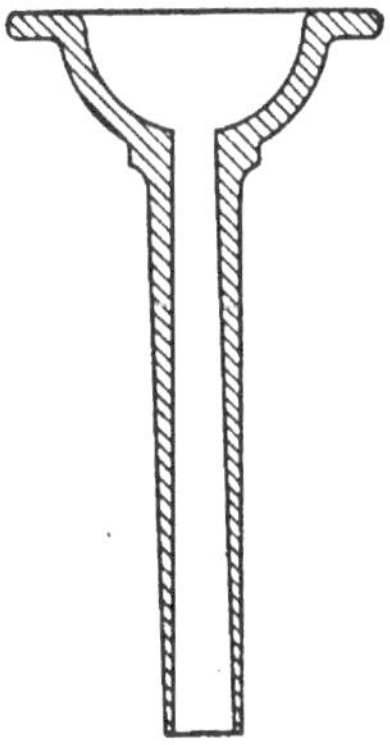

FIG. 147

When the performer blows upon this mouthpiece, a pulse of condensation enters the tube and at the same time the reed closes the opening. This pulse after reflection from the open end of the clarinet as a rarefaction, and a subsequent reflection at the mouthpiece (closed by the reed), also as a rarefaction, is again reflected at the open end, but now as a condensation; and therefore, after traveling the tube four times, the original condensation returns and forces the reed open, admitting a new pulse. The overtones which may be produced in such an instrument are evidently those of a closed pipe. It is evident that the vibration frequency is independent of the reed and depends only upon the effective length of the clarinet.

In the case of instruments like the trumpet and other brass wind instruments the current of air enters a mouthpiece similar to that shown in Figure 147. The lips of the performer act as a double reed. A pulse of condensation enters, the lips closing when the reaction of its pressure equals that of the air in the mouth of the performer. This pulse, reflected as a rarefaction from the open end of the trumpet to the lips, reduces the pressure at that point and a new pulse enters. The fundamental depends then only upon the length of the instrument. The overtones are produced exactly as in an organ pipe, by blowing more suddenly and to some extent by increasing the tension of the lips. The possible overtones are those of an open pipe.

135. The musical scale. The physical basis of harmony in music lies in the simplicity of the ratios of the vibration frequencies of the notes which are sounded together. When a note and its octave are sounded together the result is recognized by the ear as agreeable, and the two notes are said to be consonant. The ratio of the frequencies of the two notes, known as the interval between them, is in this case $\frac{2}{1}$. The next most consonant interval is that between *do* and *sol* (C and G). It is found to be the next most simple physical interval, having a value $\frac{3}{2}$. It is known as a fifth because G is the fifth note above C in the sequence of eight notes which constitutes the octave of the ordinary musical scale. If the note G and the octave of C are sounded together, the interval is $\frac{3}{4}$, as is evident from the fact that $\frac{3}{2}$ of 1 is $\frac{3}{4}$ of 2. These three notes — *do, sol,* and the octave of *do* — thus utilize the three simplest frequency ratios, namely $\frac{2}{1}$, $\frac{3}{2}$, and $\frac{4}{3}$. The next most consonant interval is that between *do* and *mi* (C and E). It is known as *the third* and represents the vibration ratio $\frac{5}{4}$. The notes *do, mi, sol* (C, E, G) sounded together are known as the *major chord.* It will be seen from the above that their relative vibration frequencies are as 4 : 5 : 6.

The so-called major *diatonic scale* is made up of three major chords. The absolute vibration number taken as the starting point is wholly immaterial, but the explanation of the origin of the eight notes of the octave, commonly designated by the letters C, D, E, F, G, A, B, C,′ may be made more simple if we begin with a note of vibration number 24. The first major chord — *do, mi, sol,* or C, E, G — would then correspond to the vibration numbers

24, 30, 36, or, as explained above, the vibration ratios 4, 5, 6. The second major chord starts with C′, the octave of C, and comes down in the ratios 6, 5, 4. The corresponding vibration numbers are 48, 40, 32, the corresponding notes *do′*, *la*, *fa*, and the corresponding letters C′, A, F. The third chord starts with G as the first note and runs up in the ratios 4, 5, 6. This gives the vibration numbers 36, 45, 54, the syllables *sol*, *si*, *re*, and the letters G, B, D′. Since the note D′ does not fall within the octave, its vibration number being above 48, the note D, an octave lower and having a vibration number 27, is taken to complete the eight notes of the major diatonic scale. The chord *do-mi-sol* is called the tonic, *sol-si-re* the dominant, and *fa-la-do* the subdominant. The relations between the notes of the octave are given below in tabular form.

Notes	C	D	E	F	G	A	B	C′
Frequencies	24	27	30	32	36	40	45	48
Intervals with C . . .	1	$\frac{9}{8}$	$\frac{5}{4}$	$\frac{4}{3}$	$\frac{3}{2}$	$\frac{5}{3}$	$\frac{15}{8}$	2
Name of interval . .		major second	major third	fourth	fifth	major sixth	seventh	octave
Intervals with next note		$\frac{9}{8}$	$\frac{10}{9}$	$\frac{16}{15}$	$\frac{9}{8}$	$\frac{10}{9}$	$\frac{9}{8}$	$\frac{16}{15}$
Name of interval . . .		major second	minor second	half tone				

Any scale the notes of which are separated by these intervals is known as a scale of *just temperament*. The scale adopted by physicists starts with middle C = 256. That adopted internationally for musical purposes has A = 435. This gives a series of values slightly higher than that of the physical scale. The scale formed on C = 256 follows.

Absolute names . . .	C	D	E	F	G	A	B	C′	D′	E′	F′	G′	A′	B′	C″
Syllables	*do*	*re*	*mi*	*fa*	*sol*	*la*	*si*	*do*	*re*	*mi*	*fa*	*sol*	*la*	*si*	*do*
Relative frequencies .	1	$\frac{9}{8}$	$\frac{5}{4}$	$\frac{4}{3}$	$\frac{3}{2}$	$\frac{5}{3}$	$\frac{15}{8}$	2	$\frac{9}{4}$	$\frac{5}{2}$	$\frac{8}{3}$	3	$\frac{10}{3}$	$\frac{15}{4}$	4
Absolute frequencies	256	288	320	341.3	384	426.6	480	512	576	640	683	768	853	960	1024

It is often desirable in music to adopt some other note than C as the fundamental note in the scale, i.e. as the *keynote*. Thus the scale formed on G as a fundamental and having intervals of just temperament would be

do	*re*	*mi*	*fa*	*sol*	*la*	*si*	*do*
384	432	480	512	576	640	720	768

If, then, a piece of music is to be played in just temperament in the scale of G, the instrument used must be capable of producing two notes in addition to those found in the major scale of C, namely the notes corresponding to the vibration numbers 432 and 720. Similar investigation of the other possible scales of A, B, D, E, and F show occasion for many other notes, so that a piano which could be played in just temperament in all the keys demanded by modern music would require about 50 notes in each octave. Since, however, the introduction of all these notes would make the manipulation of the instrument a physical impossibility, there has been devised, for keyed instruments of the piano type, another scale which is known as *the scale of even temperament.* The origin of this scale may be seen as follows.

A careful comparison of all the notes necessary for the various scales (found as above in the case of G) shows that many of these notes differ so slightly that a single note may do satisfactory duty for several of almost the same frequency, provided we are willing to content ourselves with slightly imperfect intervals. Twelve notes are accordingly chosen to replace the fifty, and these twelve are, as a matter of fact, made to divide the octave into twelve exactly equal parts. It is for this reason that the scale is called the scale of even temperament. Since there are twelve equal intervals between a note and its octave, each interval is of value $\sqrt[12]{\frac{2}{1}}$, or 1.059.

This scale is written below, starting with C = 256 for the fundamental note. For purposes of comparison the frequencies of the notes in the scale of just temperament are also given.

Notation	C	C♯ or D♭	D	D♯ or E♭	E	F	F♯ or G♭	G	G♯ or A♭	A	A♯ or B♭	B	C
Even	256	271.3	287.4	304.8	322.7	341.7	362.2	383.8	406.6	430.7	456.5	483.5	512
Just	256	——	288	——	320	341	——	384	——	427	——	480	512

It is evident from the table that although only C and its octave retain their old values in the scale of just temperament, the difference between the frequencies of any note in the two scales is so small as not to be in general noticeable.*

* The difference would, of course, be at once noticeable because of the phenomenon of beats (see sect. 149, p. 234), if the same note were sounded simultaneously in the two scales. In general this would not occur, for all instruments with

EXPERIMENT 18

Object. To find the overtones possible in open and closed pipes.

Directions. Figure 148 shows an arrangement consisting of a rotating table *T*, a siren *W*, and a five-foot open pipe *P* of diameter 5 or 6 cm. A current of air is forced through the glass tube *g* by a large bellows or other arrangement. When *W* is rotated the current of air from the bellows sends a pulse of condensation into the pipe every time an opening in *W* comes over the tube *g*. When *W* is rotated with a frequency such that the number of these pulses entering the pipe is equal to the natural vibration frequency of the pipe, resonance takes place and the pipe gives forth loudly its fundamental note. In the same way, if *W* is rotated with the proper uniform speed, *P* may be made to give forth loudly any desired overtone.

Determine exactly the number of holes which pass the orifice of the pipe for one revolution of the wheel *T*. Cause the pipe to give forth its fundamental* and hold the speed constant while a second observer takes, with a stop watch, the time of 15 revolutions of *T*. Measure the pipe length. From the data thus obtained and the correction which it was found necessary in Experiment 17 to apply to the open end of a pipe, deduce the velocity of sound in air.

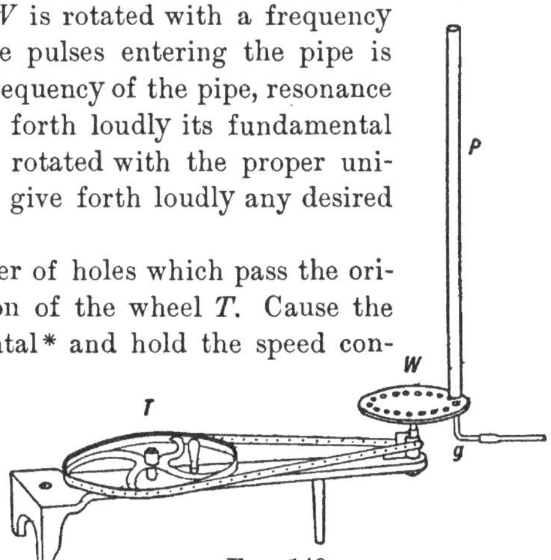

Fig. 148

Repeat the determination, using successively the first, second, third, and fourth overtones instead of the fundamental.

Transform *P* into a closed pipe by inserting a cork in the upper end, and determine as above the vibration numbers corresponding to the fundamental, the first, and the second overtones.

fixed keyboards or frets are tuned to the scale of even temperament. It is, however, worthy of note that many prominent violinists, when playing without accompaniment by an instrument tuned in even temperament, instinctively play in just temperament and so satisfy the demands of the human ear for intervals represented by the ratio of simple whole numbers.

* Difficulty may be experienced in detecting the frequency corresponding to the fundamental because of the fact that the pipe may give off its fundamental note weakly for a number of pulses smaller than that corresponding to its natural frequency (see sect. 131). This response will scarcely be noticeable except when the number of pulses is one half that corresponding to the natural period, when it may be quite pronounced. In general, however, the fundamental resonance will be so much louder that it can scarcely be mistaken. If there is doubt, multiply at once twice the pipe length by the number of pulses and see whether the product is about 34,000 cm., as of course it should be.

EXAMPLE

The plate *W* had 40 holes and made $9\frac{14}{40}$ revolutions for each revolution of *T*; hence 374 pulses entered the tube *P* for each revolution of *T*. The pipe had a length of 113 cm. and a radius of 3.8 cm.; hence the corrected length was 115.6. The average of three determinations of the number of seconds necessary for 15 revolutions of *T* when the open pipe was emitting its fundamental note was 37.4 seconds. The average found for 30 revolutions when the first overtone was sounding was 37.2 seconds; for 45 revolutions with the second overtone it was 37.6; and for 60 revolutions with the third overtone, 37.4. The corresponding frequencies were 150, 300, 450, and 600; that is, they were in the ratio of 1, 2, 3, 4. The average speed of sound from these values of the frequency and pipe length was 347.6 m. per second. The correct value as found from the temperature 21.5° C., and the relation of section 126 was 344.1 m. per second. The difference was about 1 per cent.

A cork placed in the upper end of the pipe shortened its corrected length to 113.7 cm. The average time for 15 revolutions when the fundamental of this closed pipe was sounding was 74.3 seconds. The time of 15 revolutions with the first overtone was 24.7; with the second overtone it was 14.8 seconds. The corresponding frequencies were 75.5, 227, and 380; that is, the frequencies were in the ratio of 1, 3, 5. The average value of the speed of sound from these observations was 344 m. This determination with a closed pipe differed therefore from that made with an open pipe by 1.2 per cent.

CHAPTER XIX

LONGITUDINAL VIBRATIONS OF RODS

136. Velocity of waves in thin rods of elastic material. The analysis of section 122 shows that if there is no possibility of lateral expansion of the medium, the velocity of a compressional wave depends only upon the bulk modulus of elasticity and the density of the medium. This condition is realized when a disturbance originates in the midst of an elastic medium of great extent in all directions. But when the wave travels along a thin rod there is a slight lateral expansion of that portion of the rod which is undergoing the compression. Hence if we imagine a rod of 1 sq. cm. cross section divided into centimeter cubes after the fashion of section 122, and if we imagine a small pressure dp to be applied by means of a piston p at one end (Fig. 134), then while each cube is undergoing the voluminal compression dv, the piston will move forward, not now dv, but some distance ds numerically a trifle larger than dv. From reasoning identical with that given on page 188, an equation results which differs from equation (2), page 189, namely $S^2 = dp/\rho dv$, in no respect save that ds replaces dv. We obtain, then,

$$S^2 = \frac{\frac{dp}{ds}}{\rho}. \tag{1}$$

But ds is the change in the length of the rod per unit length. Hence dp/ds is *Young's Modulus* (Y).* Thus *in thin rods compressional waves move with a velocity which is given by the equation*

$$S = \sqrt{\frac{Y}{\rho}}. \tag{2}$$

* See "Mechanics, Molecular Physics, and Heat, p. 67.

137. Natural periods of free rods. A rod surrounded by air is in every respect analogous to an open pipe, for the reflections at the ends are such as occur when a wave passes from a denser to a rarer medium. Thus such a rod will respond to a train of waves if it is of such length that a pulse 0 (Fig. 149) just entering the rod at A as a condensation will, after reflection at B, return to A and be again reflected as a condensation at the precise instant at which pulse 1 reaches A. The length $2AB$ is then the distance which a pulse 0 travels in the rod before the succeeding pulse 1 enters the rod. This is, by definition, one wave length of the note in the rod.

If one single pulse strikes the rod, the successive reflections of this pulse at A and B will cause a train of waves to be given off at each end. Thus the rod will emit a musical note the wave length of which in the rod is twice the length of the rod. The wave length in air of this note obviously bears the same relation to its wave length in the rod as the velocity of the wave in air bears to its velocity in the rod.

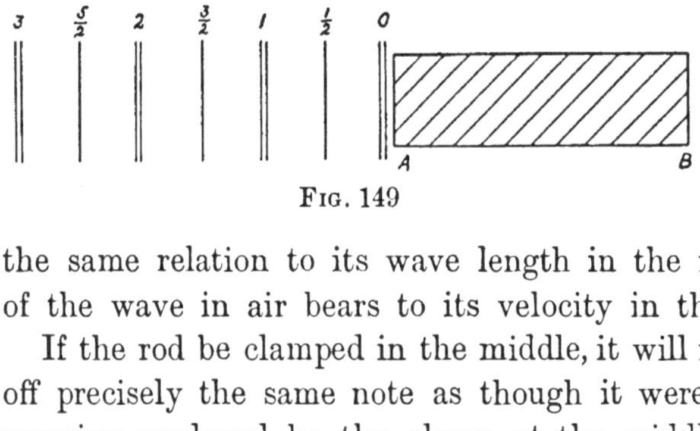

Fig. 149

If the rod be clamped in the middle, it will respond to and give off precisely the same note as though it were free, for the compression produced by the clamp at the middle produces at that point the same sort of a reflection as occurs at the boundary of a denser medium; hence the rod is equivalent to two closed pipes, each of which gives off the same note as would an open pipe (i.e. a free rod) of double the length. In order to set a rod into *longitudinal* vibrations of this sort, it is customary, instead of striking one end, to clamp it in the middle and stroke it with a rosined cloth if it is of metal, or with a wet cloth if it is of glass. The mechanism of the tone production in this case will be more fully discussed in section 140.

138. Comparison of the velocities of sound in two solids. The above theory suggests an extremely simple and satisfactory means of comparing the velocities of sound in two solids. Thus we have only to find the vibration frequencies (the pitches) of two notes produced by stroking steel and brass rods of the same length, in order to find the relative velocities of sound in steel and brass. For with rods of equal length the number of pulses communicated to the air per second by the traveling of pulses up and down the rods is obviously proportional to the velocities of sound in the two rods. Thus if S_s and S_b represent these velocities in steel and brass respectively, and n_s and n_b the corresponding frequencies produced by the rods of equal length, we have

$$\frac{S_s}{S_b} = \frac{n_s}{n_b}. \tag{3}$$

The frequencies n_s and n_b can be determined in a variety of ways; for example, by changing the length of a given sonometer wire until it is in tune, first, with the note from the steel, and then with that from the brass. Since the frequencies of the notes produced under these circumstances are inversely proportional to the lengths (see Chap. XX), we have, if l_s and l_b are the lengths of the same wire which are in tune with the steel and brass respectively,

$$\frac{S_s}{S_b} = \frac{l_b}{l_s}. \tag{4}$$

139. Nodes and loops in pipes and rods. A careful consideration of the resonance of pipes which are giving off the first or higher overtones reveals effects which have thus far been overlooked. For example, it was shown that when a pipe has its

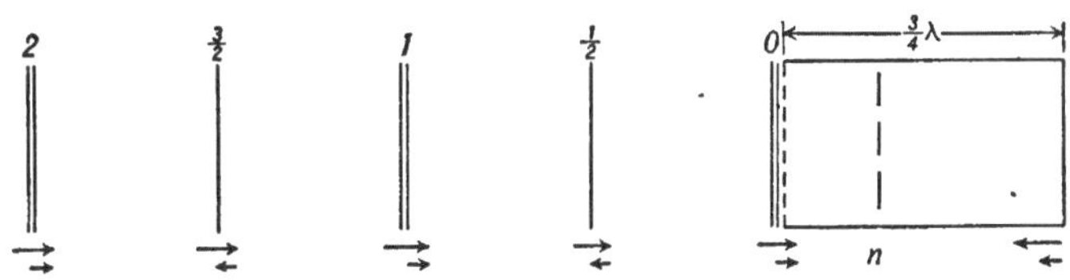

FIG. 150

second resonant length, a condensation 0 (Fig. 150) must return to the mouth of the pipe at the instant at which the rarefaction $\frac{3}{2}$ reaches the mouth; but, in the return after reflection, 0 must somewhere in the pipe collide with the advancing condensation 1. Since at the instant of the reflection of 0, 1 is one wave length behind 0, it is evident that this collision must take place just $\frac{1}{2}$ wave length from the end of the pipe, namely at n. Such a collision of two oppositely moving condensations is entirely analogous to the collision of two oppositely moving perfectly elastic balls. These are shown simply to exchange motions,* the effect being the same as though each ball passed through the other without experiencing any effect whatever from it. Thus the waves may be thought of as passing through one another, and their mutual effects may be ignored. As a matter of fact, it is of course 1 which returns to the left after the collision and unites with $\frac{3}{2}$ at the mouth, while 0 is forced back again toward the closed end of the pipe.

One half period after the collision at n (Fig. 150) of the condensations 1 and 0 ($\rightarrow\leftarrow$) there will occur at n a collision of the rarefactions $\frac{3}{2}$ and $\frac{1}{2}$ ($\leftarrow\rightarrow$). Thus the particles near n are first pushed together by opposing forces, then pulled apart by opposing forces. The result is that they do not move at all. The matter

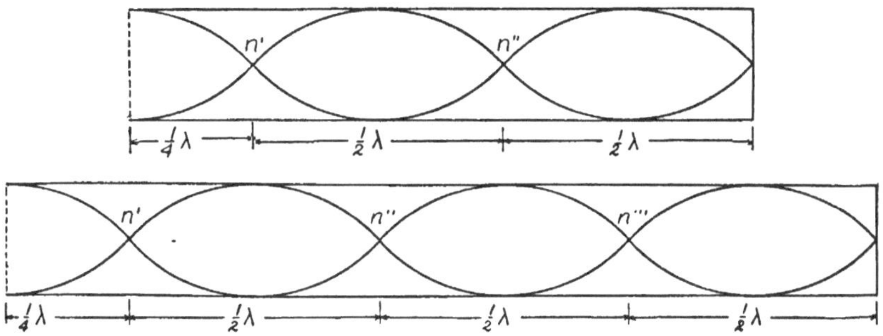

Fig. 151

about n suffers alternate compression and expansion, but the particles at n can never move either to left or to right, because they are always being urged in opposite directions by the oppositely moving waves. The point n is called a *node*. The points between the nodes where the disturbance is greatest are called *loops*.

If the length of the pipe is $\frac{5}{4}$, $\frac{7}{4}$, $\frac{9}{4}$, etc., wave length, it is evident from considerations precisely like the above that there will be nodes at n', n'', n''', etc. (Fig. 151). In other words, in any resonant closed pipe (and it is to be remembered that such a pipe is

* See "Mechanics, Molecular Physics, and Heat," Chapter VII.

resonant when and only when its length is an odd number of fourth wave lengths) the first node is one fourth wave length from the open end, and other nodes follow at intervals of one half wave length. The conventional method of representing nodes and loops in pipes is that used in Figure 151.

Since the first resonant length of an open pipe is one half wave length, and since a condensation 0 is reflected as a rarefaction, it is evident that 0 will collide in the middle of the pipe with $\frac{1}{2}$. Hence an open pipe responding to its fundamental has a node in

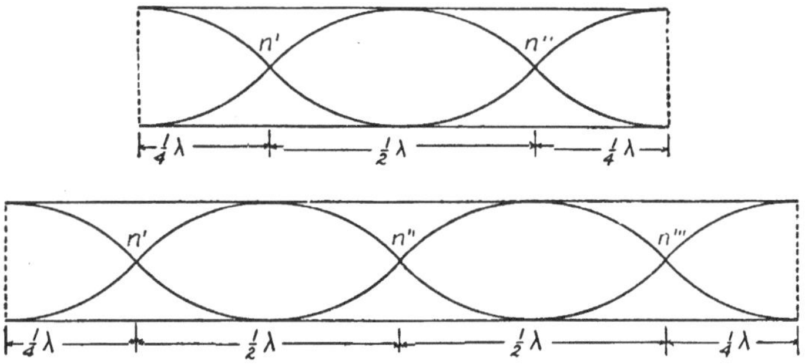

FIG. 152

the middle. Similarly an open pipe responding to its first overtone has nodes at n' and n'', each one fourth wave length from an end and one half wave length apart. Similarly for the higher overtones (see Fig. 152).

Although the preceding discussion has been limited to pipes, yet, since by section 137 a rod surrounded by air acts in every respect like an open pipe, the above conclusions hold also for rods.

140. Kundt's tube experiment. A rod *mn* (Fig. 153), to one end of which is attached a light cork piston *B*, is supported by a clamp *C* at its middle point. The piston *B* fits very loosely in a long tube *AB*, one end of which is closed by a tightly fitting piston *A*.

The rod *mn* is set into longitudinal vibration by drawing along it a cloth covered with rosin. It would seem at first thought as though the slipping of the cloth along the rod were so irregular that no musical note could be produced. As a matter of fact, however, the slipping is controlled by the natural period of the rod in much the same way as the vibrations of the air jet at the mouth of an organ pipe* are controlled by the natural period of the pipe. Thus the first slip starts a pulse down the rod which, because of the reflections at the ends, returns to the starting point at stated intervals. Of course the tendency to slip is greatest at the instant of the return of the first pulse, so that succeeding slips take place at the instants of return of succeeding pulses. Thus the rod gives off loudly the note corresponding to its natural period. The rod *mn* is clamped in the middle, but it was shown above † that the natural period in this case is precisely the same as when the rod is free. The wave length of the note produced in the material of which the rod is composed is therefore twice the length of the rod.

The piston *A* is so adjusted in position that the air column *AB* is of such a length as to be resonant to the note of the rod. Nodes are then formed which may be brought into evidence by placing

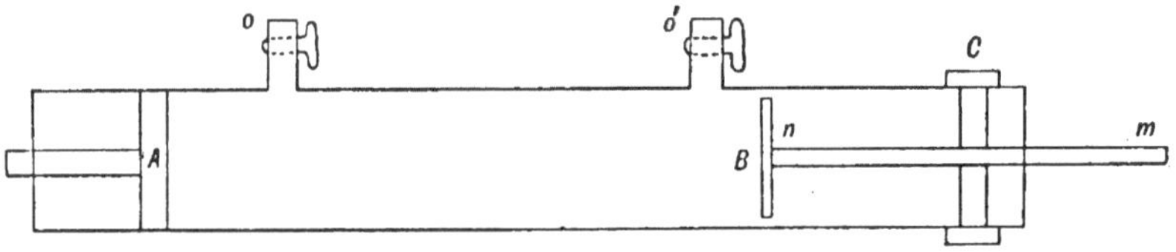

FIG. 153

along the bottom of the tube a layer of light cork filings. The cork dust will be collected into ridges at the points of maximum disturbance, i.e. at the loops. The explanation of the fact that each loop is marked not by a single ridge, but by a series of ridges, demands an analysis which is beyond the scope of this text.

The wave length of the note given off by the rod may be obtained at once by measuring the distance between two successive loops. If, then, S_r represents the velocity of the wave in the rod, λ_r its wave length in the rod, and n the frequency, we have

$$S_r = n\lambda_r.$$

Also, if the tube contains air, and S represents the velocity of sound waves in air, and λ the wave length of a note of frequency

* See section 132, page 204.
† See section 137, page 214.

n, then $S = n\lambda$. Combining these two equations we obtain as the expression for the velocity of compressional waves in the material of the rod,

$$S_r = \frac{\lambda_r}{\lambda} S, \tag{5}$$

in which λ_r is merely $2\,mn$ and λ is twice the distance between nodes in air.

By similar reasoning, if some other gas is substituted for the air in the tube and S_g and λ_g represent the velocity and wave length of the same note in that gas, then $S_g = n\lambda_g$, or

$$S_g = \frac{\lambda_g}{\lambda} S. \tag{6}$$

Since S has already been determined, and λ and λ_g may be observed, the velocity of sound in any gas may be found.

EXPERIMENT 19

(A) Object. To find the velocity of compressional waves in steel.

Directions. Following the method of the Kundt's tube experiment described in section 140, adjust carefully the sliding piston A (Fig. 153) until a maximum of agitation of the cork dust at the loops is produced when a steel rod mn is stroked with a rosined cloth. Measure the distance between A, which is a node, and the node most remote from A, say the nth, then that between A and the $(n-1)^{st}$ node, then between A and the $(n-2)^{d}$, etc. Make two vertical columns, one of measured distances, the other of the corresponding numbers of half wave lengths. The sum of the first column divided by the sum of the second gives the most accurate value of $\frac{1}{2}$ wave length which is obtainable from this sort of an observation. Shake up the cork dust and obtain a second set of readings, measuring this time first between the first and last loops, then between the second and next to the last, etc., and averaging as described above. If S_r represents the velocity of sound in steel, λ_r the wave length in steel of the note produced, and if S and λ represent the corresponding quantities when the waves have passed over into air, then S_r can be found at once from the relation $S_r = (\lambda_r/\lambda)S$. Then from S_r and λ_r the frequency n may be obtained if desired.

Compare the observed value of the velocity in steel with the theoretical value deduced from Young's Modulus, and the density, by use of the relation $S = \sqrt{Y/\rho}$.

(B) Object. To find the velocity of sound in CO_2.

Directions. Replace the air in the Kundt tube by CO_2 by permitting a gentle current from a charged cylinder to pass in at *o* (Fig. 153) and out at *o′* for two or three minutes. Then determine, precisely as above, the wave length in CO_2 of the note given forth by the steel rod. Thence deduce the velocity of sound in CO_2 and compare the result with the theoretical value obtained from the barometer height, the density of CO_2 (viz. 1.53 × the density of air), and the value of $\gamma\,(= 1.30)$.

(C) Object. To compare the velocities of sound in brass and steel.

Directions. Take two equal rods three or four meters long, one of steel and one of brass. Set them successively into longitudinal vibrations by clamping them in the middle and stroking with a rosined cloth. By means of a sliding bridge vary the length of a small wire of a sonometer (or violin) until, when picked transversely, it produces first a note in tune with that of the steel rod, then with that of the brass rod. Determine the relative velocities by the method of section 138 and compare with the relative values obtained by taking the values of Young's Modulus and the appropriate densities from a table and substituting in equation (2), page 213.

EXAMPLE

(A) The observations of the distance between nodes made according to the directions given above were as follows.

Number of ½ wave lengths	Distance in cm.
7	65.3
6	55.7
5	47.5
4	38.0
3	28.5
2	19.0
1	9.5
28	263.5

$\therefore\ \frac{1}{2}\lambda = 9.4$ cm.

Number of ½ wave lengths	Distance in cm.
6	56.1
4	36.5
2	17.7
12	110.3

$\therefore\ \frac{1}{2}\lambda = 9.2$ cm.

The average value of λ in air was thus found to be 18.6 cm. The temperature of the room was 26°C. The velocity of sound in air at this temperature is $331.27 + 26 \times .6 = 346.9$ m. per second. The length *mn* was 136.5 cm., thus making $\lambda_r = 273$ cm.

Hence S_r for steel was $\frac{273}{18.6}\cdot 346.9 = 5073$ m. per second.

Using the value of Young's Modulus previously found from experiments on a steel wire, namely 19.7×10^{11}, and taking 7.8 as the density of steel, the theoretical value of the velocity of sound in steel, as calculated from the formula $S = \sqrt{\frac{Y}{\rho}}$ was found to be 5026 m. per second. The difference between the two values is .9 per cent.

(B) When the tube was filled with CO_2 at atmospheric temperature the average wave length produced in the CO_2 by the vibrations of the steel rod was 14.5. Hence the velocity in CO_2 was $346.9 \times \frac{14.5}{18.6} = 270.4$ m. The corrected barometric pressure was 74.22 cm. The value of the velocity calculated from this pressure, together with the density of carbon dioxide, namely $.001152 \times 1.53$, and the factor $\gamma = 1.30$, was found to be 269.4. The difference was thus .4 per cent.

(C) Rods of steel and brass 290 cm. long were tuned to lengths of 21.2 cm. and 30.5 cm. respectively on a given sonometer wire.

Hence $$\frac{S_s}{S_b} = \frac{30.5}{21.2} = 1.438.$$

The theoretical value obtained by using 7.8 and 8.4 respectively for the densities of steel and brass, and 19.7×10^{11} and 10.2×10^{11} for the corresponding values of Young's Modulus, was 1.442, a difference of 0.3 per cent.

CHAPTER XX

WAVES IN STRINGS

141. General characteristics of wave motion. In the preceding sections we have discussed only compressional, or longitudinal, wave motion, and have found this to be characterized by the fact that the particles which transmit the wave move in the line of propagation of the wave itself. This is the only sort of an elastic wave which is possible in substances which do not possess rigidity.* But in substances which possess rigidity another type of elastic wave motion is possible, namely transverse wave motion. This is characterized by the fact that the particles of the medium move in paths which are *perpendicular* to the direction of propagation of the wave. The waves which travel along a rope when one end is caused to vibrate by the hand are of this sort.

Before this second type is discussed it is of importance to have clearly in mind the general characteristics of wave motion. These may be seen from a consideration of Figure 154, which represents the effect on the particles of a medium of an oscillatory motion of the piston *P*. The unit cubes of Figure 134, section 122, page 187, are here replaced by vertical lines. Thus Figure 154, *a*, represents the state of the medium before the piston has begun to move. Figure 154, *b*, represents its state when the piston has undergone its greatest displacement to the right and is ready to return, a condition which is represented by the double arrow (←→>). In the upper line of arrows each arrow represents the direction of displacement of the layer toward which it points. The small arrows below the vertical lines show the direction of the motion of the layers. The zero below any line indicates either that the medium is there at its mean (original) position, or is just changing the direction of its motion at the end of its path. The succeeding

* The large waves on the surface of a body of water are gravity waves and have nothing to do with the elasticity of matter.

figures show the progression of the initial condensation and the subsequent rarefaction for the indicated positions and directions of motion of the piston. Obviously a condensation exists, for example, in Figure 154, *b*, at 4, since the layers are there crowded together, and conversely a rarefaction exists in Figure 154, *d*, at 4, since the layers are there separated.

From the figures it is evident that the layers, or particles, of the medium are in vibration, and that at any instant these particles possess a definite configuration, for example that of particles 4 to

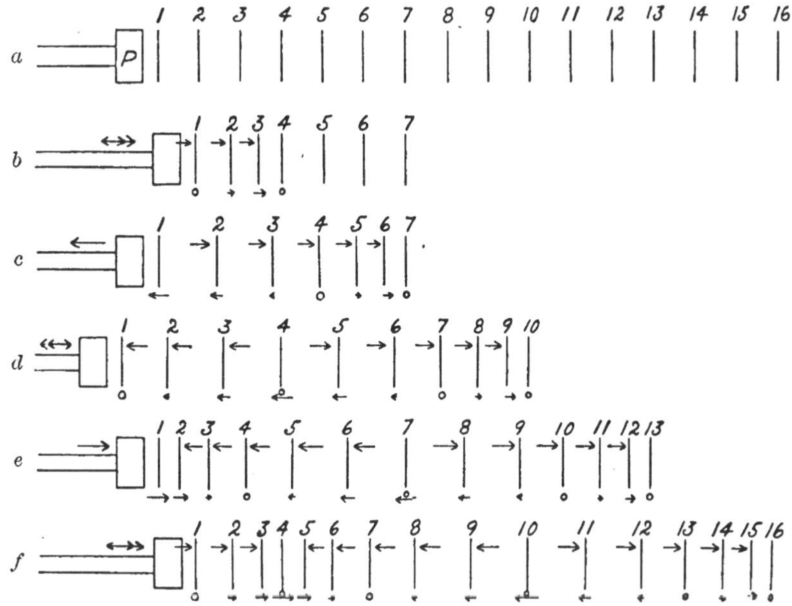

Fig. 154

16 in Figure 154, *f*. This configuration is known as the wave form, and it travels from left to right through the medium, although the particles themselves are merely vibrating back and forth across their original positions. If the motion of the piston is simple harmonic, then the motion of the particles must be simple harmonic also. As a matter of fact practically all vibrations which arise from the elasticity of matter are of this type. The *amplitude*, or maximum displacement, of the particles depends upon the amplitude of the motion of the piston, that is, upon the intensity of the

disturbance which the particles are propagating. *The difference between the times at which any two particles of the medium pass through the middle points of their paths, divided by the period of the vibration, is called the phase difference between the particles.* As has already been indicated in Chapter XVII, the distance between two successive particles which are in similar states of motion at the same time is called a *wave length.* Thus a wave length is the distance between particles 4 and 16, or 1 and 13, in Figure 154, *f*.

142. Transverse waves. The conception of wave motion just given will now be considered more analytically in connection with the transmission of transverse waves. Thus if all the particles in the line XX' (Fig. 155, *a*) are in some sort of rigid connection, and

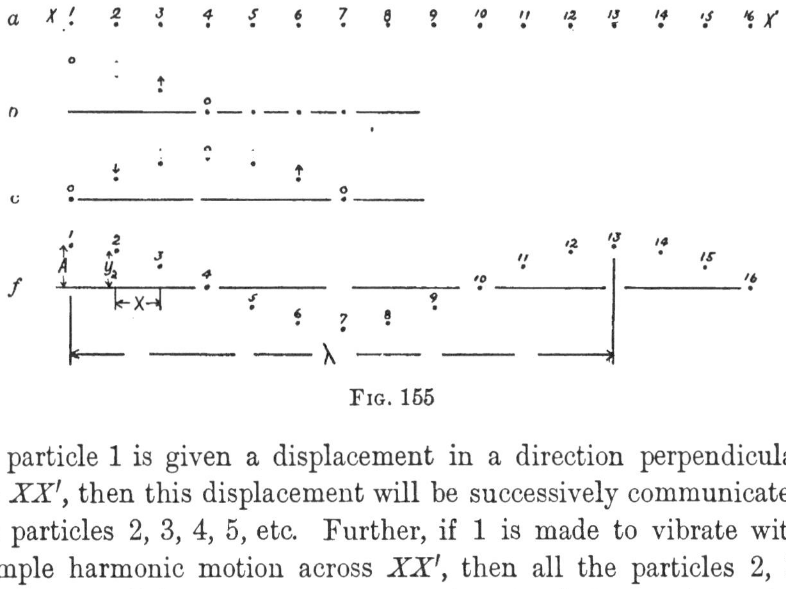

FIG. 155

if particle 1 is given a displacement in a direction perpendicular to XX', then this displacement will be successively communicated to particles 2, 3, 4, 5, etc. Further, if 1 is made to vibrate with simple harmonic motion across XX', then all the particles 2, 3, 4, 5, etc., will in succession take up this simple harmonic motion across XX'; i.e. a series of transverse waves will travel along XX'. Let the amplitude of the motion of each particle be represented by A and the period by T. Then it may be shown that the vertical displacement y_2 of any particle, such as 2, expressed in terms of A, T, and the time t since that particle left its original position, is as follows:

$$y_2 = A \sin 2\pi \frac{t}{T}. \tag{1}$$

This will be evident from a consideration of Figure 156; for a particle P moving with simple harmonic motion along the vertical path of length $2A$ has at any instant a displacement from the center O which is represented by the distance from O to the projection P upon that path of a point P' which moves uniformly in a time T about the circumference of a circle which has $2A$ as its diameter.* The angular speed of the point P' about O is $2\pi/T$ radians per second. After t seconds the particle P' has moved through an angle of $2\pi t/T$ radians. The distance OP is then A times the sine of this angle, i.e.

$$y_2 = A \sin 2\pi \frac{t}{T}.$$

The displacement at this instant of another particle such as 3 (Fig. 155, *f*) is, of course, different. Thus, suppose that this particle leaves its mean position t' seconds after particle 2 has left its mean position, i.e. let t'/T be the phase difference between the two particles. The displacement of 3 at the instant considered is obviously

$$y_3 = A \sin 2\pi \frac{t - t'}{T}. \qquad (2)$$

And similarly the displacement of any particle is represented by an expression of the form

$$y = A \sin 2\pi \frac{t - t'}{T}, \qquad (3)$$

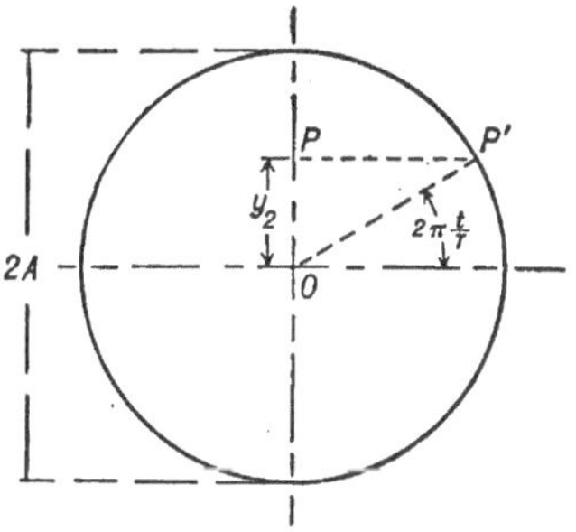

FIG. 156

where t'/T is the phase difference between the particle under consideration and the reference particle.

The particle for which $t' = T$ will be in a state of motion similar to that of the reference particle 2, and therefore distant from it by a wave length λ. It is also evident that while the first disturbance which is imparted to particle 2 is traveling forward this distance λ (for example, in Fig. 155, *f*, from 2 to 14), the particle 2

* See "Mechanics, Molecular Physics, and Heat," p. 88.

makes one complete vibration; that is, the wave travels λ centimeters in T seconds, or has a velocity S which is given by

$$S = \frac{\lambda}{T}. \tag{4}$$

Returning now to a consideration of the motion of a particle distant x from particle 2 (for example, 3, Fig. 155), it is evident that $\frac{x}{\lambda} = \frac{t'}{T}$. Substitution of the value $t' = \frac{Tx}{\lambda}$ in the expression for the vertical displacement of any particle gives

$$y = A \sin 2\pi \left(\frac{t}{T} - \frac{x}{\lambda}\right). \tag{5}$$

The several particles along the line XX' are therefore displaced from their original positions according to a sine function. The angle represented by $2\pi x/\lambda$ is called the angle of phase difference between the particle under consideration and the reference particle.

Figure 155, *b*, *c*, *f*, represents *transverse* displacements equal in amount to the *longitudinal* displacements of Figure 154, *b*, *c*, *f*. It is also evident that the confusion resulting in the figure for compressional waves from the fact that the displacements are parallel to the direction of the wave motion may be obviated by plotting those displacements vertically. With that convention in mind Figure 155, *b*, *c*, *f*, may be taken to represent the progression either of a compressional or of a transverse wave. In the discussion of nodes and loops in pipes (sect. 139, pp. 215–217) this convention has already been used.

143. Stationary waves. The phenomenon known as *stationary waves* is the result of the action upon a series of particles of two equal trains of waves traveling in opposite directions. Figure 157, *a*, shows two such trains of waves, namely A traveling toward the left, and B traveling toward the right. The waves are represented at an instant at which the crests of A are opposite to the troughs of B, and vice versa. The heavy line shows the *resultant* displacement of the series of particles. Obviously in this case each one of the particles transmitting the motion is under the action of two disturbances which tend to produce equal and opposite displacements, and as a result the particles suffer no displacement at all. In Figure 157, *b*, is shown the case in which each of the wave trains has progressed an eighth of a wave length; that is, the waves have become displaced a quarter of a wave length with respect to one another. The heavy line represents the form assumed by the row of particles at this instant as a result of the superposition of the two disturbances. Similarly, when one wave train has moved a half wave length past the other the resultant is again zero at every point.

From the figures it is evident that for any particle the resultant displacement is the algebraic sum of the displacements produced

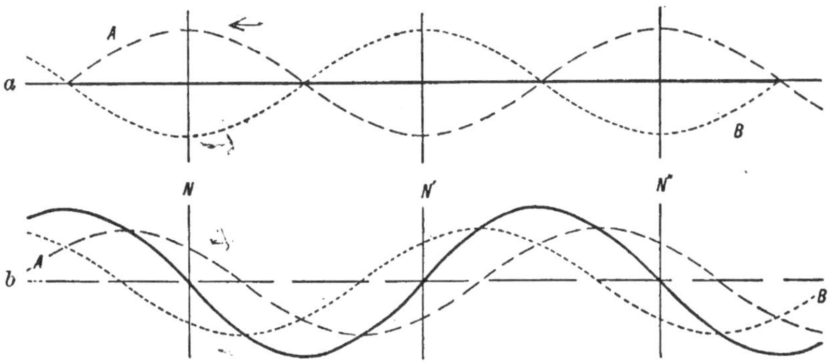

FIG. 157

at that point by the two wave motions. For certain points, e.g. N, N', N'', distant from one another by a half wave length, the resultant displacement is always zero. These points are the nodes, and correspond to the nodes for compressional waves which have been previously discussed. Between the nodes the particles are in constant vibration, but all pass through their mean positions at the same time. There is therefore no phase difference between successive particles. It is for this reason that the phenomenon of the combination of two oppositely directed trains of waves on the same particles is known as the phenomenon of *stationary waves*. The amplitudes of the vibration of successive particles vary from a maximum at the loops to zero at the nodes. On opposite sides of a node the displacements at any instant are in opposite directions. In Figure 158 are shown forms assumed by the particles between the two nodes drawn for successive instants of time. The successive positions assumed by the particles are numbered in their order.

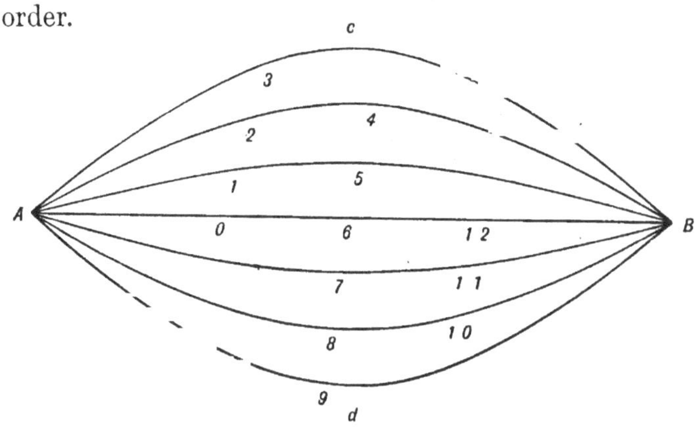

Fig. 158

144. Equation for a stationary wave. The ideas developed above from a study of the diagrams may also be obtained from a consideration of the equation of a wave motion. For let $y = A \sin 2\pi\left(\frac{t}{T} - \frac{x}{\lambda}\right)$ represent the equation for the displacement given to the successive particles by the direct wave. The reversed wave must be one for which at some given instant of time (e.g. $t = 0$) the displacements given to the same series of particles will be equal and opposite. The equation $y' = A \sin 2\pi\left(\frac{t}{T} + \frac{x}{\lambda}\right)$ satisfies this condition and represents the equation of the reverse wave. The resultant displacement Y is the sum of y and y'. That is,

$$Y = A \sin 2\pi\left(\frac{t}{T} - \frac{x}{\lambda}\right) + A \sin 2\pi\left(\frac{t}{T} + \frac{x}{\lambda}\right).$$

By expansion * and addition we have

$$Y = \left(2A \cos 2\pi\frac{x}{\lambda}\right) \sin 2\pi\frac{t}{T}. \qquad (6)$$

* For $\sin(\theta + \phi) = \sin\theta\cos\phi + \cos\theta\sin\phi,$
and $\sin(\theta - \phi) = \sin\theta\cos\phi - \cos\theta\sin\phi.$

Now consider the fundamental equation for a wave motion, namely $y = A \sin 2\pi\left(\frac{t}{T} - \frac{x}{\lambda}\right)$, and notice that it consists of two parts. The first part is the amplitude A and the second a sine function of the time t plus a phase constant x/λ. Now in the expression for Y just obtained $\sin 2\pi\frac{t}{T}$ is the sine function of the time. And evidently, since there is no phase constant in this expression, the particles must all be in the same phase of vibration. Similarly $2A\cos 2\pi\frac{x}{\lambda}$ represents the amplitude. But since x represents the distance of the particle under consideration from the reference particle, it is evident that the amplitude varies for successive particles. Also since $\cos 2\pi\frac{x}{\lambda}$ is zero when x is an odd multiple of $\lambda/4$, it follows that there are nodes, or points of zero amplitude, at points differing successively by half wave lengths. Further, the algebraic sign of $\cos 2\pi\frac{x}{\lambda}$ changes at these same points.

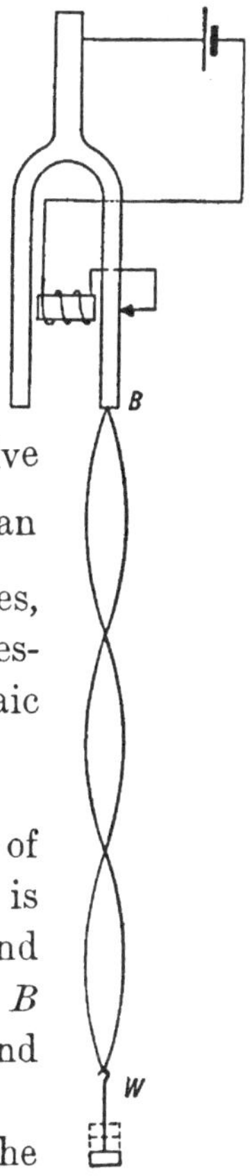

FIG. 159

145. Melde's experiment. A capital illustration of stationary waves in strings is furnished by what is commonly known as *Melde's experiment.* One end of a light cord is attached to one of the prongs B (Fig. 159) of a tuning fork, while the other end carries a weight W.

The waves which start down the cord from the vibrating fork are reflected at W, so that two trains of waves moving in opposite directions become superposed upon the cord. In accordance with the principles of the last section this condition tends to give rise to stationary waves, the positions of the nodes being at distances from W corresponding to exact multiples of a half wave length of the train sent down the cord from the fork. Since, however, the upward-moving train is again reflected at B, the condition for stationary waves in which the nodes are at distances from B corresponding to exact multiples of a half wave length is also established. It is obvious that both of these conditions can be met, and permanent stationary waves set up in the string, only if the length L of the string is an exact multiple of a half wave length.*

Instead of varying the length of the string so as to fulfill this condition, it is customary to vary the wave length by varying the load W. For the wave length λ is connected with the vibration rate n of the fork and the speed of propagation S of the train of waves along the string by means of the relation $S = n\lambda$, and the speed S is connected with the tension T in the string and its mass ρ per centimeter of length by means of a formula which is very similar to that given in the last equation on page 189. It is,

$$S = \sqrt{\frac{T}{\rho}},\dagger \tag{7}$$

* This statement is only approximately correct, since the end of the fork is not exactly at a node, but rather just as near to a node as a point near some other node which has the same amplitude of vibration as the fork.

† This formula is most satisfactorily deduced with the aid of the calculus, but it may also be obtained as follows. Let the curve *mno* (Fig. 160) represent a portion of the cord over which the deformation is being propagated. Let ee' be an element of the cord so small that it may be considered as the arc of a circle of radius R. If the string is wholly devoid of rigidity, then the only force which is urging the element toward the center c arises from the tension T in the string, and this may be regarded as a pull acting upon each end of the arc ee'.

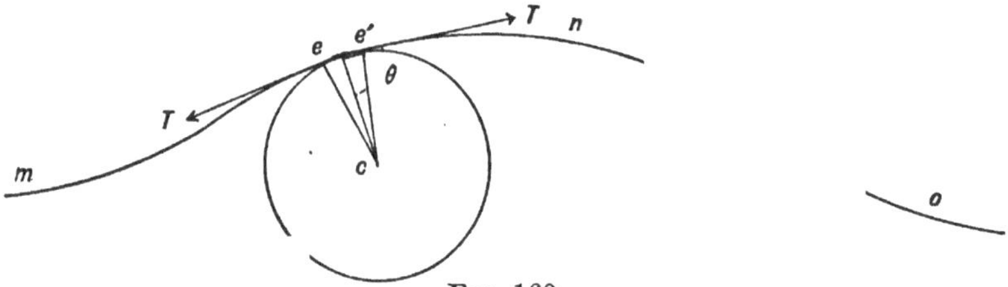

FIG. 160

The directions of these two pulls are the directions of the tangents to the curve at the points e and e' respectively (see Fig. 160). The component of each of these pulls which is urging ee' toward c is $T \sin\theta = T\frac{\frac{1}{2}ee'}{R}$. The total force f_c which is urging ee' toward c is therefore $2\,T\frac{ee'}{2\,R} = T\frac{ee'}{R}$. But since the deformation *mno* is propagating itself unchanged in character along the string, each element of the string must assume in succession the positions occupied at any instant by all the other elements of the curve *mno*. In other words, at the instant which we have been considering the element ee' is not moving at all in the so that by varying the tension T it should be possible to find a whole series of values of λ which will give rise to permanent stationary waves in the cord. Thus when $L = \lambda/2$ the string should vibrate in 1 segment, when $L = 2\,\lambda/2$ it should vibrate in 2 segments, when $L = 3\,\lambda/2$, in 3 segments, etc.

Or, in general, since by combining equation (7) with the equation $S = n\,\lambda$ we obtain

$$n\lambda = \sqrt{\frac{T}{\rho}}, \tag{10}$$

it is evident that the equations of condition for 1, 2, 3, 4, etc., segments may be written

$$n = \frac{1}{2L}\sqrt{\frac{T}{\rho}}, \quad n = \frac{1}{2L}\sqrt{\frac{4\,T}{\rho}}, \quad n = \frac{1}{2L}\sqrt{\frac{9\,T}{\rho}}, \text{ etc.} \tag{11}$$

These equations are obtained by substituting in (10) the above relations between L and $\lambda/2$. Equations (11) show, since n, L, and ρ do not change, that *the product of the stretching force by the square of the number of segments should be a constant*, and that this constant should represent the tension when the entire string is vibrating in one segment. By substituting this constant in the first of equations (11), T having been expressed, of course, in dynes, it should be possible to find the vibration rate n of the fork, provided ρ and L are known. The fact that all of these relations are found by experiment to hold, constitutes complete experimental proof of the correctness of the formula $S = \sqrt{\frac{T}{\rho}}$, for the case of strings which have no rigidity.

direction of c, but is instead moving into the position of the adjacent element on *mno;* that is, it is moving with a velocity S along the circumference of the circle which has R for its radius. Hence we may apply to its motion the law of centripetal force deduced on page 102 of "Mechanics, Molecular Physics, and Heat," namely,

$$f_c = \frac{mS^2}{R}. \tag{8}$$

But if ρ is the mass per unit length of the string, $m = ee'\rho$.

Hence (8) becomes $$f_c = \frac{ee'\rho S^2}{R}.$$

But since f_c is also equal to Tee'/R, we have

$$\frac{ee'\rho S^2}{R} = \frac{Tee'}{R}, \quad \text{or} \quad S = \sqrt{\frac{T}{\rho}}. \tag{9}$$

146. Fundamentals and overtones in strings. If a stretched string is plucked in the middle, the deformation travels in opposite directions to the two ends, is there reflected, and, since the two reflected portions returning to the middle unite in like phases at this point, the net result of the propagation of the disturbance back and forth over the string is a vibration of the string as a whole in the manner indicated in Figure 158, page 228, in which the various lines represent some of the successive positions of the string. A string vibrating in this way imparts successive condensations and rarefactions to the air in which it moves, and these, being transmitted to the ear, give rise to a note of a definite pitch which is called the fundamental note of the string.* Since the time elapsing between the instant at which the string is in the position AcB (Fig. 158) and the instant at which it assumes the position AdB is the time required for the deformation to travel over the paths cBd and cAd, it will be seen that during the time of one half vibration of the string the disturbance travels on the string a distance exactly equal to the length of the string. Hence during the period of one complete vibration of the string the disturbance travels twice the length of the string. Thus we arrive, from a wholly different point of view, at the conclusion of the preceding section, namely, that when a string is vibrating as a whole, i.e. in one segment, its length is one half the wave length of the waves which are traveling back and forth over it.

If the string is clamped in the middle as well as at the ends and plucked one fourth of its length from one end, each half vibrates precisely as the whole string vibrated in the preceding case; but since the speed of propagation is the same as before, while the distance between reflections is one half as great, the period of vibration of each half of the string must be one half as great as the preceding period. Hence the note communicated to the air is the octave of the original note, and the wave length of the note is the length of the string. The note thus produced by the string is called its

* Practically, of course, the sound thus derived is of small intensity, and in most musical instruments the greater magnitude of sound is due to synchronous vibrations which the string impresses upon its supports and through them upon sounding-boards and resonant volumes of air.

first overtone. If the string is not clamped in the middle, but is plucked one fourth of its length from one end, it still tends to vibrate as above in two segments, but this vibration is superposed upon the vibration of the string as a whole, so that the fundamental and the first overtone can be heard simultaneously. Figure 161 is an endeavor to show the appearance of a string which is vibrating so as to produce its fundamental and first overtone.

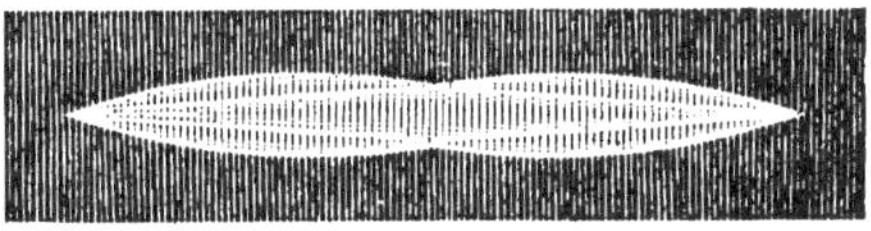

FIG. 161

Similarly, if the string is plucked one sixth of its length from one end, it tends to vibrate in three segments and the second overtone will be heard with the fundamental.

Thus the string is capable, under suitable conditions, of vibrating in any number of segments and of giving out a series of notes whose frequencies bear to the fundamental frequency the ratios 2, 3, 4, 5, 6, 7, etc. In general, in the case of the strings of musical instruments several of these overtones are produced simultaneously with the fundamental, which ones are present depending chiefly upon where the string is struck or bowed. It is to differences in the number or relative prominence of the overtones that all differences in the qualities of different notes of the same pitch are assigned.

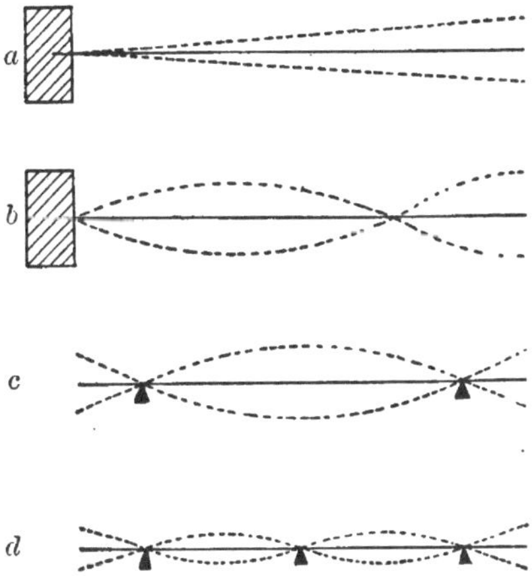

FIG. 162

147. Transverse waves in rods. In the case of rods the wave travels as the result of the rigidity of the substance of which the rod is composed. A consideration of the velocity of propagation of the wave is, however, beyond the scope of this text. The form assumed by vibrating bars may be seen from Figure 162. A bar clamped at one end gives off its fundamental note when vibrating in the form shown in Figure 162, *a*. If struck more sharply and nearer the free end, it may be made to give off its first overtone; in this case it vibrates in the form shown in Figure 162, *b*. The relation between the frequencies of the fundamental and its various overtones are not, however, simple numbers, as is the case with pipes or strings.

If the rod is supported at two points, as in Figure 162, *c*, it will vibrate in the form shown in that figure when yielding its fundamental note. The form assumed by the rod when yielding its first overtone is shown in Figure 162, *d*. In this case also the relation of the frequencies is not a simple number.

148. The tuning fork. If the rod shown in Figure 162, *c*, is bent, it is found that the nodes are brought closer together. If it has the form of Figure 163, the nodes will occur at the points marked *NN*. The higher overtones are then very difficult of production and are very much less in intensity than the fundamental. A bar bent into this form and supported at *P* is known as a tuning fork. Because of the purity of its note — that is, the absence of overtones — it has been adopted for use as a convenient standard of frequency. Any given fork must, of course, be rated first by some absolute method and then it may be used for comparison with other sources of sound.

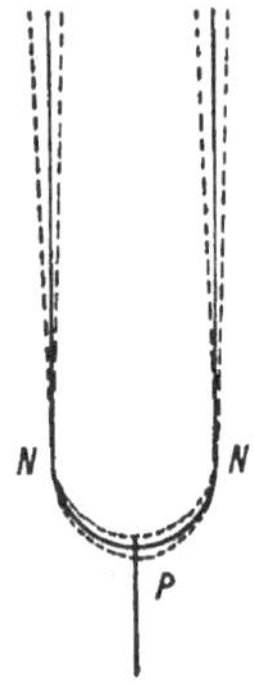

Fig. 163

149. Beats. The case of stationary waves (see sects. 143–144) is a special case of interference. Another case of especial interest in sound is the interference of the wave trains from two musical sources of sound of almost the same pitch (i.e. frequency). Obviously if a crest due to one source reaches the ear at the same time as the trough from the other source, there will be destructive interference. But since one source is vibrating slightly more rapidly than the other, an instant later two crests (or two troughs) will be in coincidence at the ear. There results then a reënforcement of the vibration. These alternations in the intensity of the sound at any point will obviously occur as many times per second as the frequency of one source exceeds that of the other. That is, if m and n represent the frequencies of the two sources, there will be $m-n$

such coincidences per second. These alternations in sound intensity are known as *beats.*

The phenomenon of beats is useful in explaining the physical basis of discords. So long as the number of beats produced by sounding two notes together is not more than five or six per second the effect is not particularly unpleasant. From this point on, however, the beats begin to become indistinguishable as separate beats and pass over into a discord. The unpleasantness becomes worst at a difference of about thirty vibrations per second. Thus the notes B and C', which differ by thirty-two vibrations per second, produce about the worst possible discord. When the difference reaches as much as seventy, the difference between C and E, the effect is again pleasing or harmonious.

But in order that two notes may harmonize it is necessary not only that they themselves should not produce an unpleasant number of beats, but that their overtones also should not do so. Thus C and B are very discordant although they differ by a large number of vibrations per second. The discord arises in this case between B (vibration number 480) and C', the first overtone of C (vibration number 512). Thus if two notes are to be consonant, neither they nor any of their overtones can fall close enough together to produce an unpleasant number of beats.

EXPERIMENT 20

(A) Object. To verify the relation $S = \sqrt{\frac{T}{\rho}}$, and to determine the rate of a given tuning fork.

Directions. Set up an electrically driven fork having from about 100 to 300 vibrations per second and connect as shown in Figure 159. Connect to the binding posts of the fork a single storage cell.* To one prong of the fork attach a light string, for example a piece of oiled fish line or linen thread about four feet in length. To the other end of this string hang a light pan for holding weights. Now vary the tension by adding weights until the string breaks up into some number of vibrating segments. The adjustment may be made exact by varying the tension until the nodes are

* If it is necessary to supply more energy to the fork in order that it may not be damped down by the tension on the string, a larger number of cells may be used. To prevent excessive sparking at the break a condenser may be shunted across it, as is done in the induction coil.

most sharply defined. It is desirable to use values of the tension for which there will result not more than seven half waves. If the tension necessary to produce this result is sufficient to stop the vibrations of the fork, follow the instructions of the footnote on the preceding page, or shorten the length of string used.

Vary the tension and note the weights corresponding to at least three different wave lengths such that the numbers of half wave lengths in the length of the string are successive numbers; for example, 3, 4, 5. See how nearly a constant number you obtain by multiplying the tension by the square of the number of segments (see sect. 145).

Weigh the string and measure its length. Calculate its linear density ρ. Express in dynes the tension necessary to cause the string to vibrate in one segment. Calculate from these data the number of vibrations per second made by the fork (see eq. (11)).

(B) Object. To find the vibration frequency of the fork used in (A) by the method of beats.

Directions. It was the conclusion of the discussion of section 149 that if two sources of sound differ slightly in their vibration frequencies, there results in a second of time a number of alternations in the intensity of the resultant sound that is equal to the difference between the frequencies of the two sources. If one of the sources is of known frequency,—for example, a standard fork,—the frequency of the second source may be determined by observing the number of these "beats." The unknown fork to be used is that of the preceding part of this experiment.*

Select by ear a standard fork of about the same note as the unknown. Always set the fork in vibration by striking it with a felt-covered hammer like a piano hammer, or with a rubber mallet. The latter is conveniently made by placing a rubber stopper on the end of a rod. Now, using a stop watch, count the number of beats for several seconds. If the number is large, it will be found easier to count the beats in groups of three or four, e.g. one, two, three; one, two, three, etc. Now make the frequency of the unknown less by attaching to one of its prongs a small piece of soft wax. Count again the number of beats. If this number is less than before, obviously the unknown has been brought nearer the standard by weighting; that is, its vibration frequency is larger than the standard by the number of beats first observed. If the act of weighting the unknown fork increases the number of beats, then the number of vibrations per second of the unknown is smaller than that of the known. Hence the number of beats per second must be added to the vibration number of the standard to obtain that of the unknown fork.

* Or it may be the fork of the falling-body apparatus, or of the inertia disk of Experiments 1 or 10 of "Mechanics, Molecular Physics, and Heat." Either of these should be rated in the supporting frame and not removed.

EXAMPLE

(A) The tension for 3 segments was 162 g. ; for 4 segments, 94 g.; for 5 segments, 60 g.; for 6 segments, 42 g.; for 7 segments, 30 g. The products of the tension in grams and the square of the number of segments were 1458, 1504, 1500, 1512, and 1470 respectively. The average product was 1488. The length of the string was 95.4 cm.; its mass, .390 g.; its linear density, .00409. Multiplying 1488 by 980 to reduce the tension to dynes and then substituting the above values of T, ρ, and L in the first of equations (11), gave $n = 99$. The fork used was marked by the maker 100 vibrations per second.

(B) Using the fork and a standard of frequency 100, one beat in two seconds was observed. Weighting the unknown increased the number of beats; hence the rate of the unknown was 99.5 per second and the per cent of error in **(A)** was .5.

CHAPTER XXI

DIFFRACTION OF SOUND AND LIGHT WAVES

150. Two theories of light. In Sir Isaac Newton's day (1642–1727) two rival theories of light were struggling for recognition. The one, the wave theory, fathered and championed by the Dutch physicist, Christian Huygens (1629–1695), regarded light, like sound, as some sort of a wave motion, the chief difference between the two being, according to this theory, that, while sound is propagated through the agency of ordinary matter, light is a wave motion in some all-pervading medium to which the name of "the ether" was given.

The rival theory, called the corpuscular theory, regarded light as due to the emission from all luminous bodies of minute corpuscles which travel in straight lines and with enormous velocities through space and produce the sensation of light when they impinge upon the retina of the eye. This theory had its most famous and most brilliant advocate in Sir Isaac Newton himself.

Newton's chief reason for rejecting the wave theory lay in the fact that he was unable to understand why, if light is a wave motion, it is always propagated in straight lines past the edges of opaque objects, instead of undergoing *diffraction*, that is, being bent around such objects, as are sound waves, water waves, and all the other types of waves with which Newton was familiar. What is commonly regarded as the decisive test between the two theories was made in the year 1800 by Thomas Young, and consisted in showing that it is possible to produce with light waves the diffraction phenomena which are to be discussed in the later sections of this chapter, and which it does not seem possible to account for from the standpoint of the corpuscular theory.

It is the object of the present discussion and of the succeeding experiments to show both theoretically and experimentally that, under suitable conditions, sound does not bend around corners, as it is commonly supposed invariably to do, and that light, on the other hand, does under suitable conditions bend around corners, as it is commonly supposed not to do. More explicitly stated, our aim will be to show that the phenomenon of straight-line propagation is characteristic of any and all types of wave motion, provided only the aperture through which the waves pass is large in comparison with the wave length of the waves. If this proposition can be proved, it will be evident that the fact of the straight-line propagation of light does not furnish any argument against the wave theory, provided the wave length of ordinary light waves is very minute in comparison with the dimensions of ordinary apertures. Before proceeding to this proposition it is necessary to consider further the nature of a wave motion in a medium of indefinite extent, and the conditions for interference in such a medium.

151. Definition of wave front. Consider S in Figure 164 to be the point source of a wave motion in an isotropic medium; that is, a medium in which the disturbance is propagated with equal speed in all directions. When the disturbance which originates at S has just reached a, it has also then just reached all other points, such as b, c, and d, which are at the same distance from S. The spherical surface passing through these points is known as the wave front of the disturbance. In general, the wave front may be defined as the surface passing through all the particles which are in the same phase of vibration.

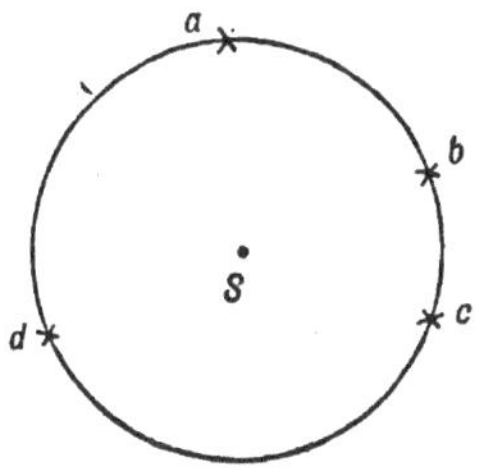

FIG. 164

The form of the wave front under the conditions just mentioned is spherical, but it will be shown later that conditions may arise in which it has not this form. Further, it will also be shown that under proper conditions a spherical wave may be converging, i.e. concave toward the direction in which it is traveling, instead of diverging, as in the case just considered.*

If the source is far enough away (rigorously, at an infinite distance), the wave front will obviously be plane.

* See section 155, page 245.

152. Construction of a wave front. Any particle in the wave front of a disturbance may be considered as a point source from which is spreading out a spherical wave. Thus consider the particles a, b, c, and d in the plane wave represented in Figure 165. A short time after the disturbance has reached these particles let the spherical wave surfaces due to them have the forms shown in the figure. If the number of these new centers is very large, it is evident from the figure, where for clearness only four have been represented, that the disturbance along the surface AD is very much greater than at any other points. In fact it may be shown by a mathematical analysis that these small spherical waves destroy each other by interference except at the surface AD. From its geometrical nature as a surface tangent to all the smaller surfaces, AD is known as their envelope. This envelope is then the new wave front of the disturbance. In general it may be said that *at any instant the wave front of a disturbance is the envelope of all the secondary-wave surfaces which are due to the action as separate sources of all the various particles that at some previous instant constituted the wave front.*

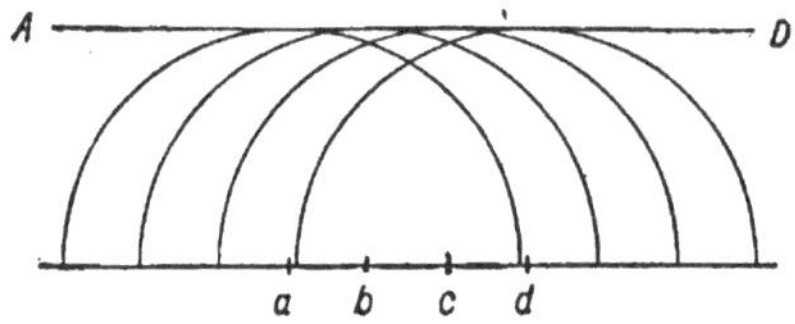

FIG. 165

153. Conditions for interference of two wave trains in a medium of indefinite extent. Let A and B (Fig. 166) be two particles vibrating in the same phase from each of which is propagated a disturbance having a spherical wave front. Let similar wave fronts be constructed for each particle. Thus the circular arcs a and a' represent the position of particles at the same distance from their respective sources and therefore in the same phase. The arcs b and b' represent the wave fronts when the disturbances have traveled one half a wave length farther; that is, each of them represents the locus of a series of particles which are exactly opposite in the phase of their vibration to the particles of a and a'. The arcs c and c' represent the wave fronts when they have traveled a whole wave length beyond a and a'. Their particles are in similar phase to those of a and a' and opposite to those of b and b'.

The particles in the line determined by the points marked x, x_2, x_3 have superimposed upon them vibratory motions of the same phase from both sources. Along this line there is therefore a reënforcement, or a maximum disturbance. Along the line determined by the points marked o_1, o_2, o_3, on the other hand, the vibrations superimposed are opposite in phase, and there is interference, or a minimum disturbance. Further, along the line determined by the points x_1, x_4, x_5 there is again reënforcement. From the construction of the figure it is evident that the condition for a maximum at any point is the existence of a difference in length of path between the point and the sources A and B respectively of some integral multiple of a whole wave length. Thus at x the difference in path is zero wave lengths, at x_1 it is one wave length, etc. Similarly, for a minimum the difference in distance must be an odd multiple of a half wave length. At o_1, o_2, o_3, etc., it is $\frac{1}{2}$ wave length. Additional maxima and minima may be found by extending the lines a, a', b, b', etc.

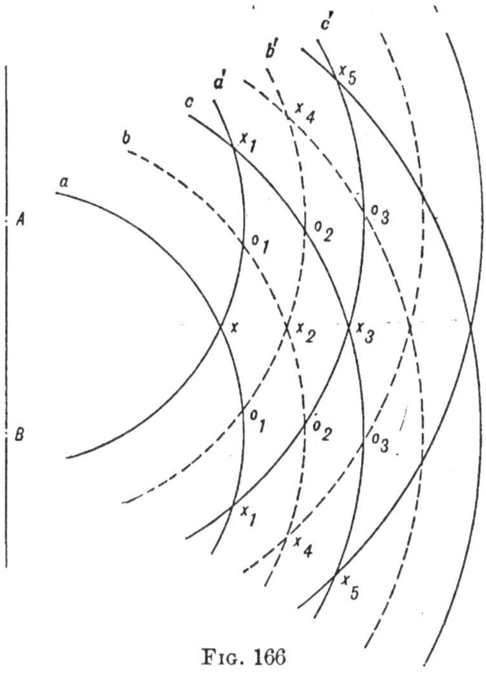

Fig. 166

It is important to notice that the lines of minimum disturbance o_1, o_2, o_3, etc., move farther and farther away from the central line of maximum disturbance x, x_2, x_3, the smaller the distance AB becomes in comparison with a wave length. Thus if AB is very large in comparison with a wave length, the line o_1, o_2, o_3 is very close to the line x, x_2, x_3, and similarly the line x_1, x_4, x_5 is close to the line o_1, o_2, o_3. But as AB becomes smaller and smaller these lines diverge more and more. When AB is just equal to a wave length the line x_1, x_4, x_5 is in the prolongation of *AB*, since it is only points in this line which can then differ by one wave length in their distances from *A* and *B* respectively. When *AB* is equal to a half wave length the line o_1, o_2, o_3 is in the prolongation of *AB*, and there are then no points of quiescence at all to the right of *AB*. When *AB* is less than a half wave length there are no points of quiescence anywhere. These considerations will now be applied to the discussion of the rectilinear propagation of wave disturbances through openings in screens.

154. The propagation of wave motions through apertures. Consider the case of a train of short waves, which, proceeding from a distant source, pass through an opening *ac* (Fig. 167) and fall upon a screen *mn*. Assume that the length *ac* is large as compared with the wave length of the train of waves. A distant source is chosen so that the wave front of the disturbance which reaches the

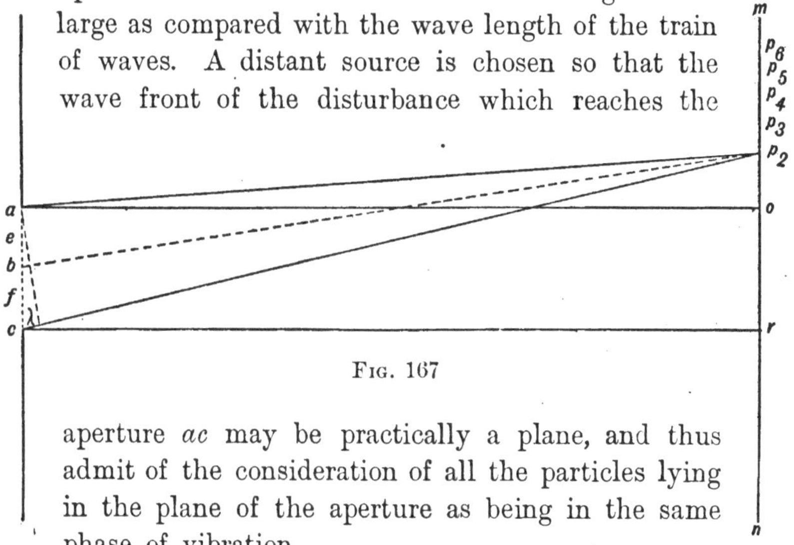

FIG. 167

aperture *ac* may be practically a plane, and thus admit of the consideration of all the particles lying in the plane of the aperture as being in the same phase of vibration.

The lines *ao* and *cr* are drawn from the source, assumed to be a point, past the edges of the opening *ac* to the screen; i.e. they are the lines which mark the limits of the *geometrical beam.* Suppose that the wave length and the opening *ac* are so related that the point p_2 on the screen, for which the distance cp_2 is exactly one wave length greater than the distance ap_2, falls outside the limits of the geometrical beam, i.e. above the point *o*. Then the particles *a* and *b* will differ in distance to p_2 by a half wave length. Hence the vibrations produced at p_2 by these two particles mutually neutralize each other. Similarly the disturbance originating in the first particle below a will at p_2 be just one half wave length ahead of the disturbance coming from the first particle below b. Thus every particle between a and b may be paired off with a corresponding particle between b and c such that the effects of the two particles neutralize each other at p_2. Hence the total effect at p_2 of the disturbances coming from the portion ab of the opening is completely neutralized by the effect of the disturbances coming from the portion bc of the opening.

Consider next a point p_4 which is so situated that the distance cp_4 is two wave lengths more than the distance ap_4. The opening ac may now be divided into four parts, ae, eb, bf, fc, such that eb neutralizes at p_4 the effect of ae, since ep_4 is one half wave length more than ap_4, and fc neutralizes the effect of bf, since fp_4 is one half wave length more than bp_4. There is therefore no disturbance at all at p_4.

At some point p_3, between p_2 and p_4, the distance cp_3 will be one and a half wave lengths more than ap_3. If we now divide ac into three equal parts, the effect of the upper third will be completely neutralized at p_3 by that of the next lower third, but the effect of the lowest third has nothing to neutralize it at p_3; hence there is a disturbance at p_3 which is due simply to one third of the particles between a and c, and even the effects of the particles in this third partially neutralize one another at p_3, since they differ somewhat in phase. It is obvious that between p_2 and p_4 the disturbance increases from zero at p_2 to a maximum at p_3, and then falls gradually to zero at p_4; that, further, there are other points of zero disturbance, p_6, etc., so situated that the distance from c to the point in question is any even number of half wave lengths more than the distance from a to this point; and that between these points of zero disturbance are points of maximum disturbance, p_5, etc., so situated that the distance from c to the point in question is any odd number of half wave lengths more than the distance from a to this point. But it will also be noticed that *the successive maxima, p_3, p_5, etc., diminish rapidly in intensity, since, while but two thirds of the particles between a and c completely neutralize one another's effects at p_3, four fifths of these particles neutralize* *one another's effects at* p_5, *six sevenths at* p_7, *etc.* Hence it is not necessary to go a great distance above o in order to reach a region in which there are no points at which there is any appreciable disturbance. Further, if we consider wave lengths which are shorter and shorter in comparison with ac, the points of maximum disturbance p_3, p_5, etc., draw closer and closer together, and soon some of them begin to fall inside the limits of the geometrical beam, i.e. below the point o. Hence those that are left above o are weaker and weaker members of the series. It follows, therefore, that when the wave length becomes very short in comparison with ac, the disturbance will have become practically zero at a very short distance above the point o. In other words, *a wave motion should be propagated in straight lines through an opening, or past an obstacle, and should not bend around appreciably into the region of the geometrical shadow, when and only when the wave length is very minute in comparison with the size of the opening;* for in this case the disturbances from the various elements of the opening must interfere in such a way as completely to destroy one another at practically all points outside the limits of the geometrical beam. The analysis of the conditions which exist inside the limits of the geometrical beam when p_2, p_4, etc., fall below o will not here be taken up, since we are not concerned at this point with showing what happens inside of or so much as with proving that practically nothing happens outside of or. Suffice it to say that experiment and theory both show that under the conditions assumed there is practically uniform disturbance within the region or.

Now since ordinary sound waves have a wave length of from 1 to 8 feet, it will be seen from the above analysis that in passing through a window or any ordinary opening they may be expected to spread out in all directions beyond the opening, as in fact we know that they do. Indeed, if the aperture is less than one wave length in width, it should be impossible to find any point of quiescence whatever on the side of the screen which is away from the source. It is clear, then, that we must produce extremely short sound waves, if we are to hope to observe with any ordinary openings the diffraction phenomena presented in the above theory.

In *light*, however, since, as we shall presently see, the average wave length according to the wave theory is only about .00005 cm., we should expect that with ordinary openings the maxima p_3, p_5, etc., would lie so near to the edges o and r of the geometrical beam as not to be easily discernible, so that, in order to bring them into evidence at all, we should expect to be obliged to work with exceedingly small openings. There is a still further condition, however, which must be met in order to bring out with maximum clearness, in the case of either light or sound, the diffraction bands p_2, p_3, etc., of the preceding theory. It is that the wave experimented upon be converging instead of plane or diverging, as assumed in the preceding discussion. Before presenting, therefore, experiments on diffraction it is desirable to consider the methods by which the form of a wave may be altered.

155. Formation of images by changes in wave form. For the sake of clearness let us take a concrete case and imagine, with Huygens, that light is a wave motion, and that a light wave originating in a point S (Fig. 168) beneath a surface of water spreads from that point as a spherical wave. Let us imagine further that light travels faster in air than it does in water. Then it can easily be seen that the wave will undergo a change in curvature in passing through the interface

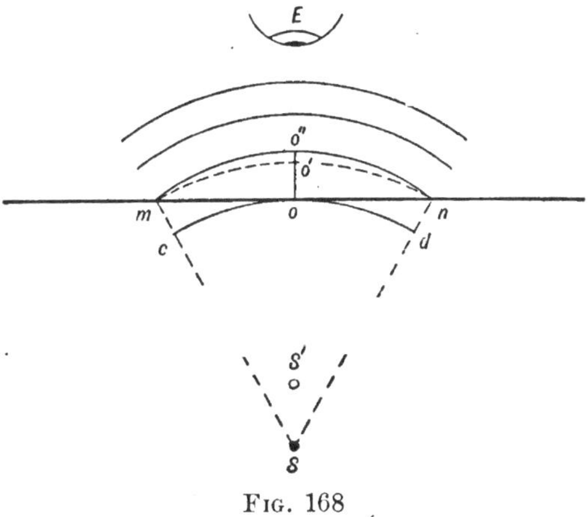

FIG. 168

mn into air. For if there were no change in the velocity of propagation in going from water into air, then the wave front which at one instant had reached the position cod would an instant later have reached the position $mo'n$, so drawn that $cm = oo' = dn$. But if oo'' represents the distance which light travels in air while it is traveling the distance oo' in water, then the wave, upon emergence into air, should occupy some position $mo''n$ instead of $mo'n$. In other words, the waves should *bulge upward* upon passing from the water to the air, so that an eye placed at E should receive a portion of a wave which actually has its center at some point S' instead of at S. This means, of course, that the point S should appear to be, not at its real position, but at some higher position S'. This is precisely what everyday experience teaches is the case; for objects under water always appear closer to the surface than they actually are. The point S' is called the *image* of the point S. It is further called a *virtual* rather than a *real* image, because, although the center of the wave which reaches the eye is at S', the actual center of disturbance is at some other point, namely at S.

Again, when a spherical wave from a point S strikes the boundary mn (Fig. 169) of a new medium, it is to be expected that a portion of it will undergo reflection, and that, at the instant at which the advancing wave front would have been in the position $mo'n$, had there been no reflection, the reflected wave will actually be in the position of the circular arc $mo''n$; so drawn that $oo'' = oo'$. This means, of course, that an eye at E, or at any point above mn, would receive a wave which has its center at S' instead of at S. As above, S' is called the image of S. In general, *when for any reason a spherical wave originating at a point in space is so modified that the wave appears to have originated at a different point, this second point is called the image of the first point.* It will be obvious that in this case, as in that preceding, S' is a virtual image of S.

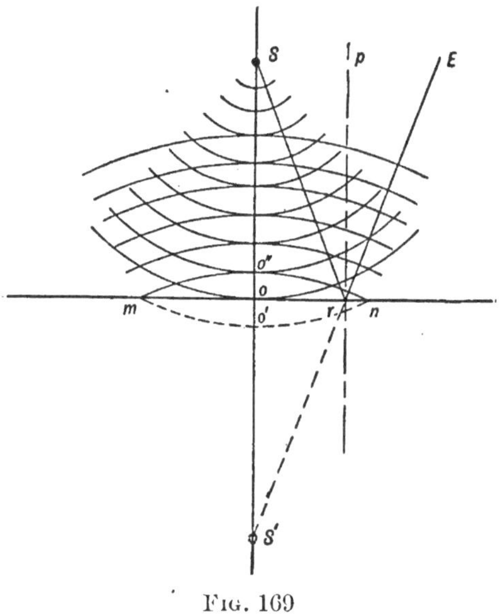

FIG. 169

Since in Figure 169 the chord mn is common to the two arcs $mo'n$ and $mo''n$, it is evident that *the image of a point in a mirror is on the perpendicular drawn from the point to the mirror, and as far behind the mirror as the point is in front of it.* The figure also shows that the light which comes to the eye by reflection from the mirror must follow the law, *angle of incidence* $Srp =$ *angle of reflection* prE; for since $oS = oS'$, $\angle rSo = \angle rS'o = \angle prS = \angle prE$.

These theoretical deductions from the wave theory are precisely the laws which experiment shows to be those which govern the reflection of light from plane mirrors.

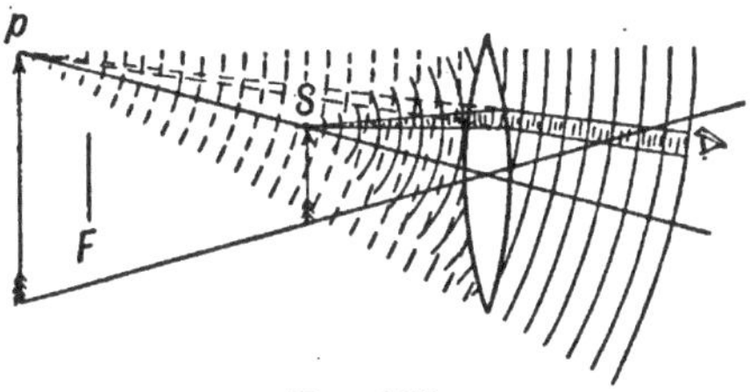

FIG. 170

A third method of modifying the shape of a wave, and one which finds large application in optical instruments, consists in causing the wave to pass through a lens in the manner shown in Figure 170. If the speed of propagation of the wave is less in the material of the lens, for example glass, than it is in air, then, since the portion of the wave which passes through the middle of the lens is retarded more than the portions which pass through the edges, it is obvious that the lens will tend to reverse the direction of curvature of the wave. If the source S is close to the lens, the curvature of the wave front which has passed through the lens may not be reversed, but it will be diminished; that is, the wave will actually be flattened so that its center will lie to the left of S. In this case an eye placed to the right of the lens and looking toward S will see a virtual image of S at the point p at which this center is located (see Fig. 170). If the distance of S from the lens is that of the so-called principal focal plane F, the emerging wave will be plane (Fig. 171); but if S is to the left of the principal focal plane F, the emerging wave will be reversed; that is, it will be concave toward the direction in which it is traveling (Fig. 172). It cannot, in general, be assumed that the emerging wave front will be strictly spherical, but if a diaphragm ac is introduced, as

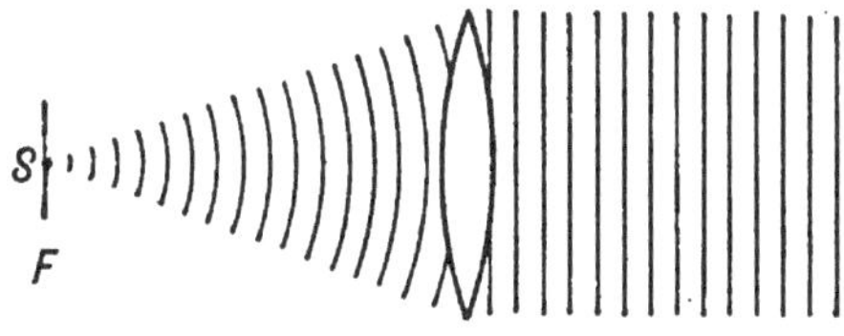

FIG. 171

in Figure 172, so as to cut out those portions of the wave which have passed through the edges of the lens, the remaining wave front may, with a suitable lens, be made practically spherical. In this case the disturbances starting from every point on the spherical surface reach the center *p* of the sphere at precisely the same instant. Since these disturbances will all be in the same phase,

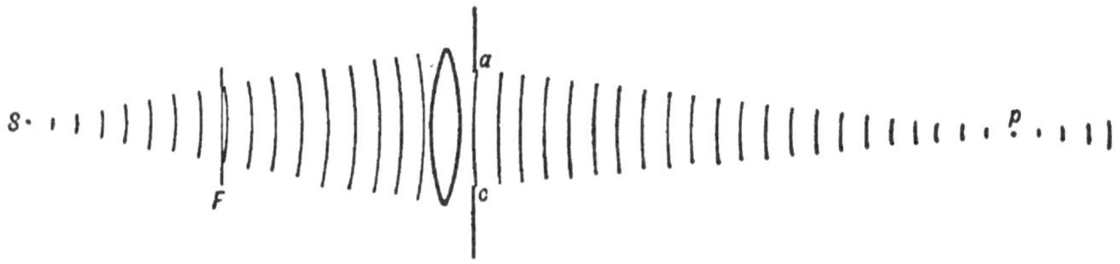

FIG. 172

they will unite to produce a disturbance of very great intensity at *p*. In ordinary language the light from *S* will be focused at *p*. In technical terms a *real image* of *S* will be formed at *p*. Such an image is distinguished from the virtual images which have thus far been discussed in that the point *p* now becomes an actual

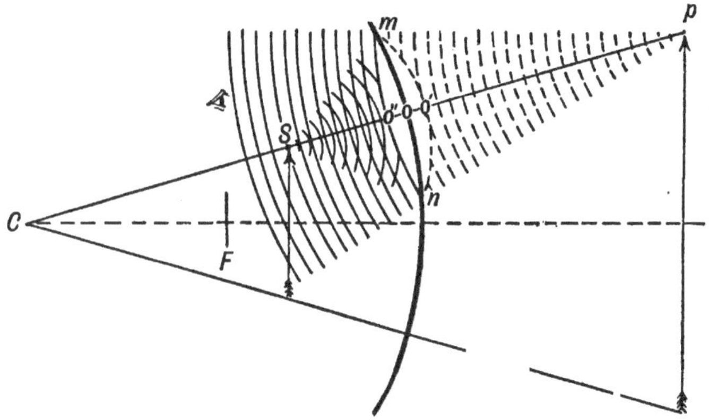

FIG. 173

center of disturbance from which waves spread out to the right of *p* as though the source *S* were itself at this point (see Fig. 172).

A second method of producing a wave which is concave toward the direction in which it is traveling is to allow a diverging wave of suitable curvature to be reflected from a concave mirror. If the source *S* (Fig. 173) is closer to the mirror than the principal focal plane *F*, the reflected wave will still be convex toward the direction in which it is traveling, as is shown in the figure which is constructed precisely as was Figure 169, oo'' being made equal to oo'. But if *S* is farther from the mirror than the principal focal plane, then the reflected wave will be concave, and a real image will be formed at *p* (see Fig. 174). In the cases illustrated in Figures 170, 172, 173, and 174, the points *S* and *p* are called *conjugate foci.* In Figure 171 the conjugate focus is obviously at infinity.

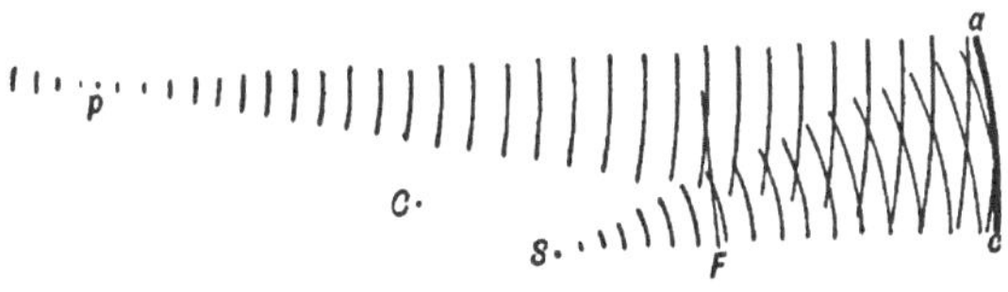

FIG. 174

156. The nature of a real image of a point source. An application of the reasoning of section 154 to the case of a converging wave shows that the image of a point source of waves should not be a single point of disturbance, but, instead, a series of maxima and minima of disturbance. For consider, as in section 154, a point p_2 (Fig. 175) far enough to one side of *p* so that ap_2 is one wave length more than cp_2. The disturbance from the portion

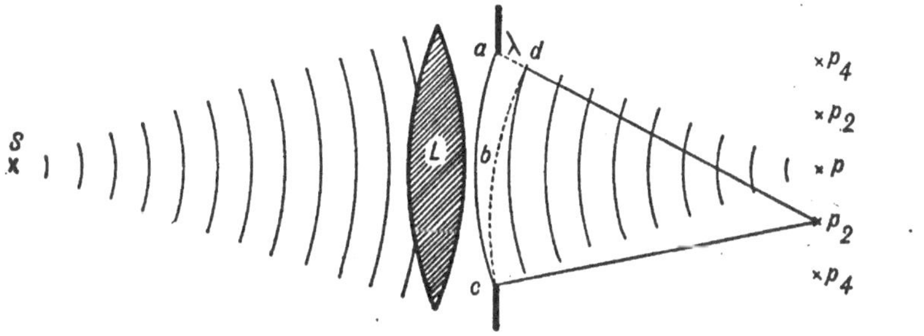

FIG. 175

ab of the wave front *abc* will completely neutralize at p_2 the disturbance from the portion *bc*. Furthermore, the contrast between the disturbance at *p* and the absence of disturbance at p_2 will be much more pronounced in this case than it was in the case illustrated in Figure 167, since in the case of Figure 175 all of the disturbances from the wave front *ac* are concentrated at the center *p*, thus making it a point of extreme brightness, while in the case illustrated in Figure 167 no such concentration occurs.

Precisely as in Figure 167, at p_3, p_5, and other points for which the distance from a to the point in question is any odd multiple of a half wave length greater than the distance from c to this point, there will be maxima of disturbance; but these maxima will decrease rapidly in intensity, p_3 being about a sixtieth as intense as p, p_5 one two hundredth as intense as p, etc. Moreover, p_3, p_5, etc., will draw very close together and very close to p as the wave length $\lambda (= ad)$ becomes small in comparison with the aperture ac. In other words, the distance apart of the maxima p, p_3, etc., *will depend simply upon the ratio between the wave length* λ *and the aperture* ac.

If, then, as the wave theory demands, the wave length of light is very minute as compared with the diameters of ordinary lenses or mirrors, the maxima p_3, p_5, etc., will be so close to p as to be indistinguishable from it; so that for practical purposes common lenses or mirrors should form, in the case of light, essentially point images of point sources, and at all points outside the limits of the geometrical beam, that is at all points not included within the region inclosed by the lines ap and cp, the light waves should mutually destroy one another.

Nevertheless, that the image of a point is not in reality a point even in the case of light waves, but consists of a series of maxima and minima, as the preceding theory demands, may be easily demonstrated by reducing the opening ac until it becomes more nearly comparable with a light wave. Thus a pin hole in a piece of cardboard held immediately in front of a bright flame and viewed at a distance of two or three feet through a small pin hole in another card held very close to the eye will appear, not as a point, but as a central bright disk surrounded by one, two, or even more black rings which correspond to the points p_2, p_4, etc., of Figure 175. Again, if a slit, say a half millimeter wide, be made to replace the first pin hole, and if it be viewed at a distance of a few feet through another slit, say one tenth millimeter wide, which is held very close to the eye, the first slit will appear as a central bright band flanked by a series of dark bands which correspond to the points p_2, p_4, etc., of Figure 175. In these experiments the remote pin hole or slit corresponds to the point S, the lens of the eye to L, the retina of the eye to the screen upon which the points p, p_1, p_2, p_3, etc., are observed, and the pin hole or slit held very close to the eye to the aperture ac, which is indeed in this case on the other side of the lens, but this fact does not alter in any way the theory of the phenomenon. The experiment may be made more striking by throwing a beam of direct sunlight through a half-millimeter slit covered with red glass, and then observing the slit through a telescope placed in the path of the beam. So long as the lens has its normal aperture the image of the slit as seen in the telescope will be sharply defined, if the telescope is properly focused.* But when the aperture is made small by slipping a second slit over the objective of the telescope parallel to the first, the series of light and dark bands p, p_2, p_3, etc., will at once appear. Furthermore, as the width of the slit over the objective is made larger and larger these bands will draw closer

* The way in which a lens or mirror forms an image of an *extended object*, such as a slit, may be readily seen by regarding the object as an assemblage of points. If the diameter of the lens or mirror is very large in comparison with a wave length, each point A or B (Fig. 176) of the object will have what is practically a point image of itself formed at a or b. Further, the points a and b will lie, approximately at least, in the prolongations of the lines drawn from A and B respectively to the center C of the lens, as is evident from the approximate symmetry of the lens, and hence of the retardation produced in the incident wave by it, about the lines AC and BC. Similarly, *the image of any point on the object will be in the prolongation of the line connecting this point with the center of the lens or mirror* (see also Figs. 170 and 173).

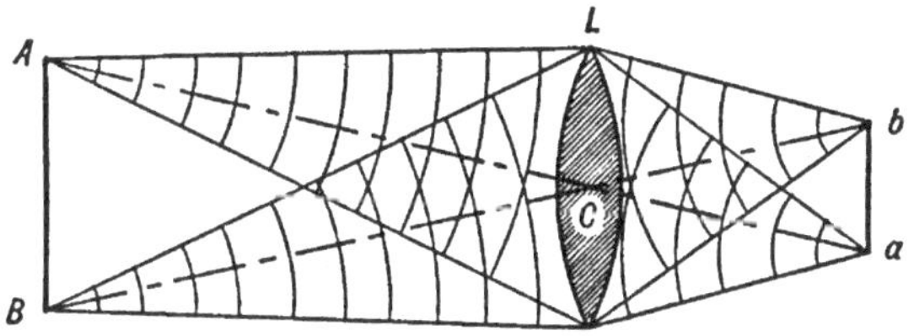

FIG. 176

When, however, we consider the clearness with which *detail* in the object is brought out in the image, we must even here remember that the images of points are not points but, instead, bright centers, each of which is surrounded by its series of bright and dark rings. If two of these centers are closer together than the radius of the first dark ring about each, they will evidently not be separated in the image by a dark region, and will therefore not be distinguishable as separate points. That is, a lens cannot produce an image which will make two close bright points appear as separate points if the angle subtended at the lens by the two points is less than λ/ac. This expression λ/ac, or the wave length divided by the aperture, is therefore known as the *limit of resolution* of the lens.

and closer together and finally be lost in the central image *p* of the first slit, precisely as the above theory demands that they should.

In order to demonstrate the existence of the same phenomena with sound waves it is obvious that, if the mirrors or lenses are to be of ordinary size, we must find a way of producing sound waves of unusual shortness. Such waves may be produced by the whistle described in the next section. They must be detected and measured, however, by a special appliance called a *sensitive flame.*

157. The sensitive-flame apparatus. The sensitive flame is a very high but very narrow flame produced by igniting a jet of gas as it issues from a tank under high pressure. The burner has a pin-hole opening and may be conveniently constructed by drawing out a piece of glass tubing with a long taper until the orifice has a mean diameter of about half a millimeter. The opening is in general made slightly elliptical and the flame is most sensitive when the shorter axis of the ellipse lies in the direction of the approaching disturbance. The entering gas is regulated until there results a long flame usually twenty or thirty centimeters high, which is just on the point of flaring.

Small disturbances of high frequency produced in the air surrounding the mouth of the burner cause the flame to flare and shorten very perceptibly, the reason being that very sudden variations in the pressure at the orifice of the burner, such as are produced by very rapid vibrations, cause corresponding variations in the rate of emission of gas, and hence corresponding variations in the height of the flame. The flame is very sensitive to exceedingly short waves such as are produced, for example, by jingling a bunch of keys. To the ordinary notes of audible sound it is wholly insensitive, because such notes do not produce sufficiently sudden changes in pressure at the orifice of the burner. There is used with the flame, therefore, as a source of waves, a small whistle shaped as shown in cross section in Figure 177, the distance between the two small openings being from one to three millimeters. The whistle is blown by compressed air and acts as

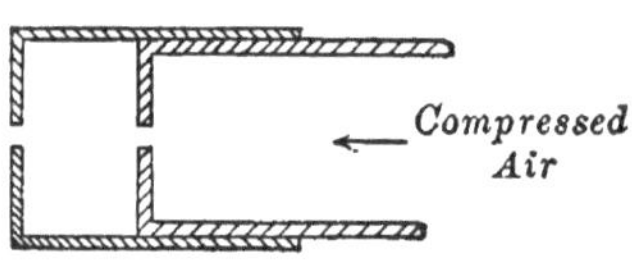

FIG. 177

a very short organ pipe,* the emission of pulses from each of the two openings being controlled by the period of the reflected pulses which bound back and forth between the two ends of the minute air chamber. Ordinarily no recognizable note is heard, for the reason that the note produced is above the limits of audition.

EXPERIMENT 21

(A) Object. To show experimentally that with sound waves of very short wave length sharp shadows are cast by ordinary objects.

Directions. Set up the sensitive flame described in section 157. Turn on the gas until the flame is from ten to fourteen inches high and just on the point of flaring, and also until a slight rattling of a bunch of keys will cause it to flare strongly. Locate the side of the flame which is most sensitive to this noise, and in all future use of the flame turn that side toward the direction from which the disturbance comes.

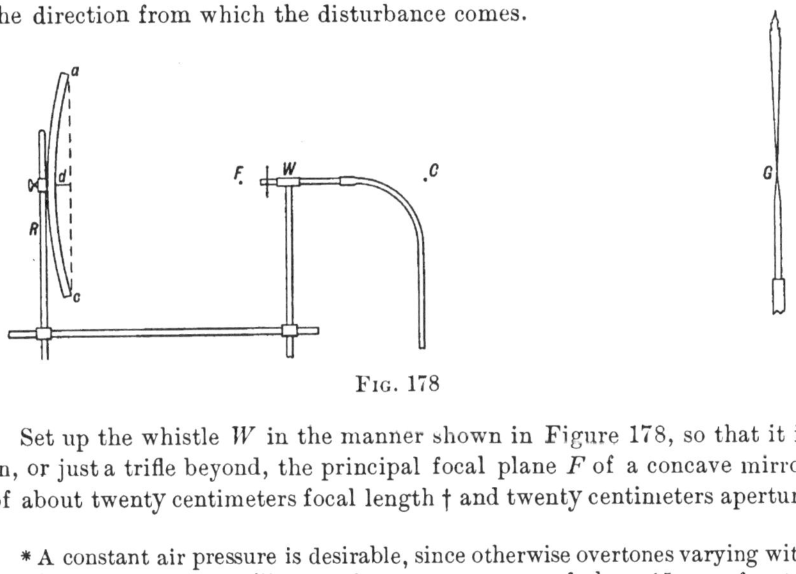

FIG. 178

Set up the whistle W in the manner shown in Figure 178, so that it is in, or just a trifle beyond, the principal focal plane F of a concave mirror of about twenty centimeters focal length † and twenty centimeters aperture

* A constant air pressure is desirable, since otherwise overtones varying with the changes of pressure will be produced. A pressure of about 15 cm. of water is usually sufficient.

† If the focal length F of the mirror is not known, and if the latter is made of plaster of Paris, or of some other material which renders a determination by optical means impossible, then determine F by measuring d and ac (Fig. 178), and substituting in the formula $F = \overline{ac}^2/16\,d$. This formula is obtained from a consideration first of the fact that $ac/2$ is a mean proportional between d and the other segment of the diameter of the circle of which the mirror is an arc; second, of the fact that, since d is very small, this other segment may be taken as the diameter itself; and third, of the fact that the focal length of the mirror is one half its radius of curvature (see Chap. XXIII).

(*ac*, Fig. 178). Place the sensitive flame *G* in the principal axis * of the mirror and at a distance from it of about eight feet. Slowly increase the air pressure with which the whistle is to be operated until the flame begins to flare strongly. Turn the tripod rod *R* which supports the mirror and whistle about a vertical axis and observe that under the conditions prescribed a sharply outlined beam of sound is produced which causes the flame to flare when it strikes it, but leaves it burning undisturbed when turned in another direction.

Turn the beam again upon the flame; then interpose between it and the whistle a strip of cardboard, or some other object a few centimeters wide, and observe that when in the right position it causes the flaring to cease, thus showing that with these short sound waves even small objects cast sharp shadows.

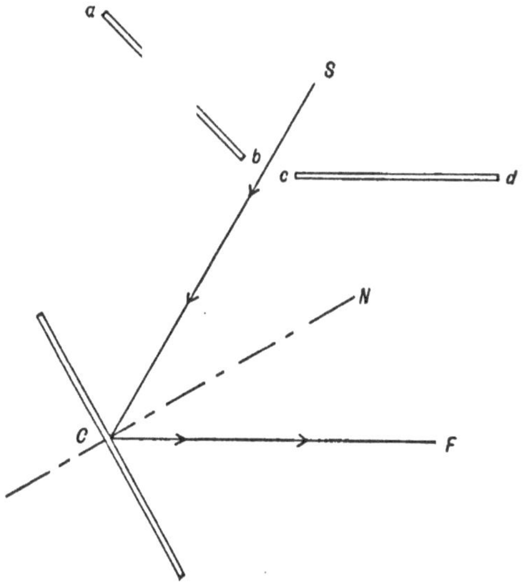

Fig. 179

Next let a beam of sound *SC* (Fig. 179) be sent from the mirror through an opening about a foot in diameter between two screens *ab* and *cd*, disposed as in the figure, and let this beam be reflected by a flat smooth board *C* to the sensitive flame placed at *F*. It will be found that the flame will flare strongly only when the board is so turned that the angle of incidence *SCN* is equal to the angle of reflection *NCF*.

(B) Object. To find the wave length of the note emitted by the whistle by locating nodes and loops in front of a reflecting wall.

Directions. Set up a smooth planed board behind the flame *G*, arranged as in Figure 178, and let the plane of the board be at right angles to the line connecting the flame and whistle. Attach a base to the board so that, as the latter is moved back, its plane remains parallel to itself. Starting with the board but a centimeter or so from the flame, move it a few millimeters forward or back until a position is found in which the flame ceases to flare. Mark the position of the board with a piece of chalk; then slide it slowly back from the flame and count the nodes, i.e. the positions of no flaring, as each is passed. Mark the position of, say, the twenty-fifth node

* The principal axis is a line drawn through the center of the mirror and its center of curvature *C*.

and find the wave length of the note given off by the whistle from a consideration of the fact that the distance between successive nodes is one half wave length.

(C) Object. To find, by means of diffraction experiments, the wave length of the note used in (**B**).

Directions. Place the whistle W (Fig. 178) in a position on the principal axis of the mirror 2 or 3 cm. from its principal focus. Then from the distance f $(= MS$, Fig. 180) from this point to the mirror, and the focal length F, compute the distance f' $(= pM$, Fig. 180) from the mirror to the conjugate focus p by means of the formula

$$\frac{1}{f} + \frac{1}{f'} = \frac{1}{F}.*$$

Place the mirror and whistle so that the orifice of the flame G (Fig. 178) is accurately at this conjugate focus p; then slowly rotate, about its own axis, the vertical rod R which carries the mirror and whistle, and thus locate, on either side of the central region p (Fig. 180) of intense disturbance, at least one point p_2 of no disturbance, and one more point p_3 of

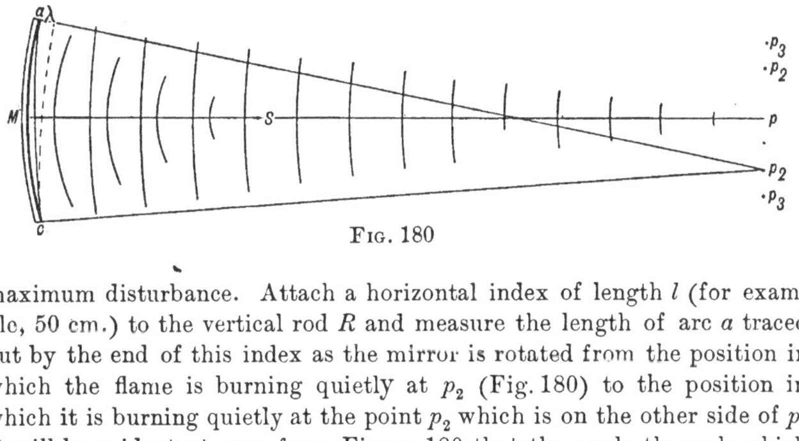

FIG. 180

maximum disturbance. Attach a horizontal index of length l (for example, 50 cm.) to the vertical rod R and measure the length of arc a traced out by the end of this index as the mirror is rotated from the position in which the flame is burning quietly at p_2 (Fig. 180) to the position in which it is burning quietly at the point p_2 which is on the other side of p. It will be evident at once from Figure 180 that the angle through which the mirror must be rotated to cause the point p to move over to the position of p_2 is pp_2/pM. But this angle is also λ/ac (see Fig. 180). Hence the equation

$$\frac{\lambda}{ac} = \frac{pp_2}{pM}, \quad \text{or} \quad \frac{2\lambda}{ac} = \frac{p_2p_2}{pM}.$$

But obviously,

$$\frac{p_2p_2}{pM} = \frac{a}{l}.$$

Hence

$$\frac{2\lambda}{ac} = \frac{a}{l}, \quad \text{or} \quad \lambda = \frac{1}{2}\frac{a}{l}ac.$$

* See Chapter **XXIII** for the derivation of this formula.

Compare the value of λ thus found with the result obtained in (**B**).*

(D) Object. To determine, by diffraction experiments, the average wave length of red light.

Directions. At a distance of about 3 m. from a vertical slit s (Fig. 181) about .5 mm. wide, set up an ordinary reading telescope T. Throw a beam of sunlight through s in the direction sT, cover s with a piece of red glass so that only red light reaches the telescope T from s; then focus T carefully upon the slit s, and thus obtain a sharply defined image of s in the eyepiece of T. Now place over the objective of T a cap containing a slit not more than .5 mm. wide, and rotate this slit until it is parallel to the slit s. We have then precisely the conditions given in Figure 175 for the formation of diffraction bands. In a manner identical with that used in (**C**) we may now turn the telescope T on its vertical support and observe the angle p_2p_2/pL (Fig. 175) through which it must pass in order to cause the cross hairs to move from one of the two dark bands p_2, adjacent to the central bright one p, to the other. In order to measure this

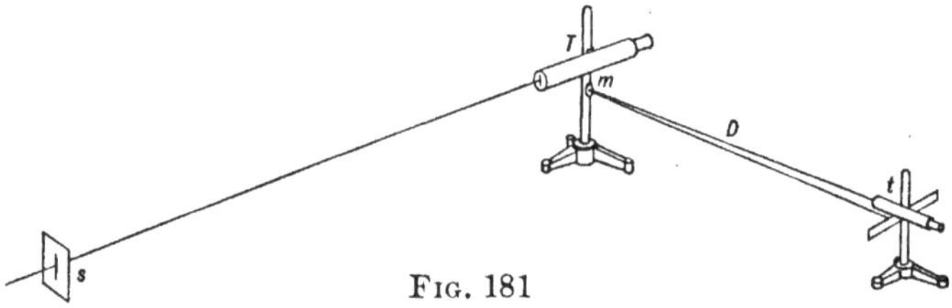

FIG. 181

angle accurately, attach to the telescope support a mirror m and observe its rotation with a second telescope and scale t (see Fig. 181). If d is the deflection observed on the scale of t, and if D is the distance from the mirror to the scale, then $d/2D$ is obviously the angle sought. To obtain as accurate a determination as possible of this angle, take the mean of the distances p_2p_2, $p_4p_4/2$ and $p_6p_6/3$. Measure with a micrometer microscope the width ac of the slit over the objective, then compute the wave length λ of red light from the relation

$$\frac{\lambda}{ac} = \frac{d}{4D},$$

and compare the result thus found with the value .000067 cm., which is about the mean value of the wave lengths of the light transmitted by ordinary red glass.

* Rigorously, the analysis given in section 156 applies only to a single plane, that is, to a rectangular aperture or slit, the dimension of which at right angles to the plane of the paper is very great in comparison with the dimension ac. The complete analysis for a circular aperture such as that here used is too complicated to be within the scope of this book. Suffice it to say that such analysis leads to the result that pp_2 for circular apertures is 1.22 times its

EXAMPLE

(A) The predictions as to sharp shadows formulated in the directions were verified by experiment.

(B) By the method of locating nodes in front of a reflecting wall twenty half wave lengths were found to be equal to 17 cm. Hence $\lambda = .85$ cm.

(C) The wave length was determined by the method of diffraction as follows. The diameter ac of the mirror was 20.4 cm., the sagitta d was 1.30 cm. Hence $R = \overline{10.2}^2/2\,(1.3) = 40$ cm. and $F = 20$ cm. The whistle was fastened to a rod attached to the vertical axis with which the mirror rotated. It was placed a distance f of 22.1 cm. from the center of the mirror along the principal axis. The sensitive flame was placed at a distance f' of 210 cm. measured along this axis from the mirror, the value of f' having been found by the relation $1/22.1 + 1/f' = 1/20$. An index 47 cm. long was fastened to the vertical rod and turned with the mirror. As the mirror turned through the angle between p_2 and p_2 the end of the index moved over 4.65 cm. Hence (see note, page 256) $\lambda = \frac{1}{2} \cdot \frac{4.65}{47} \cdot \frac{20.4}{1.22} = .83$ cm. This value agreed with that of **(B)** to within 2.4 per cent.

(D) Four dark bands on either side of the central bright image were easily observed through an aperture of width $ac = .0404$ cm. A series of readings was taken on all these bands with a telescope and scale placed 162 cm. from the mirror m attached to the first telescope T. The differences in readings on this scale for settings on the bands were as follows. From p_2 to p_2, 1.09 cm.; from p_4 to p_4, 2.16 cm.; from p_6 to p_6, 3.30 cm.; and from p_8 to p_8, 4.40 cm. Hence the average distance on the scale corresponding to the angle between p_2 and p_2 was 1.095 cm. Hence, since $\lambda = \frac{d\,(ac)}{4\,D}$, we have $\lambda = \frac{(1.095)\,(.0404)}{4\,(162)} = .0000685$. This is in agreement with the mean value of .000067 to within 1.7 per cent.

value for a rectangular opening. Hence the value of λ found as above must be divided by 1.22 in order to obtain an accurate comparison with the value found in **(B)**. Rigorously, then, we have

$$\lambda = \frac{1}{2}\,\frac{p_2p_2}{pM} \cdot \frac{ac}{1.22}.$$

CHAPTER XXII

THE DIFFRACTION GRATING

158. The principle of the grating. The phenomena of diffraction are most strikingly exhibited with the aid of an instrument devised in 1821 by the celebrated German optician Fraunhofer, and known as the *diffraction grating.* Such a grating consists essentially of an opaque screen in which are placed at regular intervals small parallel slits for the transmission or reflection of light. Thus in Figure 182 qq' represents a cross section of the grating, and the openings m, p, s, etc., represent the slits, which are thought of as extending at right angles to the plane of the page. For the sake of convenience in analysis these slits will at first be regarded as exceedingly narrow in comparison with their distance apart, that is, each opening will be thought of as a mere line. Let the source of light be so far distant that the wave surface ww' which falls upon the grating is practically plane. It has been shown in the preceding chapter that if no grating were present, a lens L interposed in the path of the wave ww' would form an image of the distant source S at some point a, while at all other points on the screen uv there would be destructive interference and therefore total darkness. Let us see how this conclusion would be modified if we take out certain portions of the wave

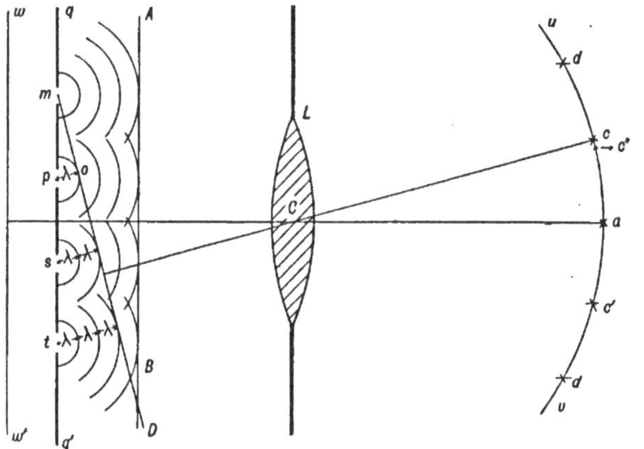

FIG. 182

front ww' by means of the grating. When the wave ww' reaches the grating qq' the points m, p, s become new sources of spherical waves, and if we draw the envelope to all these waves after the disturbance has traveled a small distance forward, we shall obtain precisely the same surface AB which we should have had if the grating had not been present, the only difference being that the intensity of disturbance in the plane AB is much less than before, since now but a few points, namely m, p, s, etc., are sending out spherical waves to AB, while before all the points in qq' were so doing. As was shown in section 155, the lens will take this plane wave AB, consisting of vibrations all of which are in the same phase, and bring it to a focus at a, so that an enfeebled image of the distant source S will be formed at this point. Thus far, then, the only effect of the grating has been to diminish the intensity of the image at a.

But AB is not now the only surface which can be drawn to the right of the grating so as to touch points all of which are in the same phase of vibration, for a surface mD, so taken that the distance from p to it is one wave length, that from s two wave lengths, and so on, satisfies this condition quite as well as does the surface AB. Hence mD may be regarded as another plane wave, which, after passage through the lens, will be brought to a focus at some point c in the line drawn perpendicular to mD through the center C of the lens,* in precisely the same way in which the plane wave AB was brought to a focus at a. It follows, then, that an image of the source should be formed at c as well as at a. Precisely the same line of reasoning will show that another image of the distant source should be formed at c' as far below a as c is above it. But a, c, and c' are not the only points at which images of the source will be formed, for it is possible to pass a plane through m such that the distance from p to this plane is 2λ instead of λ, that from s, 3λ, etc. It is obvious that all points in this plane will be in the same phase of vibration, and hence that the resulting plane wave will be brought to a focus at some point d on the perpendicular drawn from the plane through C, and at the same distance from C as are a and c.

* See footnote on page 251.

Similarly, there will be other images whose direction from C is determined by the simple condition that the successive distances from the slits to the wave front differ by a whole multiple of a wave length; thus $po = \lambda, 2\lambda, 3\lambda, 4\lambda, 5\lambda$, etc. The first image, namely that at c, where this difference is one wave length, is called the image of the first order. Similarly, that at d is the image of the second order, and so on. In a word, then, *a lens and grating disposed as in Figure 182 should produce a whole series of equidistant images of any distant source of light.* This means, of course, that under these conditions light waves will bend far around into the region of the geometrical shadow and be discernible at a large number of different points instead of simply at a.

These theoretical deductions from the wave theory of light are completely confirmed by experiment. Furthermore, the experiments illustrating them are so simple and so much a part of everyday experience that the wonder is that they escaped detection and explanation for so long a time. Thus if one looks through a handkerchief held close to the eye at a distant arc light, gas flame, or bright star, one can always see nine and sometimes as many as eighteen or more images of the light. These are due to the two sets of gratings formed by the two sets of threads which run at right angles to each other. It is usually possible to see as many as three distinct images by simply squinting at a distant light through the eyelashes, which act in this case as a very imperfect grating. In these experiments the retina of the eye takes the place of the screen uv, and the lens of the eye the place of L.

In its simplest practical form the grating consists of a plane piece of glass upon which are ruled with a diamond point, say, a thousand lines to the centimeter. The grooves cut by the diamond point constitute the opaque spaces in the grating, for the light which falls upon these grooves is scattered in all directions, so that a negligible part of it passes through in the direction in which the light is traveling. The clear glass between the rulings corresponds to the openings m, p, s in the screen of Figure 182. If such a grating is held immediately before the eye and a source of monochromatic light viewed through it, the series of images formed at a, c, d, etc., on the retina are apprehended by the observer as a series of images of the source lying in the prolongations of the lines aC, cC, dC, etc. The images may be thrown upon a screen, if screen, lens, grating, and source are given the relative positions shown in Figure 182.

159. The determination of wave length by means of the grating. If a grating is held immediately in front of the eye and a source of monochromatic light viewed through it, the wave length of the light vibration may be determined as follows. In Figure 183 let the angle between the grating and the

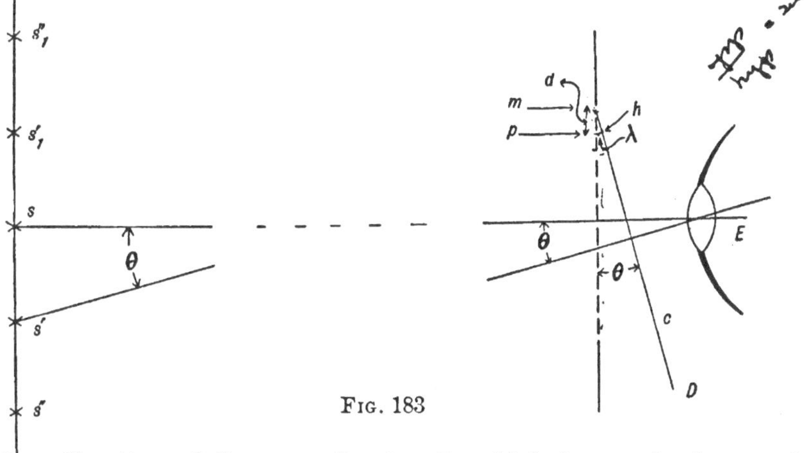

FIG. 183

direction of the wave front mD, which forms the image of the first order, be denoted by θ. Let the wave length be denoted by λ, and let d represent the distance between successive openings,—the grating space, as it is called. Now in the triangle pmh the obvious relation exists,

$$\sin\theta = \frac{\lambda}{d}. \qquad (1)$$

Similarly, for the image of the second order,

$$\sin\theta' = \frac{2\lambda}{d}. \qquad (2)$$

And, in general, if n represents the order of any image and θ the angle between the grating and the direction of the wave forming that image, then

$$\sin\theta = \frac{n\lambda}{d}. \qquad (3)$$

The determination of the angle θ may be easily made as follows. An illuminated slit is placed at s (see Fig. 183) in front of a scale fg two or three meters away from the grating and the eye. The positions s', s'', s'_1, s''_1, etc., at which the successive images formed by the grating appear, are then marked in some way upon the scale fg. If the grating has been placed so that the line of sight sE is perpendicular to it, then, since $s'E$ is perpendicular to the wave front mD, we have the angle $sEs' = \theta$, and $\sin\theta = ss'/s'E$. Therefore

$$\frac{\lambda}{d} = \frac{ss'}{s'E}, \quad \text{or} \quad \lambda = d\frac{ss'}{s'E}. \tag{4}$$

The distance $s'E$ can easily be measured with a tape. To obtain ss' it is customary to measure $s's'$ and divide it by 2. If it is the images of the second or third order which are located on the scale instead of those of the first order, then the relation evidently becomes

$$2\lambda = d\frac{ss''}{s''E}, \quad \text{or} \quad 3\lambda = d\frac{ss'''}{s'''E}. \tag{5}$$

The distance d is always obtainable from the maker of the grating.

160. The grating spectrum. If the source sends out, not monochromatic light, but, instead, white light, the series of sharply defined images of the source is found to be replaced by a single central image of the source in white light at s, bordered on either side by broad bands of colored light. In the first band the end farther from the source is red. From the red the color grades into orange, yellow, green, blue-green, blue, and finally into violet at the end nearer to the source. The band of light thus produced is called a *spectrum*, and the phenomenon of its production is known as dispersion.

The explanation of the dispersion produced by a grating is as follows. Just as in sound the peculiar tone quality of any note depends upon the combinations of wave lengths (i.e. the various overtones) which enter into its composition, so in light the tint or color quality of a light depends upon the wave lengths of the light vibrations which compose it. Thus pure white light contains vibrations of all the wave lengths which are capable of exciting the nerves of the eye. Color is then entirely dependent upon wave length and corresponds to pitch in sound. There are in fact as many pure colors as there are wave lengths for visible vibrations, and, in addition, there are an infinite number of combinations of color, just as there are of combinations of tones.

The action of the grating in producing dispersion is then easily seen; for since the position of every image except the central one is determined by the condition $\sin\theta = n\lambda/d$, it is evident that there is a different value of θ corresponding to each value of λ. Now the wave lengths which compose white light vary from about .000076 in the red to about .000039 in the violet. Hence when the source is white light the image of each order as it appears with monochromatic light is replaced by a series of adjacent images in different colors, each image corresponding to a particular wave length or color. This series of adjacent images constitutes the colored band or spectrum of each particular order. The central image is white and sharply defined because the wave front AB (Fig. 182) which gives rise to this image is at the same distance from each of the openings, and in consequence this wave front is the same for all wave lengths.

The spectrum of the first order is the only pure spectrum which a grating can produce, for it can be shown that the spectra of higher orders overlap. Thus, since for the red of the second order $\sin\theta = 2 \times .00007/d$, approximately, and since for the violet of the third order $\sin\theta' = 3 \times .00004/d$, it will be seen that $\sin\theta$ is greater than $\sin\theta'$, and hence that a part of the third violet overlaps a part of the second red. It is on account of this overlapping that one never sees more than two or three spectra on a side, for in the higher orders the overlapping is so complete as to reproduce white light.

A grating spectrum is called a normal spectrum, because the distance of each color from the central image is directly proportional to its wave length, so long as θ is small.

161. The dispersive power of a grating. It will be evident at once from Figure 182 that the smaller the distance between openings, the farther apart will be the successive images a, c, d; in other words, that the angular separation of different orders produced by a grating, and hence, also, the angular separation of different colors in the same order, increases as the distance between the lines of the grating decreases. This may be seen even more clearly by considering the equation $\sin\theta = n\lambda/d$; for if we subtract from this equation, as applied to one color of wave length λ_1, namely $\sin\theta_1 = n\lambda_1/d$, the same equation as it appears when it is applied to another color of wave length λ_2, namely $\sin\theta_2 = n\lambda_2/d$, we obtain

$$\sin\theta_1 - \sin\theta_2 = \frac{n}{d}(\lambda_1 - \lambda_2).$$

If we consider that the angles θ_1 and θ_2, being in general small, are approximately proportional to their sines, we see that the angular separation $\theta_1 - \theta_2$ of any two colors λ_1 and λ_2 is inversely proportional to the grating space d and directly proportional to the order n of the spectrum. This last result means, of course, that the spreading apart of the colors is twice as great in the second spectrum as in the first, three times as great in the third as in the first, etc.

162. The resolving power of a grating. When, however, we consider the *sharpness* with which the outline of any particular image in a particular color is formed, we find that this depends, for a given order of spectrum, not at all upon the closeness of the lines, but solely upon the number of lines constituting the grating. Thus, if there are but few openings, the individual images, *a*, *c*, *d* (Fig. 182), are found to be very indistinct in outline, but if there are a large number of openings these images become very sharply defined. In order to appreciate the reason for this let us consider a plane drawn through *m* (Fig. 182), so as to make a very slight angle with *mD*, for example so that *po*, instead of being exactly equal to λ, is just a trifle less than λ, say $\frac{99}{100}\lambda$. The disturbances which at any given instant have reached this plane will an instant later have passed through the lens and been brought together to the point c'' on the perpendicular to this plane through *C*. Now, although the disturbance which comes from *p* differs but very slightly in phase from that which comes from *m*, the disturbance from *s* differs twice as much from that which comes from *m*, etc., so that the disturbance from the fiftieth opening is just one half wave length behind that from

m, and therefore completely neutralizes it at c''. If, then, there are 100 openings, there will be complete interference at c'', although c'' is a point whose distance from c is but $\frac{1}{100}$ of ac. Just below c'' there will be a succession of maxima and minima in precisely the same way in which there are maxima and minima about a, as explained on page 249; but, as has been already shown, these maxima will decrease rapidly in intensity, so that we may say that practically there is complete interference between c'' and a point as far above a as c'' is below c. If there are 1000 openings, then the first point below c at which complete interference occurs, will be but $\frac{1}{1000}$ of the distance ac, instead of $\frac{1}{100}$ of this distance. Thus the larger the number of openings the closer does the first point of complete interference approach to c, and therefore the more nearly does the grating become able to produce at c a point image of a point source; i.e. in technical terms, the higher becomes the resolving power of the grating.

If to the successive orders c, d, etc., of a given grating (Fig. 182) we apply precisely the same reasoning as that given above, we see that, since the difference in path po corresponding to the image d is 2λ instead of λ, the first point of complete interference near to d is but half as far from d as is the case with the corresponding point in the neighborhood of c. Hence we see that with a given grating the resolving power is proportional to the order of the spectrum observed. All of the efforts, therefore, which have been made toward increasing the resolving power of gratings have been directed either toward increasing the number of lines, or else toward making it possible to work in spectra of a very high order.

The object of producing gratings of a very high resolving power is, in general, to obtain spectra of very great purity, so that colors of but the slightest difference in wave length may yet stand out as distinct colors, that is, as separate spectral lines. The largest and most perfect gratings thus far made, namely those ruled by Professor Michelson in 1908 at the University of Chicago, contain nine inches of ruled surface, the length of the lines being four and a half inches, and the number of lines to the inch 12,700. The total number of lines on these gratings is therefore 114,300.

163. Effects of finite width of slit on series of images produced in monochromatic light. In the use of a grating one often observes that the image of the third order, for example, will be brighter than that of the second, that of the fifth brighter than that of the fourth, etc. The cause of this lies in the finite width of the open spaces which have heretofore been considered to be mere lines.

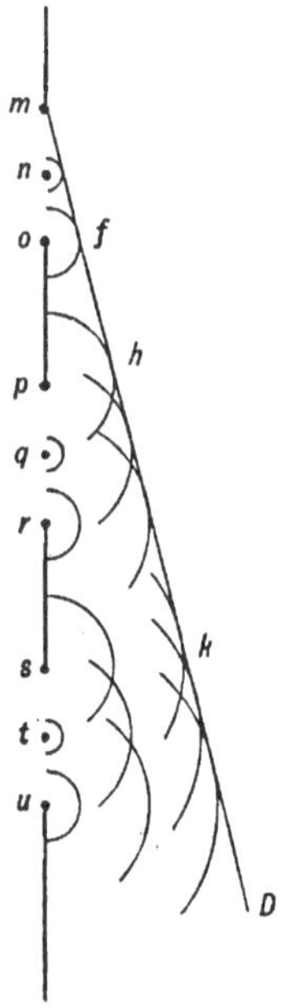

FIG. 184

In order to understand the effect of a finite width in the openings upon the relative brightness of the successive images, consider Figure 184, in which is shown on a large scale a section of a portion of a practical grating, *mo*, *pr*, and *su* representing the finite openings, and *op*, *rs*, etc., the opaque spaces. The points *m*, *p*, and *s* correspond to the line openings of the preceding discussion (Fig. 182). The line *mD* represents the wave front which gives the image of the first order. It is drawn as the envelope to the spherical waves due to the particles *m*, *p*, and *s*. The disturbances produced by these particles at the points *m*, *h*, and *k* are all in the same phase of vibration, for each point differs from the next in its distance from its respective slit by a whole wave length. The disturbances due to these points will then reënforce each other at the image *c* formed by the lens (Fig. 182). Similarly, particles such as *n*, *q*, and *t*, which bear the same relation in position to *m*, *p*, and *s* respectively, will also produce disturbances on the line *mD* at points which differ successively in their distances from these particles by a whole wave length. The disturbances due to these points will therefore reënforce each other at *c*. And similarly, for all the other points in the opening *mo* there will be corresponding points in the other openings which will reënforce these vibrations. But it is now to be noticed that the disturbances which start from the different parts *m*, *n*, and *o* of the same opening are not in quite the same phase of vibration when they reach the plane *mD*, and further that the wider the openings the greater becomes this phase difference. Suppose, then, that the open spaces *mo*, etc., are just equal to the opaque spaces *op*, etc., and that we are considering the image of the second order. We have seen that the condition which must hold for this image is that $ph = 2\lambda$, $sk - ph = 2\lambda$, etc. Since, then, $mo = op$, we have $of = \lambda$. But when this condition exists the disturbance from *n* is one half wave length behind that from *m*, and therefore completely destroys it at *c*. Similarly, the disturbances from all the points between *n* and *o* destroy at *c* the disturbances from the points between *m* and *n*.* Similarly for all the other openings. Thus the image of the second order will be entirely missing, and also, for exactly the same reason, the images of the fourth, sixth, etc., orders. If the opening is one third of the grating space instead of one half, the missing images will be those of the third, sixth, ninth, etc., orders. If conditions of this sort are only *approximately* fulfilled, as is usually the case, the images considered will be simply weakened but not entirely cut out.

164. Reflection gratings. By far the greater part of spectroscopic work is now done with the aid of gratings, but in actual work reflection gratings are much more common than transmission gratings like that which has been studied. These reflection gratings are made by ruling very fine lines on a reflecting metal surface, rather than on a transmitting glass surface. The grooves destroy the light, while the spaces between them reflect it regularly. The light from any white source which is reflected from such a grating and then brought to a focus by means of a lens shows a central white image at a position such that the angle of incidence equals the angle of reflection (see sect. 155). On either side of this central image are found spectra of the first, second, and third orders, precisely as in the transmission grating. The theory of the two gratings is in all respects identical, for it obviously makes no difference how the lines *m*, *p*, *s*, etc., of Figure 182 become sources of disturbance, whether by reflecting or by transmitting a disturbance from some other source.

A form of grating which has rendered possible some of the most important of recent advances in spectroscopy is the *concave* grating invented by the late Professor Henry A. Rowland of Johns

* See section 154, page 242.

Hopkins University. The essential difference between this and other gratings is that the lines are ruled upon the surface of a concave spherical mirror of large radius of curvature, for example 20 feet. Under such conditions the mirror itself forms the series of images corresponding to *a*, *c*, *d*, etc. (see Fig. 182, and also Fig. 180), so that it is not necessary to interpose a lens. This eliminates all difficulties arising from the absorption of the waves by the lens, difficulties which are especially pronounced in the ultra-violet and infra-red regions of the spectrum.

EXPERIMENT 22

Object. To determine the wave length of sodium light by means of a diffraction grating.

Directions. Over a Bunsen burner place a sheet-iron chimney C (Fig. 185) in which is cut a narrow vertical slit s, triangular in form. Immediately behind this place a horizontal scale at the same height as the slit. Set up a transmission grating in a clamp three or more meters away from the scale, and adjust its position until it is parallel to the scale and lies on the per-

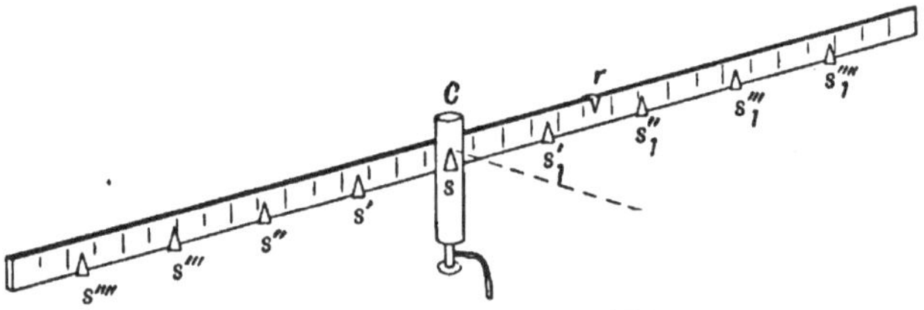

FIG. 185

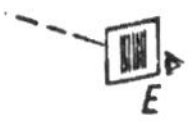

pendicular drawn to the scale through the slit. Exclude the air from the burner so that it burns with a white flame, and view the slit through the grating. We then have the conditions described in section 160. Note the overlapping of colors in spectra of higher order than the first.

Replace the white light by sodium light by readmitting the air to the burner and placing a bit of asbestos soaked in salt water in the colorless flame. Keep the eye very close to the grating in order that distances like $s'E$ may be measured to the grating rather than to the eye without introducing an appreciable error. Slide a narrow and pointed paper rider r over the scale and locate with it on the scale the positions of the images s', s'_1, s'', s''_1, and so on. Take the grating space d either from the data furnished by the maker of the grating or from a direct determination with a micrometer microscope. Measure the distances Es', Es'_1, and so on, with a steel tape. From the mean of Es' and Es'_1, and from one half of $s's'_1$, calculate λ by the use of the relations developed in section 159. Compute λ also from measurements on the brightest image of the higher orders observed.

Obtain λ from a similar set of observations upon the brightest image formed by a grating which has a different value of d.

Compare the mean of all the results with the accepted value for sodium light, namely .0000589 cm.

EXAMPLE

Using a grating with a space d of .002 cm., the distance $s's'_1$ was found to be 12.35 cm. The distances Es' and Es'_1 were 209.2 and 209.8 cm. respectively. Hence

$$\lambda = \frac{(6.17)(.002)}{(209.5)} = .00005890.$$

The images of the fourth order were bright and easily located. For these $s'''' s''''_1$ was found to be 49.7, $Es'''' = 209.6$, and $Es''''_1 = 212.3$. Hence

$$\lambda = \frac{(24.85)(.002)}{4(210.95)} = .00005890.$$

Using a second grating for which d was .001, the images of the second order were selected for the calculation ; $s''s''_1$ was 24.75, Es'' was 209.2, and Es''_1 was 210.3. Hence λ was .00005900. These results all agreed to within .2% with the value .00005890.

CHAPTER XXIII

THE REFRACTION OF LIGHT

165. Cause of refraction. It is the change in the wave front which occurs when a wave enters a new medium that accounts for the bending which light undergoes when it passes, at any other than normal incidence, from one medium into another. Thus, if a wave which originates at p (Fig. 186) has its form so changed by passage into a new medium that the center of the wave front which reaches the eye at E is at p' instead of at p, then obviously the light which has come to E from p has come over the broken path pqE. That is, *light which passes obliquely from a medium of slower speed to one of higher speed is bent away from the perpendicular qn drawn into the second medium.*

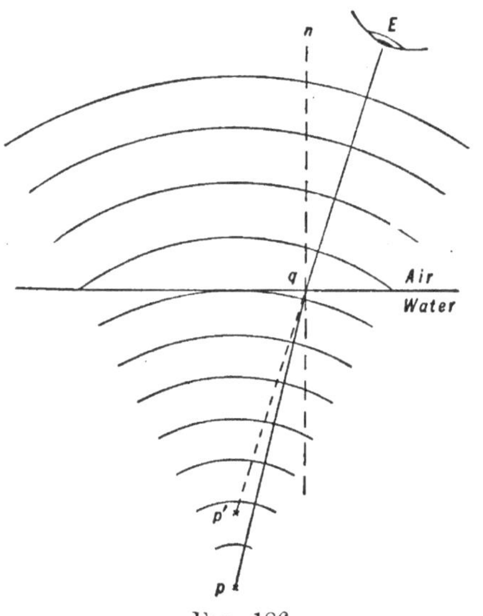

FIG. 186

If the speed in the second medium had been less than in the first, then obviously the new center p' would have been below p instead of above it. This means that *light when passing from a medium of greater to one of lesser speed is bent toward the perpendicular drawn into the second medium.*

This change in direction, due to change in wave form, which light undergoes in passing from one medium to another is known as *refraction.* In order to be in position to make a quantitative study of the phenomena of refraction it is first desirable to obtain a quantitative expression for the curvature of a wave front.

166. Measure of curvature. A circle has obviously the same curvature at every point. In Figure 187 are shown two circles, one of twice the radius of the other. It is evident from the figure that in moving along the larger circle from O to a the curve departs from the tangent (that is, curves) much less than it does in moving an equal distance from O to a' along the smaller circle; that is, the greater curvature accompanies the smaller radius. It is further evident from the figure that as Oa and Oa' become smaller and smaller, the distance from a to the tangent becomes more and more nearly exactly one half of the distance from a' to the tangent. At O therefore the *rate* at which the smaller circle is curving away from the tangent line is just twice the rate at which the larger circle is curving away from this tangent. In other words, *the curvatures of two circles are inversely proportional to their radii.* Hence *the reciprocal of the radius of any circle is taken as the measure of the curvature of that circle.* Thus if C denote the curvature of any curve at any point, and R the radius of curvature of the curve at this point, we have by definition

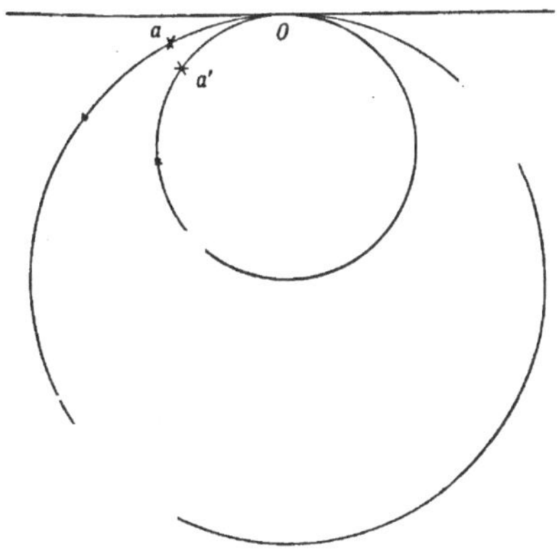

FIG. 187

$$C = \frac{1}{R}. \tag{1}$$

167. Ratio of the velocities of light in two media. If it may be regarded as established by the experiments which have preceded that light is a wave motion, then it follows from the fact that objects under water, for example, appear nearer to the surface than they actually are, that light travels faster in air than it does in water, for it is only in view of an increase in speed in going from water to air that we can account for such an upward bulging of the emerging wave as would bring nearer to the eye the center from which the waves appear to proceed (see Fig. 186).

But this phenomenon of the apparent elevation of objects under water does more than merely to establish the fact of a difference in the speed of light in the two media: it enables us to determine fairly accurately the ratio of the two speeds. For let *cod* (Fig. 188), drawn with S as a center, represent a section of the wave front at the instant it touches the water surface ww' at o; let $mo'n$ represent a section of the wave front as it would have been an instant later if there had been no change in medium, and let $mo''n$, drawn with S' as a center, represent the wave front as it actually is at this instant.

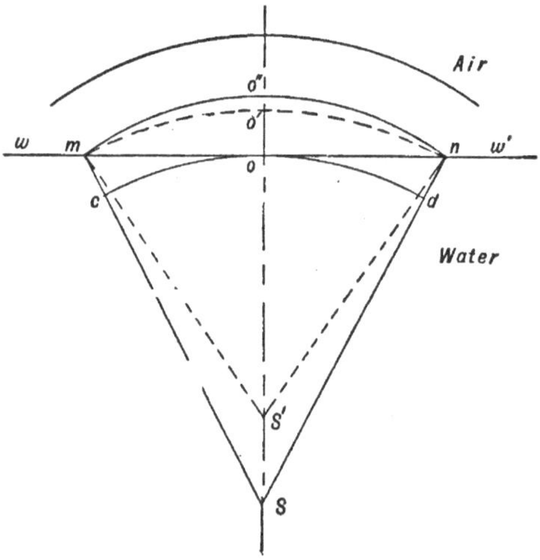

FIG. 188

Then obviously the speed of light in air divided by its speed in water is oo''/oo'. But oo'' and oo' measure respectively the curvatures of the arcs $mo''n$ and $mo'n$, so long as these arcs are small, for then they are the amounts by which these curved lines depart from the straight line ww'. Hence, since curvatures are also measured by the reciprocals of the radii of curvature (eq. 1), we have

$$\frac{\text{velocity of light in air}}{\text{velocity of light in water}} = \frac{oo''}{oo'} = \frac{Sm}{S'm}.$$

Now as one looks down vertically upon a surface of water, the section of the wave taken in by the two eyes is so small that Sm and $S'm$ are practically the distances from S and S' respectively to the surface. Hence, to measure the ratio of the velocities of light in air and water we have only to look down into a tall vessel of water, as in Figure 189, place a finger on the outside of the vessel at the apparent level of the bottom, and divide the actual depth by this apparent depth.

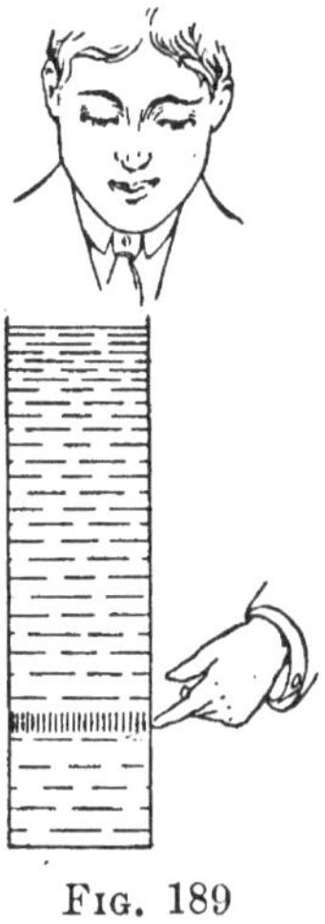

FIG. 189

With the aid of a microscope this method is often extended to the determination of the ratio of the speed of light in air and in any transparent medium which is bounded by parallel planes. Thus the microscope is first focused upon a point p (Fig. 190) when air only intervenes between p and the objective O of the microscope. Then the medium with the parallel faces ab and cd is introduced between p and O. This causes the center of the wave which reaches O to change from p to some point p'. Hence in order to see p distinctly it is now necessary to raise the microscope a distance pp'. The amount of this elevation is carefully measured. Then the microscope is again raised until the surface ab is in focus, i.e. it is raised the distance $p'n$. The ratio of the velocities of light in air and in the given medium is then $pn/p'n$.

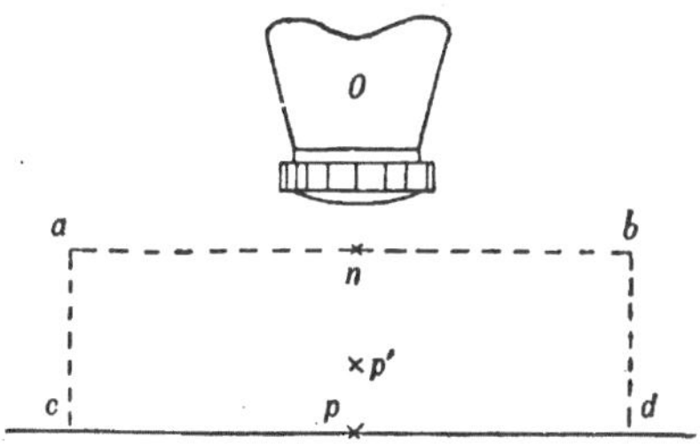

FIG. 190

168. Elementary lens and mirror formulas. In section 155 we have already discussed qualitatively the refraction produced by lenses. In the light of section 166 it now becomes very easy to deduce the formula by means of which the distance f' (i.e. Lp, Fig. 175) of the center of a transmitted wave from the lens is given in terms of the focal length F of the lens and the distance f (i.e. SL, Fig. 175) of the source from the lens. For, since the focal length F is defined as the distance from the lens of the center of a transmitted wave which before incidence was plane, i.e. had zero curvature, it follows from the definition of curvature (eq. 1) that the curvature which a lens imparts to a wave passing through it is $1/F$. It follows further from this definition, and from the fact that a convex lens always tends to reverse the direction of curvature of a wave approaching it from any source (see sect. 155), that the curvature $1/f'$ of any transmitted wave must be the difference between the curvature $1/f$ of the incident wave and the curvature $1/F$ imparted by the lens. That is,

$$\frac{1}{f'}=\frac{1}{F}-\frac{1}{f}, \qquad \text{or} \qquad \frac{1}{f'}+\frac{1}{f}=\frac{1}{F}. \tag{2}$$

This formula applies also to concave mirrors, since these also tend to reverse the curvature of an incident wave (see Figs. 173 and 174, pp. 248–249). The formula shows at once that the focal length F of a mirror is half its radius of curvature R; for when a source S (Fig. 174) is at the center of curvature of a concave mirror we have $f = f' = R$. Hence from (2)

$$\frac{1}{F} = \frac{2}{R}, \quad \text{or} \quad R = 2F. \tag{3}$$

Since, as constructions similar to those given in section 155 will show, concave lenses and convex mirrors tend to *increase* the curvature of incident waves, the corresponding formula which holds for either of these is obviously

$$\frac{1}{f'} = \frac{1}{F} + \frac{1}{f}, \quad \text{or} \quad \frac{1}{f'} - \frac{1}{f} = \frac{1}{F}. \tag{4}$$

169. Index of refraction. By considering a very small section of the wave front at q (Fig. 191) it is easy to find a relation between the amount of bending which light undergoes in passing from one medium to another and the speeds of propagation of light in the two media. Thus let qt represent so minute an element of the wave front at q that it may be considered as plane. For convenience of representation let it be greatly magnified, as in the figure. When the left edge of this element meets the new medium at q, this point becomes the center of a new disturbance which spreads as a circular wave into the new medium. If the speed of light in this medium is greater than is its speed in the lower medium, then while the disturbance from t travels a distance tv in the lower medium, that from q travels a greater distance qu in the upper medium. The disturbances, then, which at a given instant started from the various points along the line qt will an instant later have reached points which lie along the line

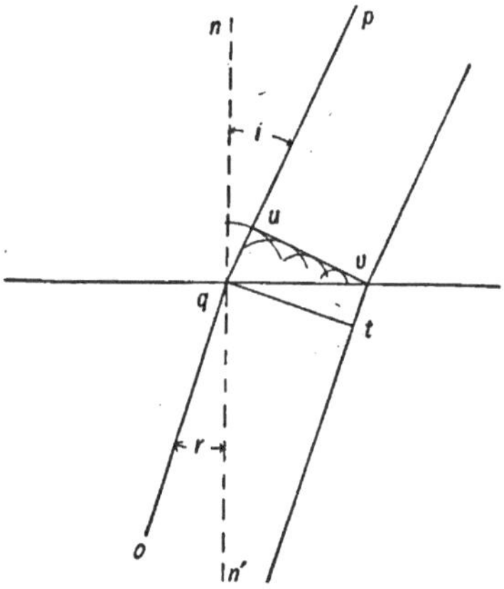

FIG. 191

uv, that is, *uv* is the new wave front in the upper medium. Whereas, then, the light approached the interface between the media traveling in the direction *oq*, it recedes from it traveling in the direction *qp*. Now if we designate by i the angle *nqp* which the direction of propagation of the light in the medium of greater speed makes with the normal drawn into this medium, and by r the angle *oqn′* which the direction of propagation of the light in the medium of lesser speed makes with the normal drawn into this medium, and if we designate by V_i the speed of propagation in the former medium and by V_r that in the latter medium, then we have $\frac{V_i}{V_r}=\frac{qu}{tv}$. But $\frac{qu}{qv}=\sin i$, and $\frac{tv}{qv}=\sin r$. Therefore $\frac{qu}{tv}=\frac{\sin i}{\sin r}$.

Hence

$$\frac{V_i}{V_r}=\frac{\sin i}{\sin r}. \tag{5}$$

It is customary to call this ratio *the relative index of refraction of the two media.* If the upper medium is a vacuum, then this ratio is called the absolute index of refraction of the lower medium, and is denoted by the letter n. The speed of propagation in air is so nearly that in a vacuum that for most purposes it is sufficiently accurate to consider the index of refraction of any medium as the ratio of the velocities of light in air and in the medium. We have, then,

$$n=\frac{V_i}{V_r}=\frac{\sin i}{\sin r}. \tag{6}$$

170. Refraction through a prism. Consider a prism composed of some refracting substance such as glass (Fig. 192) and surrounded by air. A beam of light OP is incident upon one of its faces at an angle i, and is refracted at an angle of r. This refracted beam, incident upon the second face at an angle r', undergoes further deviation, emerging

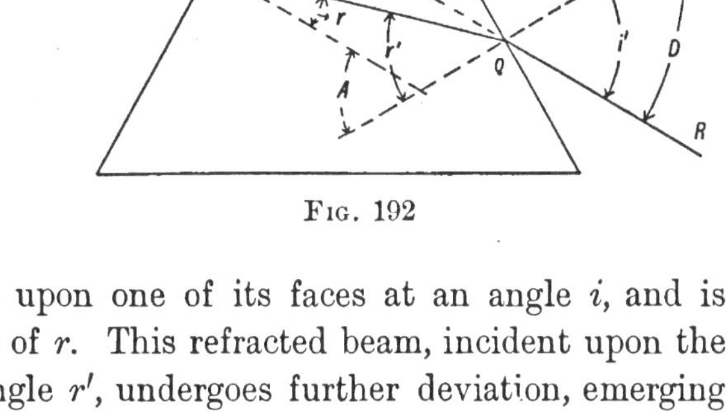

FIG. 192

into the air at an angle i'. The angle D between OP, the original direction of the beam of light, and QR, its direction after passing through the prism, represents the *deviation* produced in the direction of the beam. This angle is seen from the figure to be expressible in terms of i, r, i', and r' as follows:

$$D = (i - r) + (i' - r') = (i + i') - (r + r').$$

If A is the angle of the prism, we have, since $r + r' = A$ (see sect. 172),

$$D = i + i' - A. \tag{7}$$

171. Angle of minimum deviation. It is evident that if the beam of light in Figure 192 is caused to enter the prism in the reverse direction, namely RQ, it will travel over exactly the same path and suffer the same deviation. That is, there are two values of the angle of incidence of the entering beam for which the deviation is the same, namely i and i'. Consequently, if the angle of incidence be caused to vary from a value i to a value i', it follows, since the deviation D changes continuously between these values, and has the same value for an incidence of i as for one of i', that it must pass through a maximum or a minimum. A simple experiment to be described later shows that it passes through a minimum. Further, since this minimum value of D must occur for an angle of incidence which lies between i and i', no matter how slightly i and i' differ, it follows that it must occur when i equals i'. Hence when the deviation is a minimum we have

$$D = 2i - A. \tag{8}$$

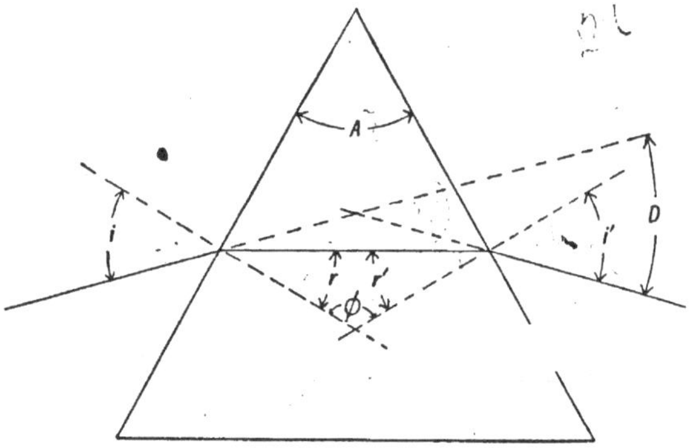

Fig. 193

172. Index of refraction from measurements upon the angle *A* of a prism and the angle of minimum deviation *D*. The index of refraction has been shown to be given by the expression $n = \frac{\sin i}{\sin r}$. For the condition of minimum deviation as given in equation (8), we have $i=(A+D)/2$. To obtain r for this case, consider Figure 193, from which it is evident that $\phi+A=180°$ and $\phi+2r=180°$; hence $r=A/2$. Substituting these values in the expression for the index of refraction, we obtain

$$n=\frac{\sin\frac{1}{2}(A+D)}{\sin\frac{1}{2}A}. \qquad (9)$$

173. The spectrometer. The spectrometer is an instrument of the form shown in Figure 194. Its essential features are represented diagrammatically in Figure 195. A circular table K, the

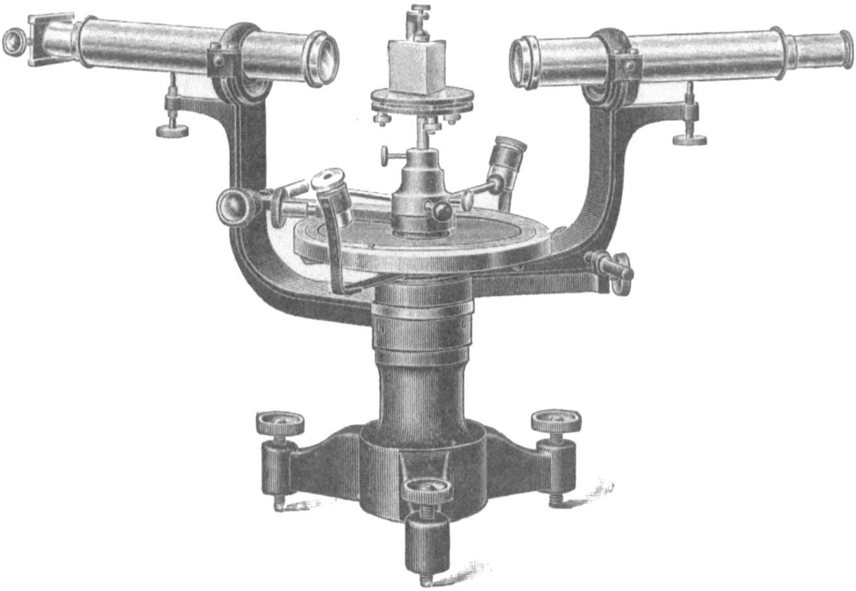

FIG. 194

edge of which is graduated in degrees, is supported upon a mounting which carries also a telescope T and a so-called collimator C. The latter consists merely of a tube carrying a slit s so mounted that it may be placed in the principal focal plane of a lens L'. The object of this arrangement is to make it possible to regard s as an infinitely distant source of light, for waves which originate at s become plane waves after passing through the lens L'. The telescope T is mounted so as to rotate about the axis of the table K. The angular position of the telescope with reference to the graduations on the table may be read with the aid of a vernier V attached to the telescope. Attached to the circular table is a second smaller circular plate called the prism table, which may be leveled by means of the leveling screws E (see also Fig. 198). This table carries the prism P. The telescope may be clamped to the mounting, and the circular table, with the attached prism table, rotated. The rotation with reference to the fixed telescope may then be read with the aid of the vernier V. A small piece of plane transparent glass m is inserted in the eyepiece e so as to make an angle of 45° with the axis of the telescope. The purpose of this arrangement is to make it possible to illuminate the cross hairs at × by throwing a beam of light from a flame or other source f into the eyepiece through the circular opening O, and thence, after reflection from the surface of m, down the axis of the telescope tube. An eyepiece provided with the opening O and the glass plate m is called a *Gauss eyepiece.*

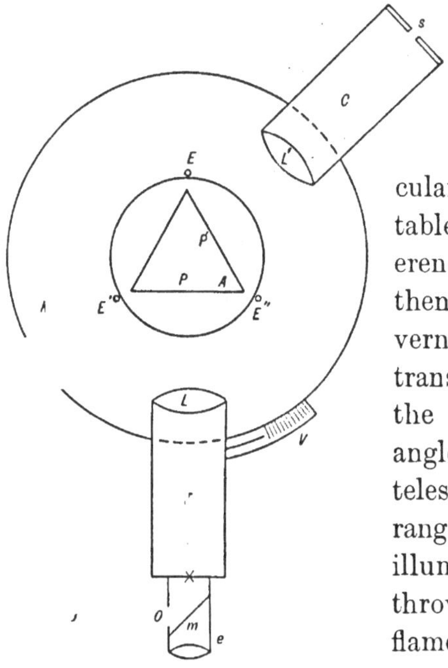

FIG. 195

174. Measurement of the angle of a prism. The prism whose angle A is to be measured is placed upon the prism table as shown in Figure 195. Now if the illuminated cross hairs × are in the focal plane of the lens L, then the light which passes from them down the tube may be reflected from the prism face P so as to return into the telescope, in which case it will be brought again to a focus in the focal plane of the lens L. Hence an eye looking into the eyepiece should see side by side two images of the cross hairs, one due to light which comes directly from the cross hairs at × to the eye, the other due to light which has passed down the tube and been reflected back again into the tube from the face P. If the prism table is turned until these two images of the cross hairs are brought into exact coincidence, we may know that the axis of the telescope is strictly perpendicular to the face P. If now the prism table is rotated until a reflected image of the cross hairs is obtained from the face P', and if this image is brought into coincidence with the direct image, the axis of the telescope will be perpendicular to P'. The angle through which the prism has been turned, as read upon the vernier and scale, is the supplement of the angle of the prism, that is, it is $180° - A$.

The second method of measuring the angle of a prism is one which makes use of the collimator. Suppose that the slit s has been placed in the focal plane of the lens L', and that the cross hairs × have been placed in the focal plane of the lens L. Then, when the light from the slit s is reflected from a prism face into the telescope as in Figure 196, an image of s will be formed in the telescope in the plane of the cross hairs. Between L' and L the light waves which form this image are plane, or, to express the same idea in different terms, the beam of light between L' and L consists of a bundle of *parallel rays*. If now the prism is set so that a part of this beam is reflected from one face of the angle A into the telescope when the latter is in the position M, and another part is reflected from the other face of the angle A into the telescope when the latter is turned into the dotted position M', then the image of the slit s can be seen in the telescope when the latter is in either of these positions. If the cross hairs of the telescope are set upon the middle of this image at M, and the reading of the vernier and scale taken, and if then the telescope is rotated into position M' and a similar setting and

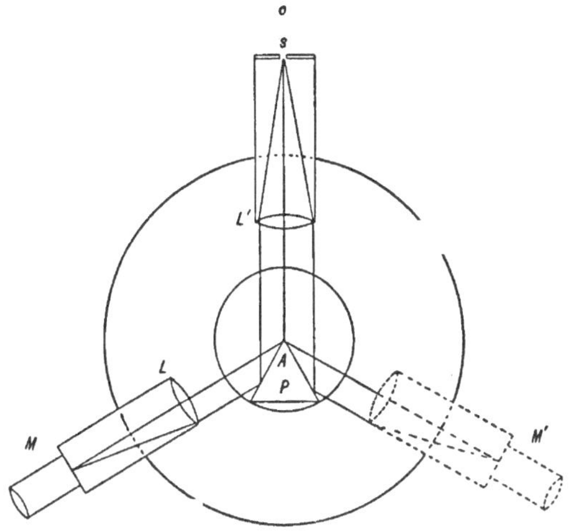

FIG. 196

reading taken, a little consideration will show that the angle A is one half of the angle through which the telescope has been rotated. This follows at once from the fact that all of the lines drawn from L' to the prism are parallel, and that "angle of incidence equals angle of reflection."

175. Measurement of the angle of minimum deviation. Since the passage of white light through a prism breaks it up into a band of colored light, the angle of deviation must be different for different colors. To obtain the angle of minimum deviation for a particular color, we first set the cross hairs in the focal plane of L, and the slit in the focal plane of L', and then set up at o a source of monochromatic light, for example, a sodium flame. We next dispose the prism, telescope, and collimator in the manner shown in Figure 197, and observe in the telescope the image of the slit formed by light which has undergone refraction in passing through the prism. To make sure that the image seen is formed by refraction rather than by reflection from the face BC, for example, we have only to place a source of white light at o and see whether the observed image of the slit is replaced by a broad band of color. Having found the refracted image, we rotate the prism table slightly, observing meantime the image of the slit in the telescope; and if this image moves in the direction of decreasing deviation, i.e. toward R, we follow it with the telescope until, without changing the direction of rotation of the prism, the image begins to return toward L. If the image first begins to move toward L, we at once reverse the direction of rotation and follow with the telescope until the position of minimum deviation has been reached, i.e. until the beam emerging from the face AC of the prism makes the smallest angle which it is found possible for it to make with

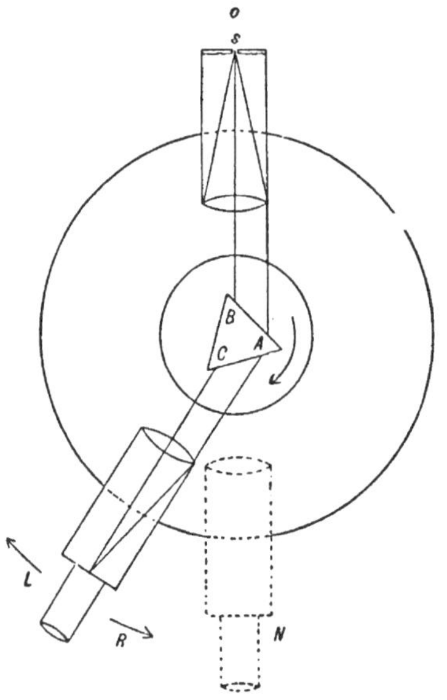

FIG. 197

the direction of the beam incident upon the face *AB*. We then set the cross hairs upon the middle of the slit, the latter being now in this position of minimum deviation, and read the vernier and scale. We next remove the prism, rotate the telescope into the position *N*, and set the cross hairs upon the image of the slit formed by this undeviated beam.* The difference between the two settings gives the angle of minimum deviation *D*.

EXPERIMENT 23

Object. To find the index of refraction of glass for sodium light by a measurement (I) of the angle of the prism, and (II) of the angle of minimum deviation.

Directions. I. Set the prism *P* (Fig. 198) upon the prism table in such a position that each of the edges is immediately above one of the leveling

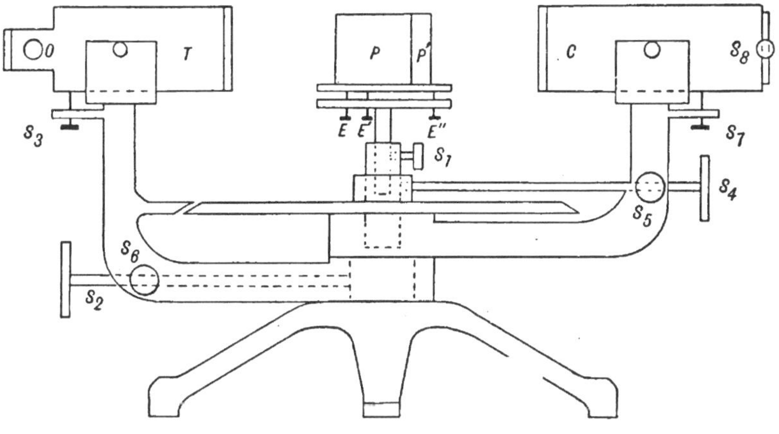

FIG. 198

screws *E*, *E′*, *E.″* Loosen the set screw S_1, so that the prism table can rotate by itself. Set up an electric light behind a ground-glass screen about a foot away from the opening *O* in the Gauss eyepiece, and on the

* It is frequently undesirable to disturb the adjustments of the spectrometer by removing the prism. In that case enough light will generally pass the edges of the prism to make possible the setting in the position *N* without removing the prism. If this is not the case the telescope may be turned into a position corresponding with that already shown in the figure, but on the other side of *N*, the prism reversed, and the position of minimum deviation determined. The position *N* then lies halfway between these two points, or, in other words, the angle through which the telescope has turned is 2 *D*.

perpendicular drawn through O to the axis of the telescope (see also Fig. 195). Clamp the telescope to the frame by tightening the set screw S_2. If the cross hairs are not perfectly distinct, slip the eyepiece alone forward or back until they are in sharp focus. Then rotate the prism table until the light which enters O from the source f (Fig. 195), and is then reflected down the telescope tube by the plane glass m, is reflected back into the tube from the face P of the prism. If the prism is being rotated somewhat rapidly, this reflection will appear simply as a flash of light across the field of view when the prism face P passes through the position in which it is at right angles to the axis of the telescope. If this flash cannot be found at first, change the inclination of the telescope by means of the screw S_3 until it appears. Keeping this reflected light in the field of view, move forward or back the draw tube which carries the eyepiece and cross hairs until a second image of the cross hairs is seen in this field of reflected light.* The cross-hairs must then be in the focal plane of the objective lens L (Fig. 195), since light waves originating in the plane of the cross hairs return to a focus in this same plane after passing through L and being reflected by a plane surface back through L. To make the adjustment perfect, turn the prism so as to bring the two images of the vertical cross hair very nearly into coincidence, and move the head from side to side to test for parallax. If this motion causes the two images to appear to move with reference to each other, focus again until such apparent movement is eliminated. The telescope is now said to be *focused for parallel rays.* This adjustment might also have been made by merely turning the telescope upon some very distant object and focusing until there was no parallax between the cross hairs and points on the object.

To make the second adjustment, which consists in setting the instrument so that the axis of the telescope is strictly perpendicular to the axis of rotation of the prism, turn the thumbscrew S_3 until the two images of the horizontal cross hair, found as above, coincide; then loosen the set screw S_4, clamp S_1, and rotate the graduated circle, with the prism, until one of the other faces, say P' (Fig. 198) instead of P, reflects the light back into the telescope. Unless the axis of rotation is already perpendicular to the axis of the telescope the horizontal cross hairs will now no longer coincide. Restore coincidence by taking up one half of the distance between the two images of the cross hairs with S_3, and the other half with the prism table screw E which is opposite to the face P'. Then rotate the graduated circle until P is again perpendicular to the telescope, and again take up half the

* If the prism has three polished faces, there will, in general, be three reflected images of the cross hairs instead of one, two of them being due to double reflections from the rear faces of the prism. To avoid the confusion arising from these double reflections, it is recommended that one face be temporarily painted with a mixture of, say, whiting and alcohol, or lampblack and alcohol. The three polished faces are needed for Experiment 24.

difference in coincidence by means of S_3 and half by means of E'', the prism table screw which is opposite to P. After two or three repetitions of this process the two images of the horizontal cross hairs should coincide, no matter which face of the prism is turned so as to reflect light back into the telescope. If exact coincidence cannot be obtained, it is because the edges of the prism itself are not quite parallel. In this case it is customary to distribute the error in coincidence between the faces.

Next mark carefully the particular angle of the prism which you wish to measure; then obtain the reflected image of the cross hairs from one of the faces of this angle. See that the clamp S_2 is tight, so that the telescope will henceforth remain quite stationary, rotate the graduated circle until the two sets of vertical cross hairs are in approximate coincidence, and, to make this adjustment perfect, clamp S_4 and complete the coincidence by means of the fine-adjustment, slow-motion screw, S_5. Take the exact reading, in degrees, minutes, and seconds, of the vernier and scale; then loosen S_4, rotate the graduated circle with the prism until the other face of the angle to be measured is perpendicular to the telescope, clamp S_4, make a final setting, as before, with S_5, and then take a second reading of the vernier and scale. The difference between the two readings should be the supplement of the angle of the prism.

As a check upon this determination, measure the same angle by the second method of section 174. To do this, first see that the telescope is still focused as above for parallel rays; then remove the prism, loosen S_2, and rotate the telescope so as to see through it into the collimator. Level the collimator approximately by means of S_7 until the cross hairs are in the middle of the slit; then loosen the tube which carries the slit of the collimator, and slip it in or out until the image of the slit observed through the telescope is in sharp focus. To make the adjustment perfect, move the slit until there is no parallax between the cross hairs and the image of the slit. Now complete the leveling of the collimator by adjusting S_7 again. Then set the prism as shown in Figure 196, clamp S_4, loosen S_2, and place the telescope first in position M, then in position M', and see if a sharp reflected image of the slit can be seen in both positions. If not, adjust the position of the prism until both images appear; then make the slit very narrow by means of S_8, turn the telescope into position M, clamp S_2, and by means of the fine-adjustment screw S_6 set the vertical cross hair upon the middle of the slit, and then take the reading of the vernier and scale. Loosen S_2, turn the telescope to position M', and repeat. The difference between the two readings should be twice the angle A of the prism.

II. To measure the angle of minimum deviation, first clamp S_4 and loosen both S_1 and S_2. Then set up a sodium flame at o (Fig. 197), and, by rotating both the prism and the telescope, find the refracted image of the slit in the telescope when the known angle A is placed as in the diagram. If the image cannot be found at first, it is probable that the prism needs to be rotated somewhat farther in the direction of the arrow. Having found the refracted image, proceed as indicated in section 175, setting first approximately upon the position of minimum deviation, then clamping both S_1 and S_2, and setting the cross hairs exactly on the middle of a very narrow slit by means of S_6. This done, rotate the prism by turning S_5, and see if the slit can be moved by it at all in the direction of decreasing deviation. If so, move the cross hair up to the new position by means of S_6.

Finally, compute the index of refraction from equation 9.

EXAMPLE

I. The scale of the graduated circle of the spectrometer used was divided into degrees and $\frac{1}{3}$ degrees; that is, the smallest scale division was 20 minutes. The vernier was divided by long lines into 20 divisions, each of which was divided into two parts. These 40 vernier divisions corresponded to 13 degrees, that is, to 39 scale divisions; hence one vernier division was equal to $\frac{39}{40}$ of a scale division, that is, each vernier division lacked $\frac{1}{40}$ of being a whole scale division. Hence the vernier read to $\frac{1}{40}$ of 20 minutes, or to 30 seconds, the long lines on the vernier indicating minutes.

The readings for the setting on the face P were 315° 46′ 30″ on vernier a, and 135° 46′ 30″ on vernier b. The readings for the setting on face P' were 75° 44′ on a, and 255° 44′ on vernier b. The prism table was therefore rotated through 119° 57′ 30″. Hence A = 60° 2′ 30″.

The reading on vernier a at M (Fig. 196) was 330° 36′ 30″, and that at M' 210° 32′ 30″. The difference is 120° 4′. Hence A = 60° 2′. The mean value of A was therefore 60° 2′ 15″.

II. The readings for the angle of minimum deviation gave $D = 53° 6'$.

Hence $$i = \frac{A + D}{2} = \frac{60^\circ\, 2'\, 15'' + 53^\circ\, 6'}{2} = 56^\circ\, 34'\, 8'',$$

and $$r = \frac{A}{2} = \frac{60^\circ\, 2'\, 15''}{2} = 30^\circ\, 1'\, 8''.$$

Therefore $$n = \frac{\sin i}{\sin r} = 1.6679.$$

CHAPTER XXIV

TOTAL REFLECTION

176. Form of the wave front in the second medium. In connection with the discussion of the formation of an image by a lens (see sect. 155), the statement was made that, in general, a wave which is spherical in one medium is no longer spherical after it has passed into a second medium. The correctness of this assertion

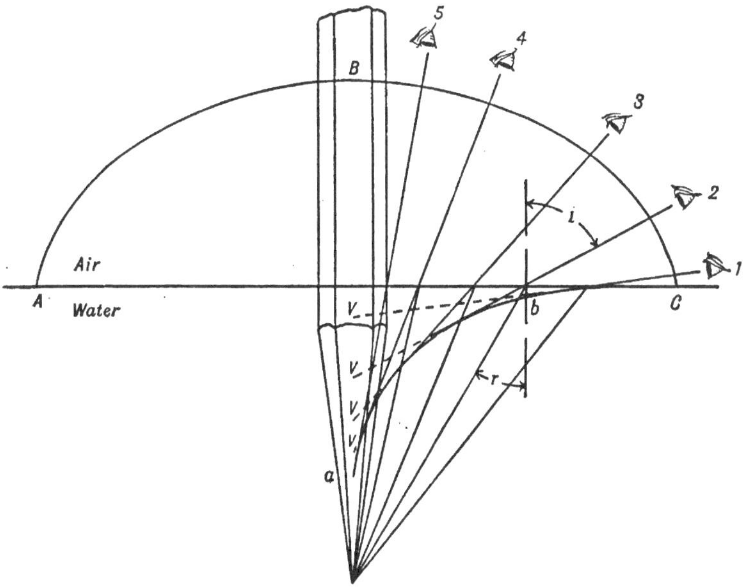

FIG. 199

can be convincingly shown by the following experiment. If a pencil is thrust normally into a body of water and viewed by an eye which looks nearly along the surface (Fig. 199), the portion of the pencil which is beneath the surface will appear extremely short, while, if the eye is slowly raised from position 1 to position 5, the length of the submerged portion will appear to become greater and greater. This means, of course, that the center of curvature of the wave front ABC is different for different points on ABC; that is, the curvature is not constant. In other words, the wave front ABC is not spherical.

The same conclusion is reached by a consideration of the law of refraction deduced in the last chapter, namely,

$$n = \frac{\text{velocity in air}}{\text{velocity in water}} = \frac{\sin i}{\sin r}.$$

This equation shows that whenever light travels from a medium of lesser speed to one of greater, sin i is greater than sin r, that is, the ray emerging into the medium of greater speed bends away from the perpendicular. Hence, as r is increased, i must reach 90°, and the emerging ray lie flat along the surface before r reaches 90°. If, then, this last ray enters an eye in the surface, the point beneath the surface from which the ray comes must, of course, appear to lie in the surface itself. That is, as the eye is raised from a position in the surface through positions 1, 2, 3, and 4 (Fig. 199), the point from which the light comes must appear to sink from the surface to a greater and greater depth beneath it, precisely as it was observed to do in the preceding experiment.

177. Caustics. This fact of the modification of a wave front by reflection or refraction so that it is no longer spherical, and therefore has not a single definite focus, or center, after such modification, is the cause of a group of phenomena known as *caustics*. These can be observed most easily with lenses or mirrors of large curvature. One of the most familiar of such caustics is that formed by a cylindrical lens such as a tumbler of water. When the sun's rays fall upon such a vessel, the rays which pass through the outer portion are brought together at points much nearer the tumbler than are those which pass through the central portion. In other words, the changes in

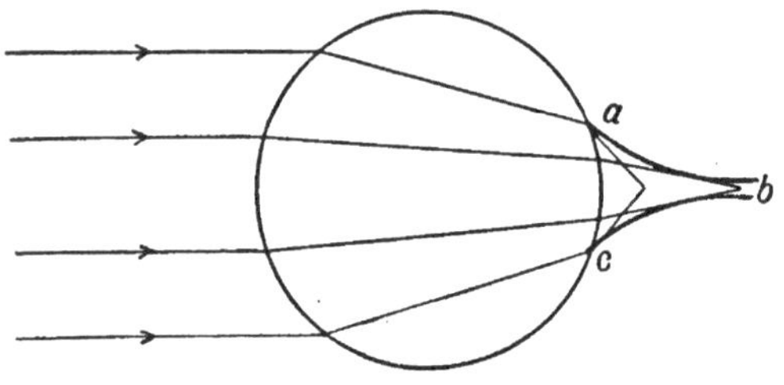

FIG. 200

the form of the wave front which take place, both when the wave enters the lens and when it leaves it, are such that the outer edges of the wave which has passed through the tumbler have their centers of curvature relatively close to the tumbler, while the center of curvature of the central portion of the wave is more remote (see Fig. 200). The caustics are the curved lines *ab*, *cb*, which represent the envelopes of the radii of curvature of the different portions of the wave (see also *ab*, Fig. 199). Figure 201 shows a caustic formed by reflection from a spherical mirror of very large aperture. It will be seen from the figure that a spherical mirror cannot be said to have a definite focus unless its aperture is small, that is, unless the angle subtended at *F* by the mirror does not exceed fifteen or twenty degrees.

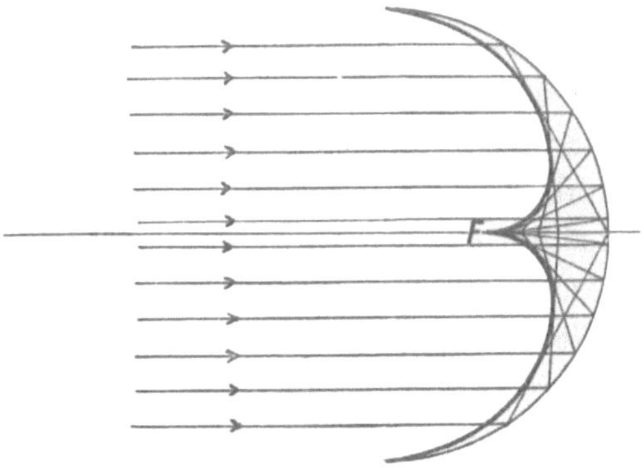

FIG. 201

The only general case of the modification of a spherical wave by reflection or refraction, in which the modified wave has a strictly spherical form, is the case of reflection from a plane surface (Fig. 169, p. 246). Nevertheless, for a given form of incident wave, for example a plane wave, it is always possible to give such a shape to the reflecting or refracting surface as to make the resulting wave front spherical. Thus when a plane wave falls upon a *parabolic* mirror the reflected wave front is strictly spherical, and hence no caustic is formed (see Fig. 202). Conversely, when a wave originates at the focus of a parabolic mirror it is reflected as a rigorously plane wave, that is, as a parallel beam. Hence the use of parabolic mirrors in search lights.

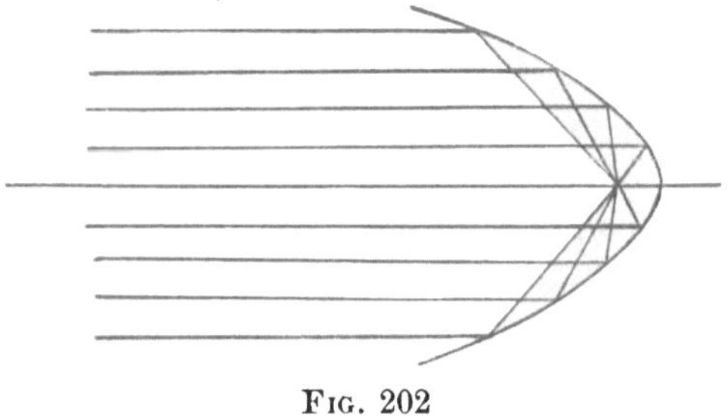
FIG. 202

The lack of sphericity in the waves which have passed through ordinary lenses bounded by spherical surfaces is seen in the fact that it is impossible with such lenses to obtain images in which there is sharp definition; for the rays which pass through the edges of the lens are brought to a focus nearer to the lens than are the rays which pass through the middle portion. This phenomenon of caustics, as shown by ordinary lenses, is usually known as *spherical aberration.* It may be reduced by decreasing the aperture of the lens, but this also decreases its resolving power (see note *, p. 251). It is climinated entirely by special combinations of convex and concave surfaces, or by grinding the surfaces of a single lens into special form.

178. Total reflection. From the law of refraction stated in section 176, it will be seen that when a ray of light travels from a medium of lesser speed to one of greater, if we continue to increase r after i has become 90°, there can be no refracted ray at all. Experiment completely confirms this conclusion and shows that light which is incident upon the upper medium (Fig. 203) at an angle greater than that for which $i = 90°$ is totally reflected in accordance with the usual law of reflection, namely, the angle of incidence r equals the angle of reflection r'.

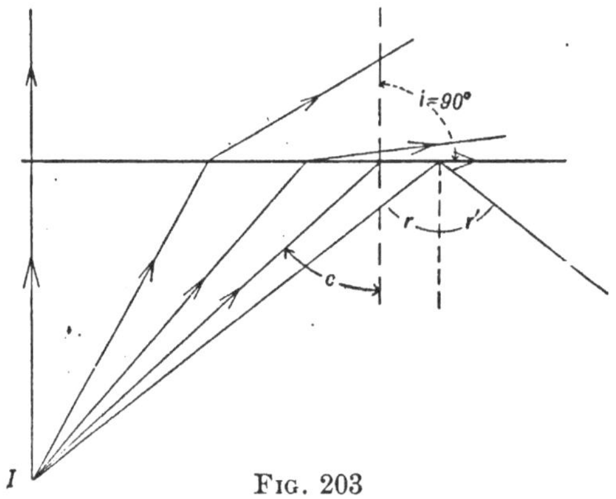

FIG. 203

Figure 203 shows the course of various rays starting at different angles from the point I beneath a surface of water. The value of r for which $i = 90°$, and hence the value of r beyond which total reflection takes place, is called the *critical angle* (see c, Fig. 203). It depends upon the relative index of refraction of the two media. Thus since

$$n = \frac{\sin i}{\sin r},$$

it is evident that, since $\sin 90° = 1$, for c, the critical angle, there is the following relation:

$$n = \frac{1}{\sin c}.$$

Thus for water, for which $n = 1.33$, $\sin c = 1/1.33$, or $c = 48.5°$. This shows that no ray of light which comes from water to a surface separating water and air so as to make with the normal an angle greater than 48.5° can pass out into the air.

179. Total-reflection prism. An application of the principle of total reflection is made in the construction and use of the so-called *total-reflection prism*, a device for changing the direction of a beam of light by 90° without sensibly diminishing its intensity or producing in it dispersion, if it is a beam of white light. The index of refraction of practically all forms of glass is more than 1.5. Now for a substance for which n is 1.5, c is 42°. Hence if a beam of light op (Fig. 204) enters a right-angled prism at normal incidence, it will strike the face AB at 45°, that is, at an angle greater than the critical angle. It will therefore be totally reflected and pass out normally through the face AC without refraction or dispersion.

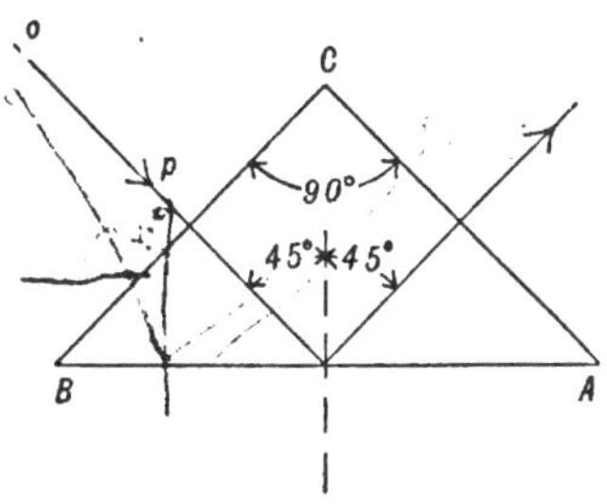

FIG. 204

180. Determination of the index of refraction from measurements upon the critical angle. One of the most satisfactory methods of demonstrating the facts of total reflection, and at the same time of determining the index of refraction of a substance which can be put into the form of a prism with three polished faces, is as follows. Let MON (Fig. 205) be any broad source of monochromatic light, CDE a prism so set that light from MON is refracted and reflected to the telescope T in the manner shown in the figure. The different rays which come to T from different points on MON will obviously strike the surface CD at different angles. Those rays which strike CD at an angle greater than the critical angle will be totally reflected, while those which strike at an angle less than the critical angle will be partially reflected and partially transmitted. If $OpqS$ is the ray which strikes exactly at the critical angle, then all of the light which comes to T from the portion Cq of the face CD will have struck CD at angles greater than the critical angle, and hence it will have undergone *total* reflection. But all of the light which comes to T after reflection at qD will have struck CD at angles less than the critical angle, and hence it will have undergone only *partial* reflection, the other part having been transmitted. Hence the surface CD should appear to consist of two parts of unequal illumination, the portion Cq, which corresponds to total reflection, appearing brighter than the portion qD, which corresponds to partial reflection. The line of junction of these two portions is quite sharply marked, so that the precise point q at which total reflection begins can be accurately determined. Suppose that the cross hairs of the telescope T are first set upon this point and the telescope then rotated until it is at right angles to the face ED, that is, suppose that the angle i is accurately measured. The index of refraction may then be obtained from i and the angle A of the prism as follows. We have

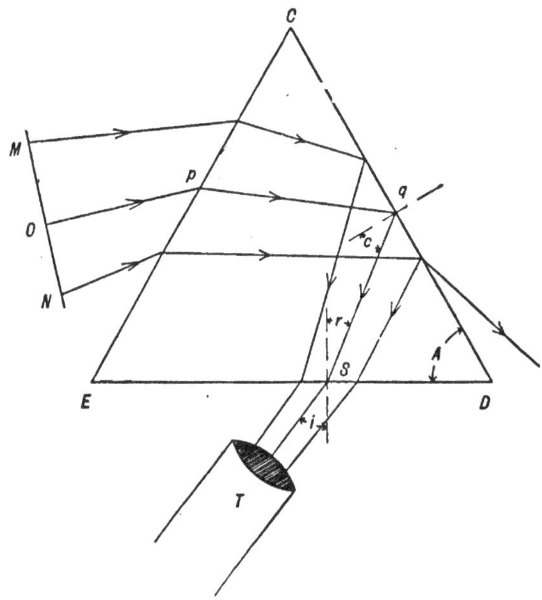

Fig. 205

$$n = \frac{\sin i}{\sin r}. \tag{1}$$

$$n = \frac{1}{\sin c}. \tag{2}$$

$$A = r + c. \tag{3}$$

From these equations we have only to eliminate c and r in order to obtain n in terms of i and A. From (1) and (3) we obtain

$$n = \frac{\sin i}{\sin (A - c)} = \frac{\sin i}{\sin A \cos c - \cos A \sin c}. \tag{4}$$

From equation (2) $\sin c = 1/n$, and also

$$\cos c = \sqrt{1-\sin^2 c} = \sqrt{\frac{n^2-1}{n^2}} = \frac{1}{n}\sqrt{n^2-1}.$$

Substitution of these values in equation (4) gives

$$n = \frac{\sin i}{\sin A \cdot \frac{1}{n}\sqrt{n^2-1} - \cos A \cdot \frac{1}{n}}, \tag{5}$$

or

$$n^2 - 1 = \left(\frac{\sin i + \cos A}{\sin A}\right)^2. \tag{6}$$

Therefore

$$n = \sqrt{\left(\frac{\sin i + \cos A}{\sin A}\right)^2 + 1}. \tag{7}$$

EXPERIMENT 24

Object. To find by a total-reflection method the index of refraction of the prism used in Experiment 23.

Directions. Focus the telescope for parallel rays and set its axis at right angles to the axis of rotation by the method of Experiment 23. Then set up a broad sodium flame *MON* (Fig. 205) on the level of the telescope of the spectrometer and at a distance from it of two or three feet. Turn the telescope and prism into about the relative positions shown in the figure, or until the image of the flame reflected from the face *CD* can be seen in the telescope. Then rotate prism and telescope together until the yellow line which divides the field into two parts of unequal intensity is seen. Set the cross hairs upon this line, and read the vernier and scale. Without moving the prism rotate the telescope into a position at right angles to the face *DE*, as determined by the coincidence of the two images of the cross hairs as seen in the Gauss eyepiece; read again the vernier and thus determine the angle i. From this value of i and from the value of A as obtained in Experiment 23, find n by equation (7). Compare the value of n thus found with that obtained in Experiment 23.

EXAMPLE

The reading of the vernier for the setting on the line of total reflection was 303° 54′. The reading for normal incidence was 262° 46′. Hence $i = 41° 8'$. The angle A as found in Experiment 23 was 60° 2′ 15″. Hence

$$n = \sqrt{\left(\frac{\sin 41° 8' + \cos 60° 2' 15''}{\sin 60° 2' 15''}\right)^2 + 1} = 1.6683.$$

The value found in Experiment 23 was 1.6679. The difference is .02 per cent.

CHAPTER XXV

PHOTOMETRY

181. Laws of illumination. When a surface is illuminated by light from a luminous point we may define the *intensity of illumination* as the quantity of luminous energy which falls upon the surface per second divided by the area of the surface, that is, as the quantity of light per unit area. It will be obvious at once from this definition that if we consider two surfaces at a given distance from the point, the one normal to the direction of propagation of the light and the other so inclined that its normal

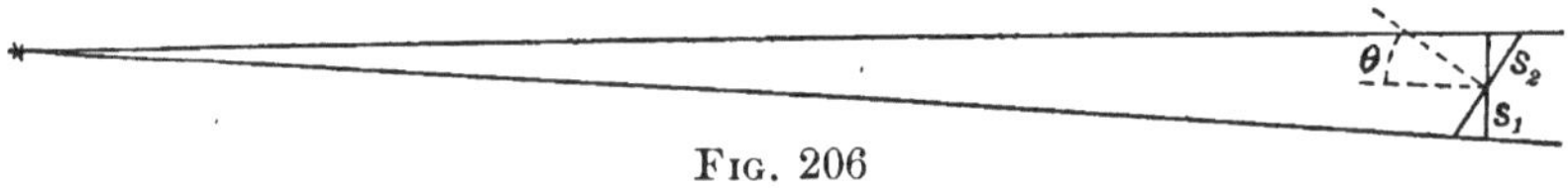

FIG. 206

makes an angle θ with this direction, then the illumination I_2 of the inclined surface is related to the illumination I_1 of the normal surface by the equation

$$I_2 = I_1 \cos\theta. \qquad (1)$$

For, since the same energy E falls upon the inclined surface s_2 (Fig. 206) as upon the normal surface s_1, we have $I_1 = E/s_1$ and $I_2 = E/s_2$; hence

$$\frac{I_2}{I_1} = \frac{s_1}{s_2} = \cos\theta.$$

Again, the illumination upon two surfaces, s_1 and s_2, at distances r_1 and r_2 respectively from a given point source, and making the same angle with the direction of propagation of the disturbance, are related by the equation

$$\frac{I_1}{I_2} = \frac{r_2^2}{r_1^2}; \qquad (2)$$

for, since the same energy E falls upon the surface s_2 (Fig. 207) as upon the surface s_1, we have $I_1 = E/s_1$ and $I_2 = E/s_2$. Hence

$$\frac{I_1}{I_2} = \frac{s_2}{s_1} = \frac{r_2^2}{r_1^2}.$$

The two results thus obtained may be stated as follows: *Intensity of illumination is inversely proportional to the square of the distance from the point source, and directly proportional to the cosine of the angle which the normal to the illuminated surface makes with the direction of the incident light.*

This conclusion has followed simply from the definition which we have given to the word *illumination.* If this definition is to have any practical value in photometry, it is necessary that two surfaces appear to the eye equally bright whenever they are illuminated with equal intensities, as here defined. Experiment shows

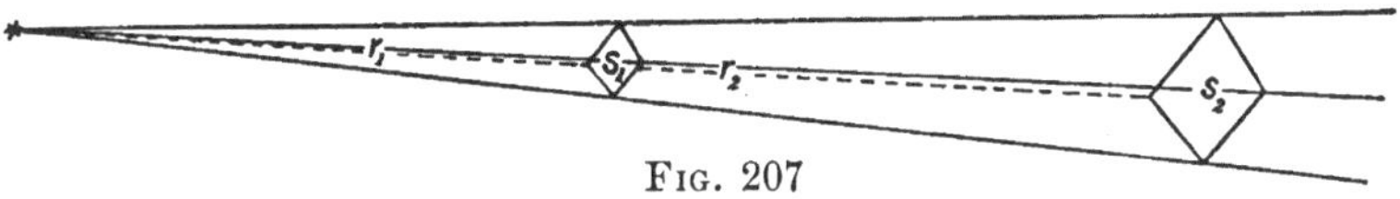

FIG. 207

that this is the case when, and only when, the two surfaces are illuminated by lights of the same, or nearly the same, color. For example, a screen placed at the distance of one meter from a single candle appears exactly as bright as a similar screen placed at a distance of two meters from four precisely similar candles placed very close together.

182. Intensities of different sources of light. The example just given suggests at once a method of comparing the quantities of light emitted by different sources of the same color. For we have only to arrange the sources so that each illuminates at the same angle one of two adjacent surfaces, and then to vary the distances of the two lights from their respective screens until these two screens appear to have equal illumination. It will then be seen at once from the example of the preceding section that the intensities of the two sources, that is, the quantities of light emitted by them, are directly proportional to the squares of their respective distances from the equally illuminated surfaces. Algebraically stated, if L_1 and L_2 represent the intensities of the sources, and d_1 and d_2 their respective distances from equally illuminated screens, then we have

$$\frac{L_1}{L_2} = \frac{d_1^2}{d_2^2}. \tag{3}$$

183. Photometric standards. The unit of light-emitting power first used for a comparison of the intensities of different sources of light was *the candle.* Thus, a light of sixteen candle power is one which produces the same intensity of illumination at a distance of four meters as does a single candle at a distance of one meter. Since candles of different composition and size, burning under different conditions, differ widely in the amounts of light which they emit, it is obviously necessary to specify the type of candle to be used as a standard. The so-called *normal candle* is a candle of paraffin, 2 cm. in diameter and burning with a flame 50 mm. high. In Great Britain the legal standard of light is a sperm candle which burns 7.776 g. of spermaceti per hour. In Germany the standard candle has been replaced by a special form of lamp invented by Hefner-Alteneck. It burns amyl acetate, and, when regulated so as to have a flame 40 mm. high, emits one so-called *Hefner* unit of light. This is equivalent to .81 of a normal candle. Another special form of oil lamp, called the *Carcel* standard, is in use in France. It is equivalent to about 9.4 normal candles. For many purposes it is very convenient to use a carbon-filament glow lamp as a standard, since with suitable precautions it emits a very constant light.

184. The Lummer-Brodhun photometer. The most approved modern instrument for comparing the intensities of different sources of light is the Lummer-Brodhun photometer. The surfaces the illumination of which by the two sources S_1 and S_2 (Fig. 208) is made the same, are the opposite sides of a white opaque screen AB. These surfaces are viewed by an eye at E with the aid of two plain mirrors M_1 and M_2. In order to bring the two sides of AB into immediate juxtaposition, as seen by the eye at E, the principle of total reflection is made use of in the construction of the prism CD. This consists of two right-angled prisms, CGH and DGH, pressed very firmly together along the faces *GH*, which are made so as to come into perfect contact in certain places, but not to come into contact in other places. Now the light which comes to *E* through the portion of the interface *GH* in which the surfaces are in perfect contact is light which comes from the left side of *AB*, undergoes reflection at M_1, and then passes without change of medium through the prism from the face *CG* to the face *HD*. On the other hand, the light which comes to *E* from the portions of the face *GH* which are not in perfect contact, that is, from places at which an air film exists between the two surfaces, is composed entirely of rays which have come from the right side of *AB* by way of the mirror M_2, and have then undergone total reflection at the surface of the air film. Hence, if the two sides of *AB* are exactly similar surfaces, and if M_1 and M_2 are exactly similar mirrors, it is only necessary to set *AB* at such a point between the sources S_1 and S_2 that the whole surface *GH*, as seen from *E*, is of uniform illumination, and then to apply equation (3). In order to eliminate any possible inequalities in the two sides of *AB*, or in the mirrors M_1 and M_2, the whole instrument is usually rotated through 180° about an axis passing through *AB*. This interchanges the two sides of *AB* and also the mirrors M_1 and M_2. The mean of the settings before and after reversal is then taken as the correct setting. Figure 209 shows a horizontal section of the instrument.

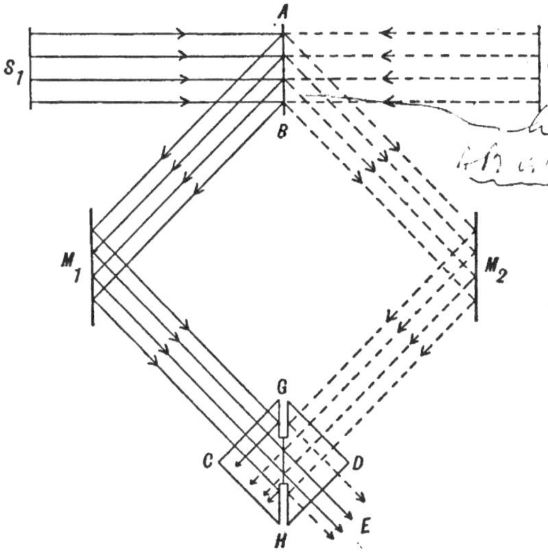

Fig. 208

185. Photometric values of lights of different colors. As stated in section 181, such an instrument as that described in the last section is capable of yielding concordant results only when the two sources are of the same color. If the colors of the sources differ, the impression of color contrast is so strong that the setting for equality of illumination becomes an estimate which even the same person cannot duplicate closely, and upon which different persons will have widely different judgments. If we are to continue to define illumination, as in section 181, as the quantity of luminous energy per unit area, it would, of course, be possible to compare the intensities of emission of lights of different color, such as red and blue, by allowing the radiations from the different sources to fall at a given distance upon equal surfaces which completely absorb them both, and then measuring the relative amounts of heat developed in these surfaces per second. This would presuppose, of course, the elimination of the heating effects due to the nonluminous radiations, which, in general, accompany luminous radiations. This might be accomplished by passing the lights from both sources through a prism, and allowing only the visible portions of the spectra to fall upon the comparison screens whose change in temperature was to be observed. Since, however, even within the limits of the visible spectrum, surfaces illuminated with equal energy do not, in general, appear equally bright, it is obvious that such a comparison would give us no information as to the relative values for the purposes of vision of different lights.

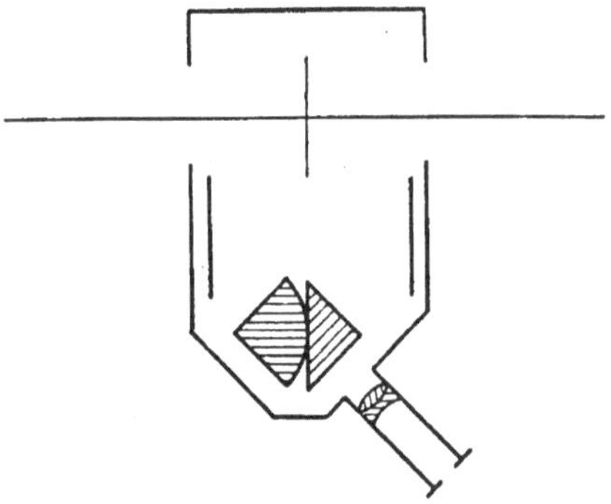

Fig. 209

Photometry, then, as an accurate science, is limited by the very nature of the eye to the comparison of lights of approximately the same color. In the case of complex lights, such as most commercial lamps produce, we must, in general, be content with rough approximations in our estimations of relative candle power, for such lamps generally differ considerably in color. Nevertheless, as suggested above, it is always possible to pass both of the lights to be compared through a prism and thus to separate each into its constituent colors, after which the relative intensities of any particular color in the two sources may be accurately compared. This process is called *spectro-photometry.*

186. The laws of radiation. The fact that even when a body is placed in the best obtainable vacuum its temperature continually falls when it is surrounded by a colder body, such, for example, as liquid air, shows that all bodies are at all temperatures continually radiating energy in the form of ether waves. It follows that when a body is at constant temperature it must be absorbing energy at precisely the same rate at which it is radiating energy. This principle is known as *Prevost's law of exchanges.*

How the total amount of radiated energy varies with the temperature, and how it is distributed among the waves of different wave length are questions which have been made the subjects of many important investigations, both experimental and theoretical. In 1879 Stefan, of Vienna, discovered experimentally that when a black body, such as a carbon filament, is heated to different temperatures *the total intensity of emission is directly proportional to the fourth power* of the absolute temperature. This is known as *Stefan's law of radiation.* It has since been deduced from purely theoretical considerations. It holds strictly only for black bodies, that is, bodies which absorb all radiations which fall upon them, but is approximately correct for most solids.

With reference to the distribution of the emitted energy among the different wave lengths some information may be obtained from very familiar experiments. It is a matter of common observation that as the temperature of any solid is continuously raised it at first emits only heat waves, that then visible waves of a very dull red color make their appearance, and that the color then changes first to orange, then to yellow, and finally to a brilliant white. This behavior shows that as the temperature is raised shorter and shorter wave lengths are added to the emitted light. It must not be supposed, however, that a white-hot body emits less red light than does a red-hot one. In general the intensity of emission of all wave lengths increases rapidly with temperature, but the rate of increase is more rapid in the case of the shorter waves. If the light is passed through a prism and the energy of radiation measured in the different portions of the spectrum by means of its heating effect upon a suitable thermometer, the wave length at which the heating is a maximum continually shifts toward shorter and shorter wave lengths as the temperature rises. A law known as *the displacement law* has been brought to light by the most careful experimental and theoretical investigation. It asserts that the wave length λ_m, at which the heating is a maximum, so shifts with rising temperature that the product of λ_m by the absolute temperature T is a constant. Thus at 820° absolute the wave length at which the radiated energy is a maximum is .00356 mm., while at 1640° absolute λ_m is .00178. In symbols the law is

$$\lambda_m T = \text{constant.}$$

The enormous increase, however, in the intensity of radiation of any particular wave length with temperature is shown by the following example. If the intensity of the red light ($\lambda = .000656$ mm.) emitted by a body at 1000° C. is called 1, at 1500° C. the intensity of the same wave length is over 130, and that at 2000° C. over 2100.

All bodies in which the radiation of heat and light is unaccompanied by permanent chemical or molecular changes of any sort are found to become visible at the same temperature, namely at about 525° C., and to become white hot at about 1200° C. Nevertheless the total intensity of emission depends not only upon the temperature, but also upon the nature of the radiating body. At a given temperature, however, no body emits any wave length in greater intensity than does a black body. These statements apply only to radiations produced by *temperature* alone. When the radiation is produced by molecular or chemical changes it is of a different type, called *luminescence.* It is illustrated by the electrical discharge in vacuum tubes, by the glowworm light, and other similar phenomena. In cases of luminescence there is often a strong emission of light with very little heat.

187. Optical efficiency. It will be seen from the preceding paragraph that if we are to use temperature radiation for the purposes of commercial lighting, then the chief requisite of the incandescent body is that it be capable of withstanding a very high temperature. Thus the chief advantage of the arc light over the incandescent electric light is that in the former the carbon is in such form that it can be raised to the extreme temperature of 3800° C., while the slender filament of the latter permits a temperature of only about 1900° C. Similarly, the high efficiencies of the Welsbach and the Nernst lights are due to the fact that in both of them the incandescent body is raised to a temperature of about 2300° C.*

The optical efficiency of a source of light is defined as the ratio of the luminous energy radiated per second to the energy required to maintain the light for this time. In the ordinary oil or gas light not more than 1 per cent of the total heat energy produced by the combustion is represented in the luminous radiations. The electric light is much more efficient. An ordinary incandescent lamp consumes about 3.5 watts per candle power. The luminous energy radiated per second by a light of one candle power is about 1.3×10^6 ergs. Since a watt is 10^7 ergs per second, the efficiency e of the incandescent lamp is given by

$$e = \frac{1.3 \times 10^6}{3.5 \times 10^7} = .037 = 3.7\%.$$

In the best arc lamps the efficiency is as high as one-third watt per candle power, or approximately ten times that of the incandescent light. This means that in such an arc light, if we neglect the consumption of the carbons,† as much as 37 per cent of the total energy expended is utilized in the production of light. These figures relate to the new flaming arc produced between carbons which are impregnated with some salt like calcium fluoride, and in which the light comes largely from the incandescent vapors in the arc itself. The efficiency of the ordinary arc, in which the light comes chiefly from the luminous center of the positive carbon, is in general not greater than 1.2 watts per mean spherical candle power.

The efficiency of the Nernst lamp is about twice that of the incandescent; that of the Cooper-Hewitt light about six times that

* It should be said, however, that the arc light is probably not an example of pure temperature radiation. The light is due in small part to luminescence.

† The neglecting of this factor obviously renders the result quite uncertain.

of the incandescent. The new tungsten and tantalum lamps have about twice the efficiency of the carbon filament, the gain being wholly due to the higher temperatures employed.

In rating the efficiency of an electric light it is not customary to reduce to absolute units, as was done above, but merely to give the ratio of the candle power produced to the watts consumed.

EXPERIMENT 25

Object. To plot the efficiency curve of an incandescent lamp.

Directions. Arrange a Lummer-Brodhun photometer P, a 16-candle-power, 50-volt lamp l, an ammeter A, reading to 5 amperes, a voltmeter V, a variable resistance R, and a battery or dynamo B, as in Figure 210. Then connect a 16-candle-power, 110-volt lamp L to the lamp socket, and, using L as a standard, find the candle power of l for a series of values of the P.D. across its terminals. Beginning with a P.D. of 40 volts, decrease R and thus increase P.D. by about 3 volt steps until 65 is reached, taking readings on the ammeter, voltmeter, and photometer at every change.

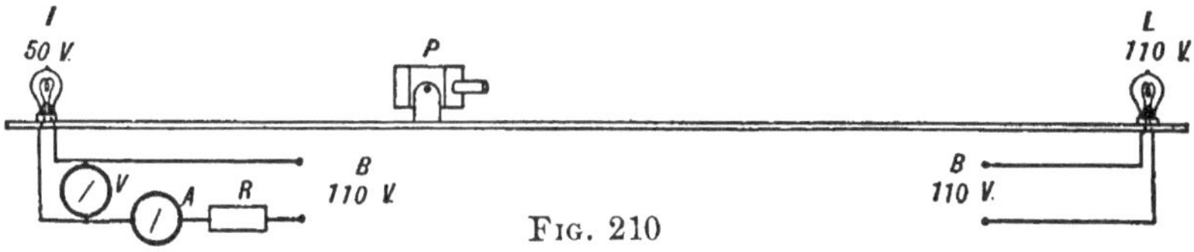

FIG. 210

Compute in every case the number of watts (volts × amperes) required to produce one candle power of illumination, and then plot a curve in which abscissas represent volts and ordinates represent optical efficiencies (candle power divided by watts). Tabulate the data in one corner of the sheet and let the graph constitute the record of the experiment.

EXAMPLE

The record of this experiment is shown in Figure 211. It is seen that the efficiency continuously increases with the P.D. and would doubtless continue to do so until the lamp burned out. The reason that the lamp is not run at a higher P.D. than that marked upon it is that the increased optical efficiency is more than offset by the decreased life of the lamp.

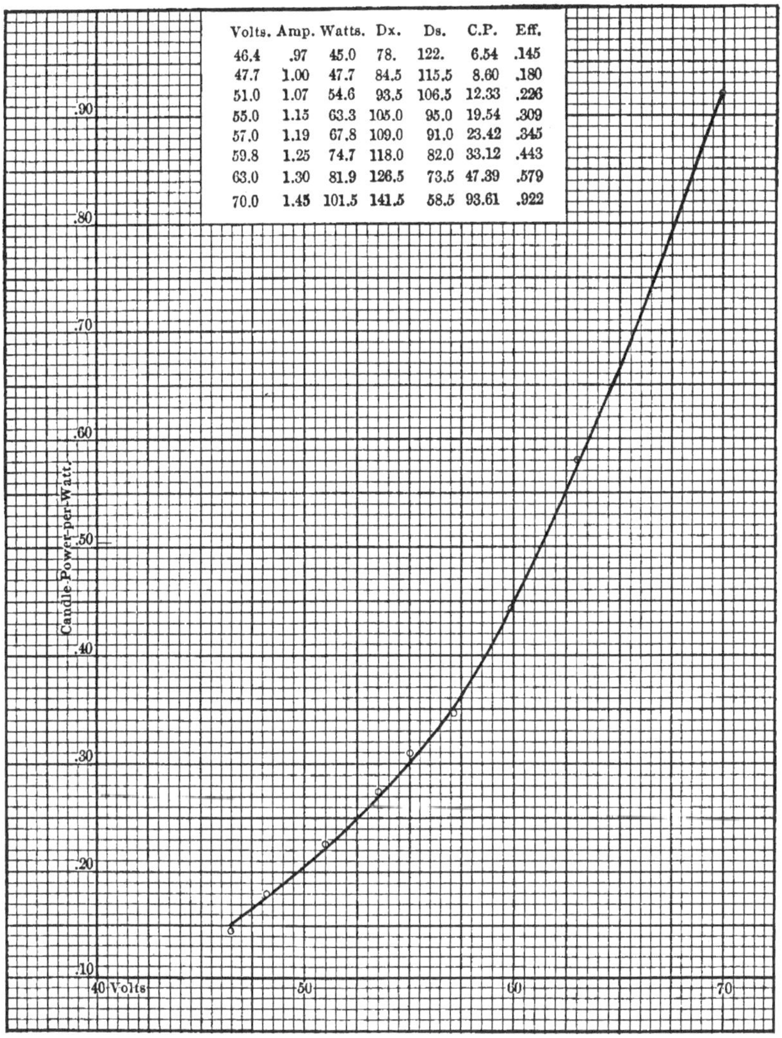

Volts.	Amp.	Watts.	Dx.	Ds.	C.P.	Eff.
46.4	.97	45.0	78.	122.	6.54	.145
47.7	1.00	47.7	84.5	115.5	8.60	.180
51.0	1.07	54.6	93.5	106.5	12.33	.226
55.0	1.15	63.3	105.0	95.0	19.54	.309
57.0	1.19	67.8	109.0	91.0	23.42	.345
59.8	1.25	74.7	118.0	82.0	33.12	.443
63.0	1.30	81.9	126.5	73.5	47.39	.579
70.0	1.45	101.5	141.5	58.5	93.61	.922

FIG. 211

CHAPTER XXVI

DISPERSION AND SPECTRA

188. Newton's experiments on dispersion. It was in the year 1669 that Newton, at the age of twenty-five, published his justly celebrated experiments on the analysis and synthesis of white light, — experiments which during more than two centuries formed the basis of all explanations of the phenomena of color.

These experiments consisted in admitting light through a small aperture *A* (Fig. 212) into a darkened room and observing that the round image *BC* of the sun which was produced on the wall, before the prism *P* was placed in the path of the beam, became replaced upon the interposition of the prism by the band of colors *RV*. This band was red at the end which corresponded to the smallest amount of refraction, and changed through yellow, green, and blue, into violet at the other end. Newton further placed a second prism in the path of the colored band in the position indicated by *P'* of the figure, and observed that the colors were perfectly recombined on the wall into white light. He also showed that it was impossible by means of a second prism to further decompose any one of the spectral colors into more elementary parts.

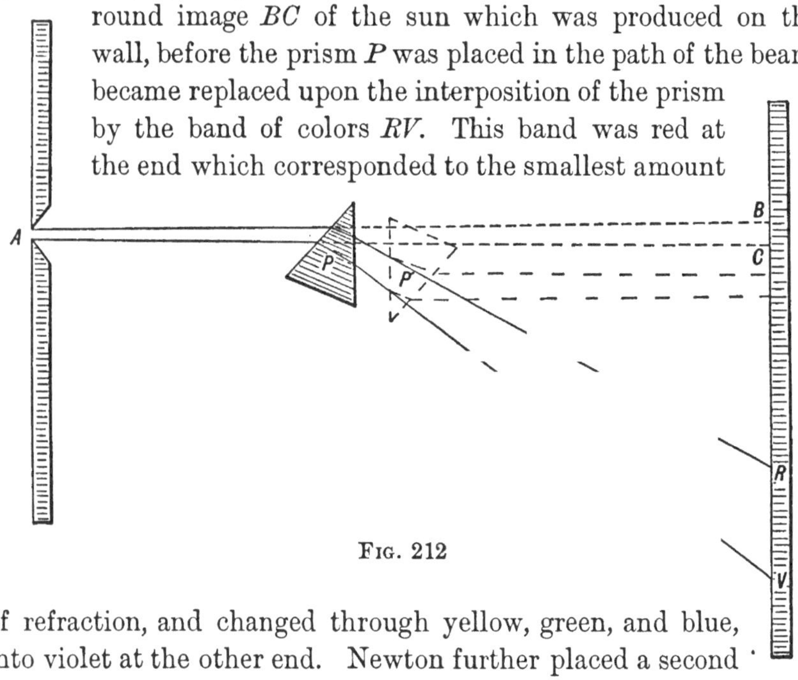

FIG. 212

In view of these experiments it has been customary, since the time of Newton, to regard white light as composed of a mixture of elementary colored lights of every conceivable wave length between that of the longest red and that of the shortest violet. As a matter of fact, Newton's experiments show, not that white light actually consists of all of these colored lights, but merely that white light is decomposed by a prism into these colored lights, and that by recombining these colors we do actually reproduce upon the retina the effect of white light. However, we are led into no conclusions which are at variance with experiment if we adopt Newton's view point as to the nature of white light, and we shall therefore make this viewpoint the basis of much of our reasoning. We shall, however, return to a more critical analysis of this subject in a later section (see sect. 195).

Since violet light is refracted more than red light, and since the amount of refraction is a measure of the change of velocity in going into a new medium, it is clear that the shorter visible waves undergo a greater change of velocity in going into glass than do the longer waves. In other words, the velocity of propagation of violet light through glass is less than that of red light. The exact ratio of these velocities is the ratio of the indices of refraction of glass for red and violet. This for flint glass is about 1.62/1.67.

189. Pure spectra. Since Newton's spectrum consisted merely of a row of circular images of the sun in different colors, and since these images overlapped, as is seen from a consideration of Figure 212, it is evident that the color at any given point of this spectrum was a combination of two or more colors; that is, this spectrum did not consist of colors each of which corresponded to one particular wave length. It was on account of this fact that Newton failed to notice some of the most interesting characteristics of the solar spectrum; such, for example, as the Fraunhofer lines (see sect. 192). In order to obtain a pure spectrum it is necessary to avoid this overlapping of the images of the aperture in different colors. But since a small aperture of any shape whatever will always produce a round image of the sun at *BC*, in order to obtain a pure spectrum two alterations must be made in Newton's arrangement. First, the aperture must be a very narrow slit; and second, a lens must be placed at such a point between the aperture and the screen as to form upon the screen, not an image of the sun, but an image of this narrow slit. In no other way can the image in any particular color be made a mere line. When, however, the slit and the screen are at conjugate foci of a lens, as in Figure 213, the spectrum becomes simply a row of adjacent line images in different colors of the line source. A spectrum formed in this way is called a *pure spectrum.*

When a spectroscope is in such adjustment that an image of the slit is formed in the focal plane of the eyepiece of the telescope, it is evident that a pure spectrum may be obtained. The best way of ascertaining whether or not this condition exists is to illuminate the slit with monochromatic light, such as that produced by incandescent sodium vapor, and then to see whether or not a sharply defined image of the slit is formed in the eyepiece of the telescope.

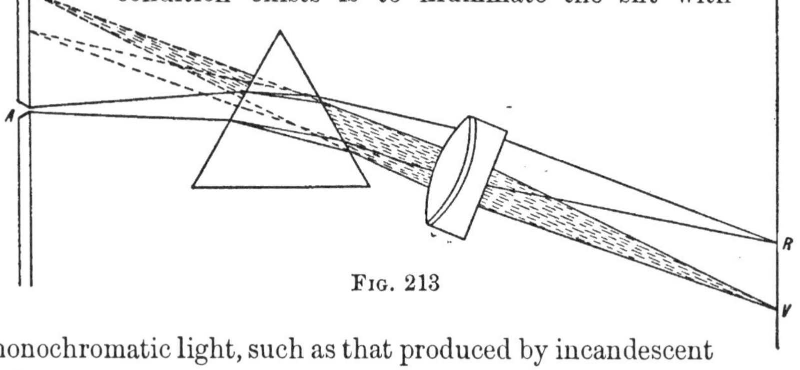

FIG. 213

190. Normal and prismatic spectra. In Chapter XXII it was shown that the spectrum produced by a grating is one in which the angular separation of any two colors is directly proportional to the difference in wave length between these colors, provided θ is small (see p. 264); for then

$$\theta_1 - \theta_2 = \frac{n}{d}(\lambda_1 - \lambda_2). \tag{1}$$

This means that two photographs made with two different gratings, but reduced to the same size, are identical in the arrangement and proportion of their colors. Because of the ease of comparison resulting from this fact the grating spectrum has been adopted as the standard, or *normal,* spectrum.

In general, the spreading of the different colors produced by the passage of light through a prism is not at all proportional to the difference in wave lengths, nor indeed does a substance which produces a large mean refraction always produce a correspondingly large spreading or *dispersion* of any two colors. In other words, dispersion is not proportional to refraction, although Newton supposed, from his early investigation of the subject, that such proportionality existed. In general, prismatic spectra differ from normal spectra in that the reds and yellows are relatively little separated, while the blues and violets are abnormally spread out. If, then, a photograph of a spectrum produced by a prism is made to the same scale as one produced by a grating, the different colors will not occupy at all the same positions or the same relative spaces. Nor, indeed, are the spectra of prisms made of different materials found to agree with one another, the red and yellow, for example, suffering a larger relative separation in one case than in another. It is on account of this so-called *irrationality* of prismatic dispersion that it is possible to construct *direct-vision spectroscopes*. These instruments consist of prisms so combined as to produce dispersion without producing any mean deviation of the beam (see Fig. 214). It is also on account of the irrationality of dispersion that it is possible to produce so-called *achromatic lenses*. These will be considered in the following section.

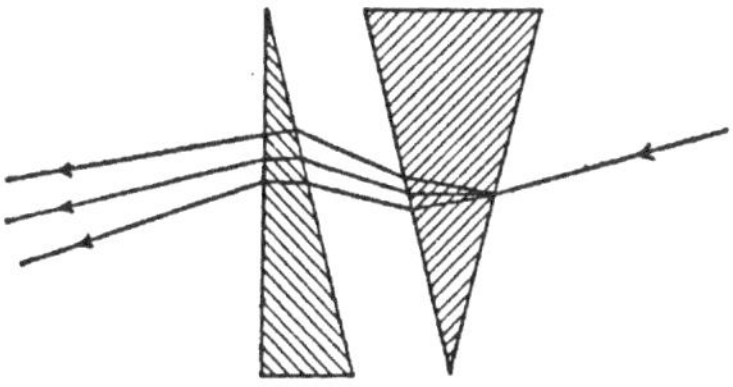

FIG. 214

191. Chromatic aberration and the achromatic lens. The fact that a glass lens produces dispersion is responsible for a phenomenon observable with all simple lenses and known as *chromatic aberration*. When white light falls upon such a lens, since the violet waves are refracted more than the red ones, the focus for the violet waves must obviously be closer to the lens than is that for the red. If v and r (Fig. 215) represent these two foci respectively, then the foci of the colors

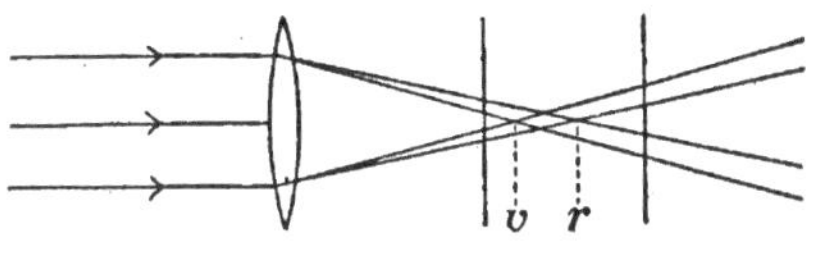

FIG. 215

intermediate between violet and red will obviously occupy positions intermediate between v and r. It is because of this phenomenon that the image formed by a simple lens is in general indistinct and fringed with color. If a card is held nearer to the lens than the mean focal plane, the outer edge of the image is fringed with red, since the red rays, being least refracted, are here on the outside, as shown in the figure. If the card is moved to a position just beyond the mean focal plane, the image is fringed with violet, since the violet rays, after crossing in the focal plane, are here on the outside.

The problem of eliminating chromatic aberration is obviously the inverse of the problem of constructing a direct-vision spectroscope, for in the latter it is necessary to produce dispersion without producing mean deviation, while in the former it is necessary to produce refraction without producing dispersion. The problem is solved by combining into one lens a convex lens of crown glass and a concave lens of flint glass in the manner shown in Figure 216. The flint-glass lens then overcomes practically all the dispersion produced by the crown-glass lens, without, however, overcoming the refraction. Such lenses are called *achromatic lenses*. They are used in the construction of all high-grade optical instruments.

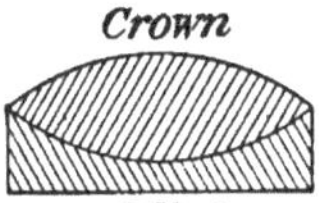

FIG. 216

192. Continuous and discontinuous spectra. In general when a pure spectrum is formed of the light emitted by an incandescent solid or liquid, it is found that there are no breaks whatever in the band of color which constitutes the spectrum, no matter how narrow the slit may be. Looked at from Newton's standpoint this means that the light emitted by a white-hot solid or liquid contains every conceivable wave length between the longest red and the shortest violet. All that we are able to assert with certainty, however, is that the light *which has passed through the prism* contains all conceivable wave lengths within these limits.

When, however, a gas or vapor is brought to incandescence, its pure spectrum is found to consist of a number of separate images of the slit, each, of course, in its own color. This means that an incandescent gas or vapor emits only radiations of certain definite wave lengths. Spectra of this type are called *bright-line spectra.* They are produced by vaporizing metallic salts in a hot flame, like that of the Bunsen burner, or by sending electrical discharges through tubes containing, in rarefied form, the gases to be examined. The characteristic spectrum of sodium, for example, is usually formed by placing common salt or some other compound containing sodium, for example an ordinary glass rod, in a Bunsen flame. It consists of two bright yellow lines very close together. In ordinary spectroscopic work it is seen simply as a single line, but if the slit of the spectrometer is made exceedingly narrow, the original broad image of a broad slit is found to separate into two narrow images of the very narrow slit, thus showing that the images were not originally distinguished as two merely because they overlapped. The spectra of other gases and vapors are not so simple as that of sodium, and the lines are in general scattered through the whole range of visible wave lengths. The fact that the spectrum of an ordinary gas flame is of the continuous rather than of the bright-line type is due to the fact that the incandescent body in a gas flame is not a gas at all, but is rather solid carbon particles suspended in a nearly colorless flame like that of the Bunsen burner.

193. Absorption spectra. In addition to the bright-line spectra discussed above there is another type of discontinuous spectrum, namely the so-called *dark-line* or *absorption spectrum.* When a pure spectrum is formed by the light from the sun after it has passed through a sufficiently narrow slit, it is found, upon examination, to consist of a continuous spectrum crossed by a large number of fine dark lines. These lines were first noticed by Wollaston in 1802, but were afterwards rediscovered in 1814 and investigated by Fraunhofer, who located about 700 of them, and after whom they were named. The existence of these lines in the solar spectrum shows clearly that certain wave lengths are either absent from the sunlight which has passed through the prism, or, if not entirely absent, are at least much weaker than are their neighbors. When the solar spectrum is compared with a sodium spectrum formed by the same spectrometer, it is found that two dark lines in the former are exactly identical in position with the two sodium lines. A similar comparison of the spectra of other elements with the solar spectrum has resulted in the identification in position of many of the dark lines of the latter with the bright lines of the former. An explanation of this phenomenon is suggested by the following experiment.

When a pure solar spectrum is formed by means of any instrument, it is found that if the intensely yellow flame which is produced by burning metallic sodium is placed anywhere in the path of the beam of sunlight which falls upon the slit of the spectrometer, the dark lines of the solar spectrum which correspond in position with the bright lines of the sodium spectrum are intensified, that is, they are very much darker than before. This seems to show clearly that the two prominent lines in the yellow part of the solar spectrum are due in some way to sodium vapor through which the sunlight has somewhere passed on its way to the spectrometer, since making it pass through more sodium vapor increases the prominence of these lines. Now we know that whenever the waves from a sounding tuning fork fall upon another fork of exactly the same pitch, the latter is set into sympathetic vibrations; in other words, the second fork absorbs the vibrations emitted by the first. We know further that this phenomenon of sympathetic vibrations cannot be produced unless the two forks have precisely the same natural periods. It is customary to assume, therefore, in explanation of the two dark lines in the yellow portion of the solar spectrum, that the extremely intense radiation of all wave lengths, due to the extremely hot solid nucleus of the sun, has had some of its wave lengths weakened by absorption as it has passed through the cooler sodium vapor in the gaseous envelope (the chromosphere) of the sun, but that the only wave lengths so weakened are those which correspond to the exact periods of vibration which the absorbing vapor itself is capable of emitting. A similar explanation holds for the other Fraunhofer lines. These lines are not then, in general, devoid of light, but merely appear dark in the solar spectrum because of the very much greater intensity of the light of adjacent wave lengths which have not been weakened by absorption. Thus the same sodium vapor which, when viewed by itself in the spectroscope, appears as two bright lines, appears as two dark lines when viewed against the brilliant background of the solar spectrum. In other words, the darkness is merely a matter of contrast. Figure 217 shows the location of the main lines of the normal (see p. 263) solar spectrum. The corresponding wave lengths are given in so-called Ångström units or 10^{-10} meters. Some of these lines, such as *A* and *B*, are known to be due to absorption which takes place in the earth's atmosphere. *A* and *B* are in fact due to the oxygen of our air. The group *b* is due to magnesium vapor in the sun. *C*, *F*, and *h* are due to hydrogen in the sun. Those of the lines which are due to absorption in the sun are distinguished from those due to absorption in the earth's atmosphere by the fact that when the light from that edge

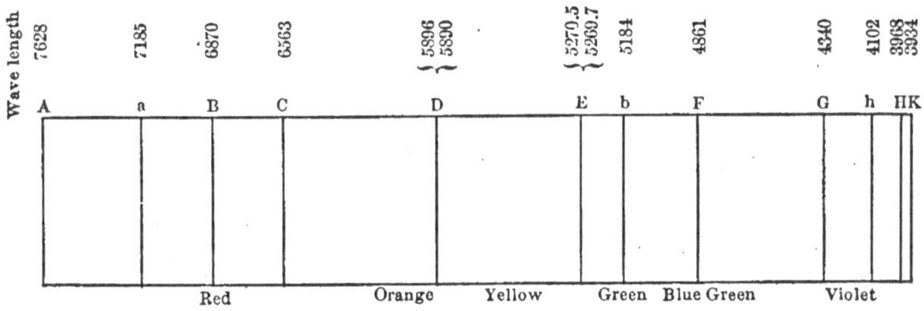

Fig. 217

of the sun which, in view of the sun's rotation, is moving toward the earth, is thrown into the slit of a spectroscope, the lines which have their origin in the sun are slightly displaced toward the violet end of the spectrum in accordance with the principle, known as Döppler's principle, that when a vibrating body moves toward an observer the wave length is shortened. Lines due to our atmosphere obviously could not show such displacement.

194. The reversal of the sodium lines. The phenomenon of the absorption of the light emitted by sodium vapor when this light is passed through cooler layers of sodium vapor may be shown as follows. If the light produced by burning metallic sodium in the Bunsen flame is observed through the spectrometer, the spectrum will at first consist simply of the two brilliant sodium lines, but as the burning continues a point is reached at which the hot sodium vapor in the interior of the flame has become surrounded by dense masses of its own vapor at a lower temperature. At this moment the absorption occurs and the centers of the two lines suddenly turn black. This phenomenon can be observed in any dense incandescent vapor. It is called the phenomenon of reversal.

195. Theory of bright-line and continuous spectra. The fact that gases and vapors give forth only certain definite wave lengths is easily explained from the standpoint of the kinetic theory of matter. For since, according to this theory, the molecules of ordinary gases are for the most part outside the range of one another's influence, it is to be expected that when their constituent parts, for example, their electrons, are once set into vibration by any cause, they will continue to vibrate in their natural periods quite undisturbed during the whole interval between two impacts. Hence these vibrating electrons will send out relatively long trains of perfectly definite wave lengths. Furthermore, if the atom is a very complex structure, it is entirely possible that a given atom might send forth a large number of different wave lengths. This picture of the mechanism of light emission by incandescent gases requires, indeed, a very great complexity in some atoms, such, for example, as that of iron, the spectrum of which contains several thousand different lines.

The explanation of the continuous spectrum of incandescent solids offers greater difficulty. A theory which was in vogue up to about 1895 was developed in view of the three following facts.

(1) If the density of a gas is increased, its spectral lines grow broader.

(2) When liquid solutions show absorption spectra the absorption bands are never fine dark lines, as in the case of gases, but are in general broad bands.

(3) Theory shows that although a body which vibrates without damping must have a perfectly definite and unchanging period, and must therefore emit a homogeneous train of waves, a vibration which occurs with considerable damping is one of constantly changing period, the limits of which are larger and larger the greater the damping. If, then, we assume that the closely packed molecules of liquids and gases are capable, on account of their mutual influences, of producing only strongly damped vibrations,

we can account for the fact that matter in the rarefied condition shows narrow absorption bands and emits bright-line spectra, while in the dense condition it has broad absorption bands and emits, with increasing density, broader and broader lines which finally run together into the continuous spectrum.

It has been pointed out, however, especially by Gouy in France and Rayleigh in England, that it is not necessary to assume any periodicity at all in the source which emits white light. For it can be shown by mathematical analysis that it is possible to resolve an irregular jumble of pulses, such as might be communicated to the ether by atomic shocks, into such a number of homogeneous vibrations, having periods which lie very close together, as is found in the continuous spectrum. Indeed, if we assume that the particles of an incandescent body do vibrate in an infinite number of different periods, it is clear that, since these vibrations must all be transmitted simultaneously to the eye by the same ether, the resultant disturbance of any particular point or particle of the ether would be very irregular, and even according to Newton's view point this irregular disturbance must be resolved by the prism or grating into the extremely close series of regular wave lengths which the continuous spectrum shows. It is therefore not necessary to assume that white light ever consisted of anything but the jumble of irregular pulses which, in any case, the prism or grating is obliged to resolve. It is then very easy to see how a body like a solid or liquid, in which the molecules make extremely short excursions between impacts, might emit such a jumble of ether pulses as this theory requires for the constitution of white light.

196. Spectroscopic analysis. Since a given substance in a gaseous condition always has a characteristic spectrum, and since, furthermore, the spectrum of an elementary substance like hydrogen is in general found to appear in the spectrum of any compound containing hydrogen, it will be seen that the observation of the character of the spectrum of a substance of unknown composition furnishes a very satisfactory method of testing for the presence of certain substances in the compound. Since gases alone have characteristic spectra, the method of spectroscopic analysis is obviously limited to the observation of the spectra of vaporized substances. There are, in general, three ways in which gaseous spectra are compared.

(1) The first is represented in Figure 218. The image of an illuminated scale *L* is formed in the focal plane of the eyepiece *O* of the telescope *T* by means of a reflection from one face of the prism *P*. At the same time some substance, for example,

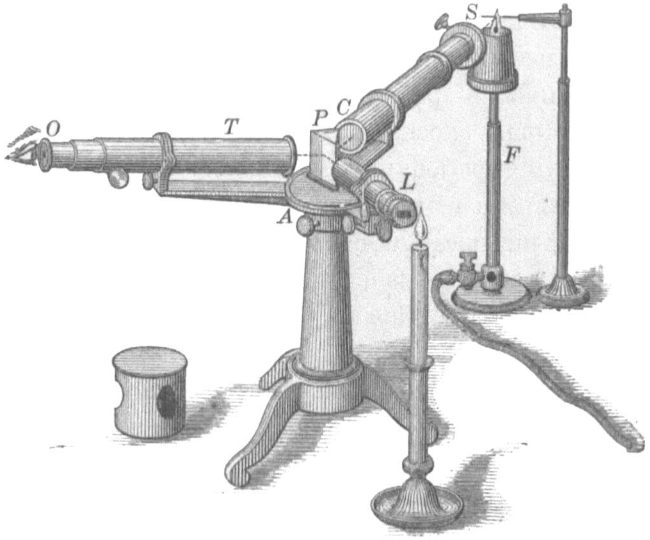

FIG. 218

lithium chloride, is vaporized in the flame *S*, and the characteristic spectrum of lithium is therefore also formed in the focal plane of the eyepiece *O*. The exact positions of the lithium lines on the scale are recorded. Some unknown compound which is to be tested for the presence of lithium is then vaporized in the flame *S*, or in a spark tube which replaces *S*, and the presence of lines on the illuminated scale in exactly the position just occupied by the lithium lines is looked for in the spectrum of this compound.

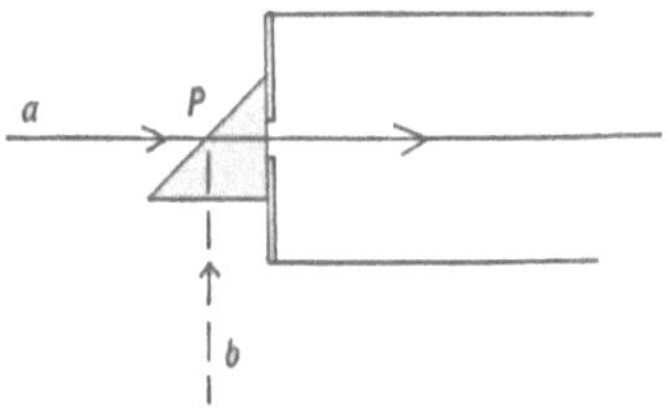

FIG. 219

(2) In the second method the slit of the spectroscope is covered through half its length by means of a small total-reflecting prism *P* (Fig. 219). Light from the

compound is then caused to enter the upper half of the slit from the flame *a*, while light from a lithium flame *b*, for example, is caused to enter the spectroscope through the lower half of the slit by means of total reflection within the prism *P*. If, then, the characteristic lines of lithium which appear in the upper half of the field of view of the telescope *T* are found to continue clear across the field of view, then lithium must be in the compound which is being vaporized at *a*, since the spectrum of this compound occupies only the lower half of the field of view. This method evidently renders the scale *L* unnecessary.

(3) The third, and perhaps the most satisfactory method, consists in replacing the telescope *T* by a photographic camera and taking a photograph of the unknown substance which is being vaporized at *a* with the photographic plate partially covered, and then replacing the unknown substance at *a* by the substance for which the test is being made, and, without altering at all the position of the plate, taking another photograph when only the previously covered portion of the plate is exposed. It is then only necessary to see whether the characteristic lines obtained by the last exposure coincide with those obtained in the first exposure. Figure 220 is a copy of a photograph so taken, the upper and lower portions representing the bright-line spectrum of iron, and the middle portion the dark-line spectrum of the sun. It is obvious that iron is present in the sun. Some of the other substances which have been identified in this way in the solar spectrum are calcium, oxygen, hydrogen, aluminum, nickel, magnesium, cobalt, silicon, carbon, copper, zinc, cadmium, silver, tin, and lead. The lines characteristic of the element now known as helium were observed in the sun before the element was known to exist on the earth.

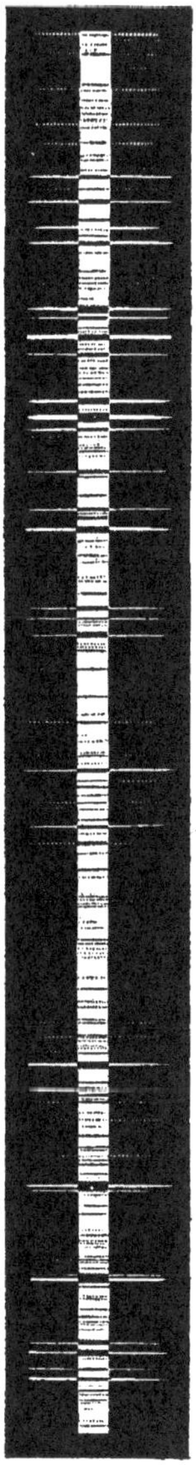

FIG. 220

EXPERIMENT 26

(A) Object. To become familiar with the spectra of different substances.

Directions. Using a spectroscope either of the form shown in Figure 194 or Figure 218, see that the prism is in approximately the position of minimum deviation for sodium light (see sect. 175, p. 280). Using a slit about half a millimeter wide, focus the telescope until the image of the slit is in sharp focus. Then replace the sodium light by a white light and make a rough chart, similar in form to that shown in Figure 222, of the distribution of the light in the spectrum, indicating on the chart by brackets or otherwise what portions of the total length of the spectrum are occupied by the red, the yellow, the green, the blue, and the violet respectively.

Replace the white light by sodium light, and reduce the width of the slit until the yellow sodium line is seen to be in reality two distinct lines very close together. If it does not appear as such at first, make the slit still narrower and focus the eyepiece more sharply. If it is not even then double, it is probable that the spectroscope has not a sufficiently high resolving power. Now make a chart of the sodium spectrum; that is, draw two fine lines, or one, if but one is seen, beneath the central portion of the region marked "yellow" in the chart above and label it "sodium." If you are using a spectroscope of the kind shown in Figure 218, indicate the exact numerical position of the sodium lines upon the scale.

Fig. 221

Using a relatively wide slit (.5 mm.), introduce successively with different platinum wires into the flame of a Bunsen burner *S* (Fig. 218) the chlorides of lithium, strontium, calcium, barium, and potassium, and make a chart of the spectrum of each. In the case of strontium notice particularly the isolated line in the blue; in the case of potassium, the line in the extreme red. Some of these lines are persistent, while others appear only for an instant after the salt is introduced into the flame.

In a similar way make charts of the spectra of nitrogen, mercury, and hydrogen, obtaining these spectra by means of an induction-coil discharge through a vacuum tube (see Fig. 221) placed in front of the slit.

(B) Object. To analyze a mixture for the presence of various elements.

Directions. By either of methods (1) or (2) in section 196 analyze an unknown mixture furnished by the instructor.

(C) Object. To observe the phenomenon of the reversal of the spectral lines of sodium.

Directions. Use again, if possible, a sufficiently narrow slit to bring out the sodium spectrum as a close double line. Burn a piece of metallic sodium about twice as large as a pea in a Bunsen flame placed a foot or so in front of the slit, and observe the reversal of the sodium lines.

(D) Object. To show the existence of sodium and hydrogen in the sun and to compare a prismatic with a normal solar spectrum.

Directions. With a lens of about the same focal length as that of the collimator throw an image of the sun upon the slit of the spectroscope. Make this slit very narrow, focus the eyepiece until the Fraunhofer lines are very distinct, then identify as many as possible of the lines *A*, *a*, *B*, *C*, *D*, *E*, *b*, *F*, and *G*, of Figure 222, which shows a prismatic spectrum. Note that the distances apart of *A* and *D*, *D* and *F*, and *F* and *G* are approximately the same, and also that the distance from *B* to *D* is about the same as that from *D* to *b*. Place a sodium flame before the slit and note that when the sunlight is cut off the yellow sodium lines appear in the exact position of the *D* lines. In the same way show that *C* and *F* are hydrogen lines.

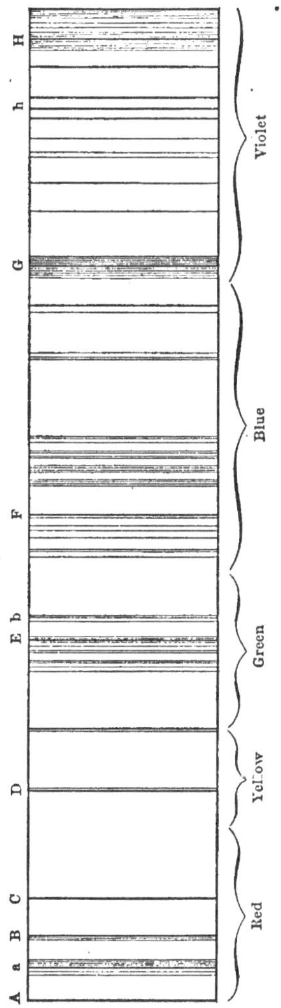

FIG. 222

Replace the prism by a reflection grating, or take another instrument in which a grating is in place, and set this grating in such a position that light coming through the slit falls upon the grating at an angle of incidence not exceeding 45 degrees. Turn the telescope so as to take in the light reflected from the grating face in accordance with the law, "angle of incidence equals angle of reflection." This light should produce in the field of the telescope an uncolored image of the slit. This image corresponds to the central uncolored image produced by the transmission grating of Experiment 22. Focusing, as above, the sun's rays upon a very narrow slit, adjust the leveling screws of the grating, or grating table, until the uncolored image of this slit is in the middle of the field of view. Then turn the telescope and observe the spectra of the first and second orders on either side of the central image. Observe and record the distances apart in this normal spectrum of *A*, *D*, *F*, and *G*, and of *B*, *D*, and *b*. Compare these distances with those shown in Figure 217. See if the distance apart of the *D* lines of the sun's spectrum, or the bright lines of the sodium spectrum, is not much greater in the spectrum of the second order than it is in that of the first (see sect. 162, p. 264).

CHAPTER XXVII

POLARIZED LIGHT

197. Polarization by reflection. All of the phenomena of light which have been thus far studied have been found to be explicable upon the basis of the same wave theory which applies to the phenomena of sound. In other words, so far as the fundamental facts of reflection, refraction, diffraction, emission, and absorption are concerned, sound and light are identical in all respects except in the lengths of their waves and in the nature of the media which act as their carriers.

There is, however, a class of phenomena, known as the phenomena of *polarization,* which differentiate light completely from sound, and show that light waves are not compressional waves at all as are sound waves, but are instead transverse waves similar to those which elastic solids are able to propagate by virtue of their rigidity. These phenomena are so far removed from ordinary observation that they will be here presented in connection with a series of qualitative experiments. The facts presented in the first experiment were discovered in 1810 by the French physicist Malus (1775–1812).

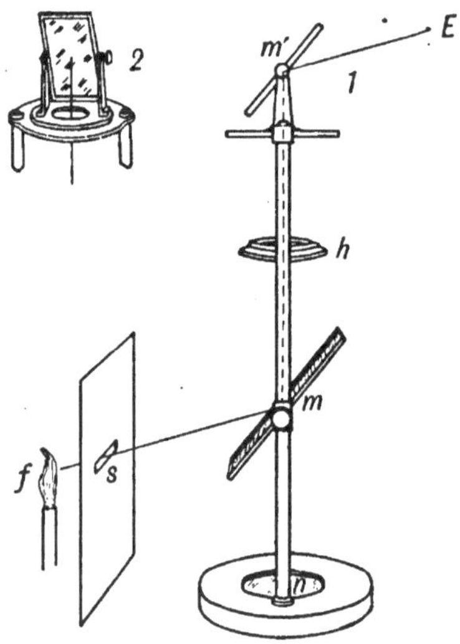

FIG. 223

Experiment 1. Set the plane glass reflector m of the so-called Nörrenberg polariscope of Figure 223 so that its plane makes an angle of about 33° with the vertical. Adjust the position of a horizontal slit s (about 5 mm. wide) and a sodium flame f so that when you remove the black glass mirror m' and look vertically down upon the middle of m you see a portion of the flame.

In this experiment the mirror n may be covered with a piece of black paper. Place m' in position and turn it so that it is exactly parallel to m, that is, so that its plane also makes an angle of 33° with the vertical. Place the eye at E in such a position that when you look at the middle of m you see the twice-reflected image of the sodium flame. Then rotate m' in its frame about a vertical axis and observe the image of the flame as you do so. When you have turned m' through 90°, that is, into the position shown in Figure 223, 2, the image of the flame will have completely disappeared.

The experiment shows that light waves cannot be longitudinal, for if the particles of the medium which transmits the light from m to m' vibrated in the direction of propagation of the light, then the conditions of symmetry would demand that the wave be reflected in precisely the same way after m' has been rotated through 90° as before. But if light consists of waves in which the direction of vibration of the particles of the medium is always transverse to the direction of propagation of the waves, and if, in a very short interval of time, the vibrating particles which give rise to light waves change their direction of vibration many times, then the above phenomena can be very easily understood. For suppose that in Figure 224 *ab*, *cd*, *ef*, *gh*, etc., represent successive directions of vibration of the particles of the medium across the path of the ray *sm* (Figs. 223 and 224). All of these vibrations can be resolved into two component vibrations, the one *perpendicular* to the plane of incidence *smm'*, that is, at right angles to the plane of the page, and represented by the dots in the line *sm*, and the other *in* the plane of incidence and represented by the straight lines drawn across the path of the ray *sm*.

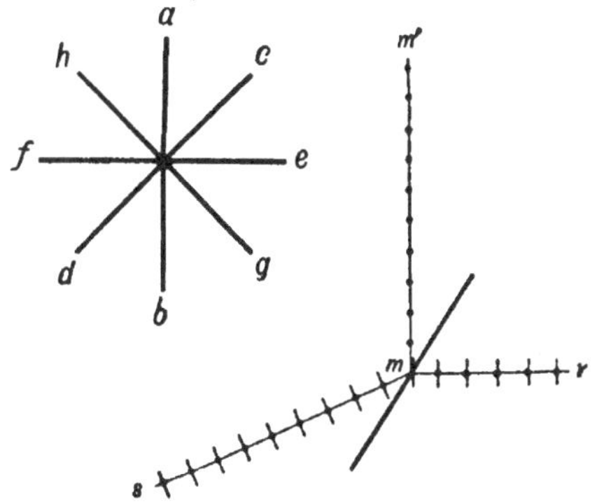

FIG. 224

Now when the ray *sm* strikes the mirror it is clear that the general law of reflection, namely angle of incidence equals angle of reflection, requires that there be some angle of incidence such that the refracted ray *mr* and the reflected ray *mm'* are at right angles to each other. But when this is the case that component vibration of the refracted ray which lies in the plane of incidence coincides with what should be the direction of the reflected ray, namely the line *mm'*; hence this component vibration obviously has no component which is perpendicular to *mm'*. But if light vibrations are always perpendicular to the direction of propagation of the light, this means that there should be one angle of incidence for which no part of the component vibration of the original light which is in the plane of incidence can be reflected.

Considerations of symmetry require, however, that that component vibration of the ray *sm* which is perpendicular to the plane of incidence, and represented by the dots in the figure, should be reflected at all angles of incidence. Now this is precisely what experiment shows to be the case. *At the angle of incidence of the ray sm for which the angle rmm' is a right angle, the reflected ray mm' consists only of vibrations which are perpendicular to the plane of incidence.* The ray *mm'* is said to be a ray of *plane polarized light,* and the angle of incidence at which the ray *sm* must fall upon the mirror in order that the reflected ray *mm'* may consist only of vibrations in this one plane is called the *polarizing angle.* That this angle is always the angle for which the reflected and refracted are at right angles was discovered in 1815 by Sir David Brewster (1781–1868), and is known as Brewster's law. It may easily be shown that another form of statement of the same law is the following: *the angle of complete polarization is the angle the tangent of which is the index of refraction of the reflecting substance.* This is the form in which Brewster announced his law. That the two forms of statement represent one and the same physical relation may be seen from the following:

If the angle *cod* (Fig. 225) is 90°, then we have $90° - i + 90° - r = 90°$, or $i + r = 90°$; hence $\sin r = \cos i$; hence the index $n (= \sin i / \sin r)$ may be written in the form

$$n = \frac{\sin i}{\cos i} = \tan i, \quad \text{or} \quad i = \tan^{-1} n. \qquad (1)$$

The reason that we originally set the mirror m so as to make an angle of 33° with the vertical was that the index of refraction of crown glass is about 1.55, and the angle the tangent of which is 1.55 is 57°. In order that the angle of incidence might be 57° it was necessary to make the angle between the plane of the mirror and the vertical 33°.

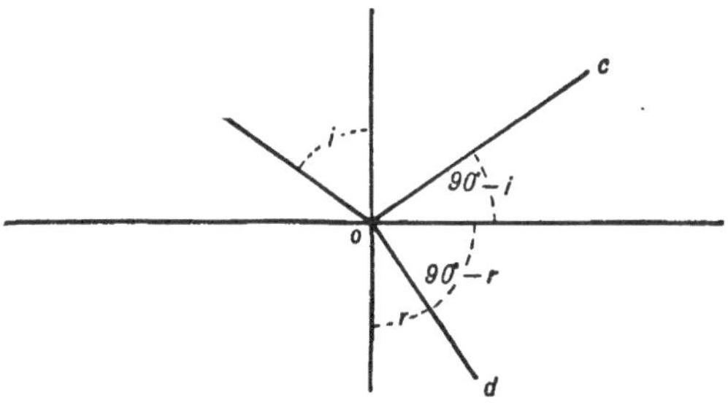

FIG. 225

It will now be obvious why we obtained no reflected light at all from m' when we had rotated it from position 1 to position 2 (Fig. 223). For in this latter position m' bore precisely the same relation to the vibration of the ray mm' as did the mirror m to the component of sm which was vibrating in the plane of incidence smm'.

Experiment 2. Now set m and m' again so that there is no light from f reflected at m' and then rotate m' about a *horizontal* axis, observing the middle of m' all the while. You will find that there is always some of the light ray mm' reflected from m' except when m' is set exactly at the polarizing angle. The amount of the light thus reflected will be found to increase rapidly as the position of the mirror departs in either direction from the polarizing angle.

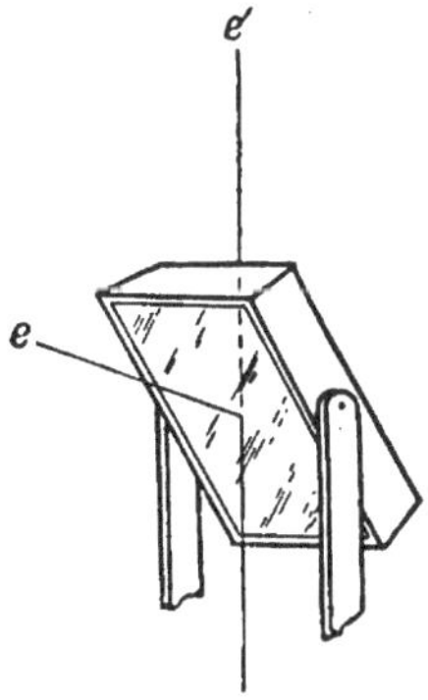

FIG. 226

Experiment 3. Replace the glass mirror m' by a pile of about fifteen thin glass plates set at the polarizing angle (see Fig. 226), and then observe not only, as above, the reflected ray e, but also the ray e' transmitted by the plates as the pile is turned about a vertical axis. You will find that when the pile of plates is in position 2 (Fig. 223), that is, in the position such that the reflected ray disappears, the transmitted ray is of maximum brightness, and when the plates are rotated into position 1 (Fig. 223), that is, into a position such that the reflected ray is of maximum brightness, the transmitted ray has almost entirely disappeared.

In explanation of these effects consider that an incident beam sm (Fig. 227) of ordinary light is resolved into two components, one vibrating in, and one normal to, the plane of incidence. Let the intensity of each of these components be represented by 50 (see Fig. 227). At the polarizing angle none of the 50 parts which vibrate in the plane of incidence are reflected, while photometric measurements show that about 16 per cent of the light which is vibrating perpendicular to the plane of incidence is reflected; that is, 8 per cent of the incident beam is reflected at the polarizing angle.

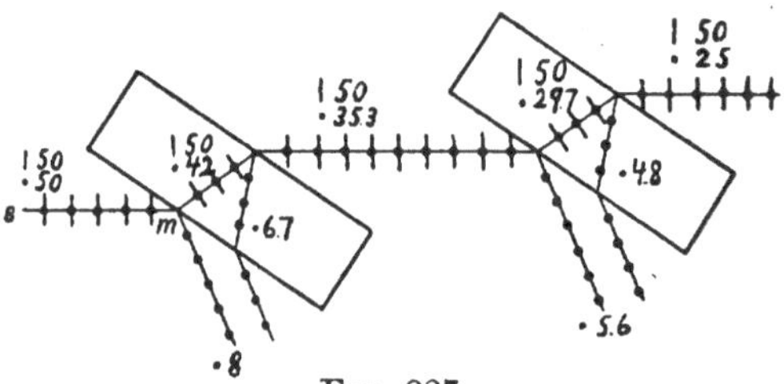

FIG. 227

Hence, after the first refraction, the transmitted light consists of 50 parts vibrating in the plane of incidence and 42 parts vibrating in the plane perpendicular to the plane of incidence. After the second refraction these numbers have become 50 and 35.3; after the third refraction, 50 and 29.7; after the fourth, 50 and 25, and so on. After passage through twelve or thirteen plates the transmitted light has become nearly plane polarized by this process, the plane of its vibrations obviously being at right angles to the plane of vibration of the reflected light. A pile of plates of this sort furnishes a very inexpensive means of obtaining plane polarized light, but it suffers from the disadvantage that the polarization is not quite complete. If no light whatever were absorbed or scattered by the glass, the transmitted ray would become more and more nearly plane polarized the larger the number of plates, but in practice there is found to be no advantage in increasing the number of plates beyond thirteen or fourteen.

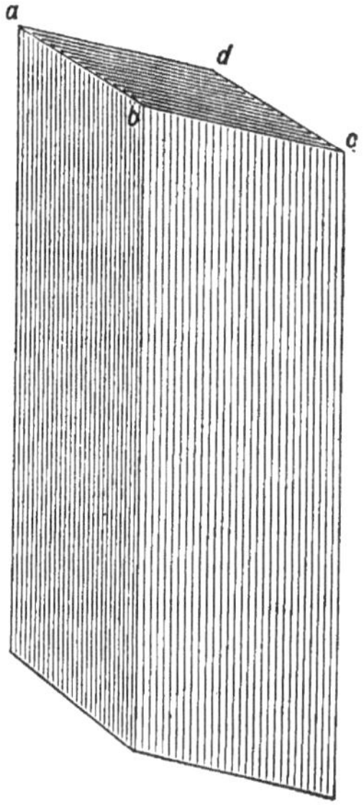

FIG. 228

Experiment 4. Replace the upper mirror *m′* by a Nicol prism (Fig. 228), the construction of which will be considered later, and looking down through this Nicol at the image of *f* reflected in *m*, rotate the Nicol about a vertical axis. You will find that the ray mm' is cut off completely by the Nicol when the latter is in a certain position, but that the light from the flame is transmitted with maximum brightness when the Nicol has been rotated through an angle of 90° from this position. From a knowledge of the plane of vibration of the ray mm' (Fig. 223) decide what must be the plane of vibration of a ray with respect to the face *abcd* of the Nicol, in order that it may be wholly transmitted by the latter, and mark the direction of this transmitting plane of the Nicol by an arrow drawn on the face of mounting containing the Nicol. Henceforth you can use the Nicol as a detector of the plane of vibration of any polarized light which you may observe.

198. Polarization by double refraction. The phenomena which will be presented in the following experiments were discovered in 1670 by the Danish physicist Erasmus Bartholinus (1625–1698), who first noticed the fact of double refraction in Iceland spar, and by Huygens (1629–1695) in 1690, who first noticed the polarization of the doubly refracted beams produced by the Iceland spar, and first offered an explanation of double refraction from the standpoint of the wave theory.

Experiment 5. Make a pinhole in a piece of black cardboard, and lay the cardboard on a piece of plane glass on the frame *h* (Fig. 223). Some inches beneath this, for example on the plate *m* turned into the horizontal position, lay a piece of white paper and illuminate it well. Then lay a crystal of Iceland spar (Fig. 229) over the hole in the cardboard. Remove m' and look vertically down upon the crystal. You will see two pin holes instead of one. Rotate the crystal about a vertical axis. One image will remain stationary, while the other will rotate about it.

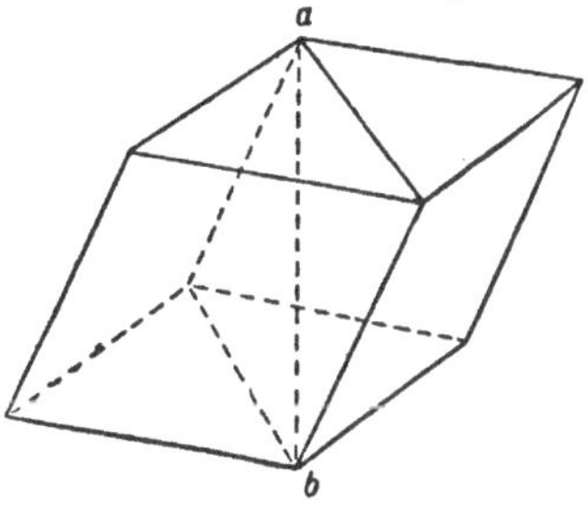

FIG. 229

That the image which remains stationary is produced by light which has followed the usual laws of refraction is evident from the fact that it behaves in all respects as it would if viewed through a glass plate. The image which rotates, however, must be produced by light which has followed some extraordinary law of refraction; for although it has passed into the crystal in a direction normal to the bottom face, and out of it in a direction normal to the top face, it must have suffered bending inside the crystal, since it emerges from the crystal at a point different from that at which the other ray emerges. We must conclude, then, that a ray of light which is incident upon the lower face of such a crystal of Iceland spar is split into two rays by the spar, and that these two rays travel in different directions through the crystal. The ray which follows the ordinary law of refraction is called the *ordinary* ray, the other the *extraordinary* ray.

Experiment 6. To find the direction in which the extraordinary ray travels, rotate the crystal about a vertical axis above the pin hole and note that the extraordinary image always lies in the line connecting the ordinary image and the solid obtuse angle of the face which is being viewed, and, further, that the extraordinary image is always on that side of the ordinary which is away from this solid obtuse angle.

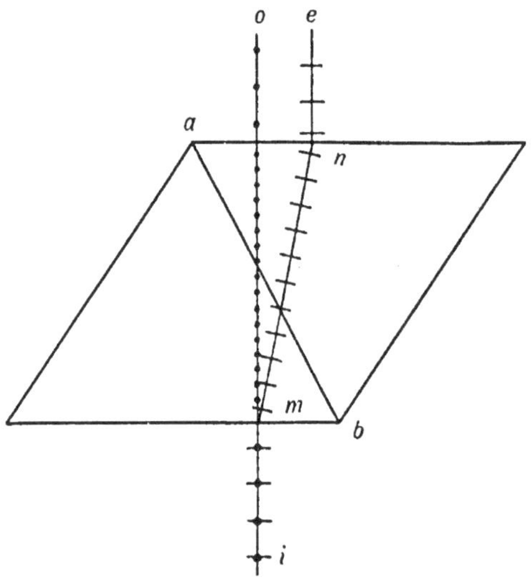

FIG. 230

It will be evident, then, from these experiments that if Figure 230 represents a section of the crystal made by passing a plane normal to the top and bottom faces and through the two solid obtuse angles of the rhomb, the line *io* will represent the path of the ordinary ray through the rhomb, while the broken line *imne* will represent the path of the extraordinary ray.

Experiment 7. In order to determine whether the ordinary or the extraordinary ray travels the faster through the rhomb, observe again the two pin holes, or, better, observe at close range a dot on a piece of white paper upon which the crystal lies, and note which image, the ordinary or the extraordinary, appears to be the nearer to the upper face.

This will evidently correspond to the ray which has suffered the largest change of velocity in emerging into the air (see p. 272); that is, it will correspond to the ray which travels more slowly in the crystal. This will be found to be the ordinary ray.

Experiment 8. If you can obtain a crystal which has been cut so that its top and bottom faces are planes which are at right angles to the line *ab* (Fig. 229) which connects the two obtuse angles of a perfect rhomb, that is, a rhombohedron having all of its faces equal, view the pin hole normally through this crystal. You will observe that there is now but one ray, and that this ray does not change position upon rotation; that is, that it behaves in the ordinary way.

The direction of the line connecting the two obtuse solid angles of a crystal all of whose sides are equal is the *optic axis* of the crystal. This axis is not a line, but rather a direction. Any ray of light which passes through the crystal in a direction parallel to the line *ab* (Fig. 229), that is, parallel to the optic axis, does not suffer double refraction.

Experiment 9. Place the Iceland spar again over the pin hole in the manner indicated in Experiment 5, and view the two images through the Nicol prism as the latter is rotated about a vertical axis. You will find that both the ordinary and the extraordinary images consist of plane polarized light, but that the planes of vibration of the waves which produce the two images are at right angles to one another.

Hence we may conclude that the Iceland spar has in some way separated the incident light into two sets of vibrations, one of which consists of all the components of the initial vibrations which were parallel to a particular plane in the crystal, while the other consists of all of the components of the initial vibrations which were perpendicular to this plane.

Experiment 10. With the aid of the Nicol, the transmitting plane of which you determined in Experiment 4, find whether the ordinary or the extraordinary ray consists of vibrations which are parallel to the plane which includes the optic axis of the crystal.

You will find that it is the extraordinary ray the vibrations of which are in this plane, while the vibrations of the ordinary ray are perpendicular to this plane (see Fig. 230).

199. Theory of double refraction. The elementary theory of double refraction is as follows:

For the sake of simplicity we shall confine attention to the wave form in a single plane in the crystal, namely the plane which is perpendicular to the upper and lower faces of the crystal and includes the optic axis. This is called the *principal plane.* It is the plane of the paper in Figures 231 and 232. As we have already seen, any incident beam of light which passes normally into the crystal through the hole in the cardboard (see Figs. 231 and 232) may be thought of as consisting of equal vibrations in two planes, one perpendicular to the plane of the paper and the other parallel to this plane. Let us consider these two vibrations as separated, so that we may treat of one in Figure 231 and the other in Figure 232. Let us suppose, further, that any vibrations which are parallel to the direction of the optic axis *ab* pass through the crystal with greater facility, that is, with greater speed, than do vibrations which are perpendicular to this direction. The component in the plane of the paper (see Fig. 231) of the incident vibrations will give rise at the boundary *mn* of the crystal to transverse disturbances which will travel outward in all directions through the crystal. The portion of the wave-front, however, which travels at right angles to the axis *ab*, that is, in the direction *md* (Fig. 231), will have its vibrations parallel to

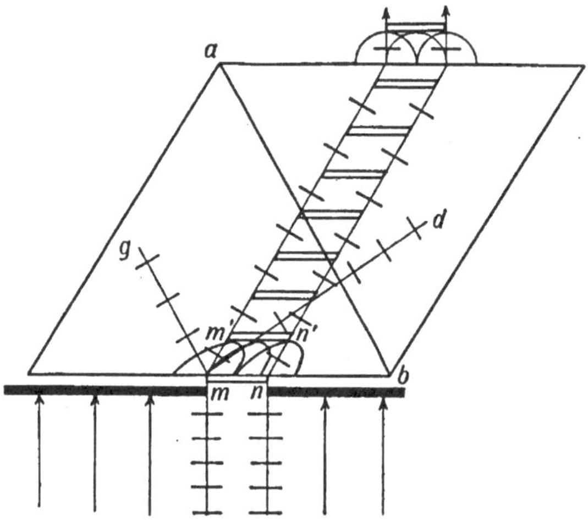

Fig. 231

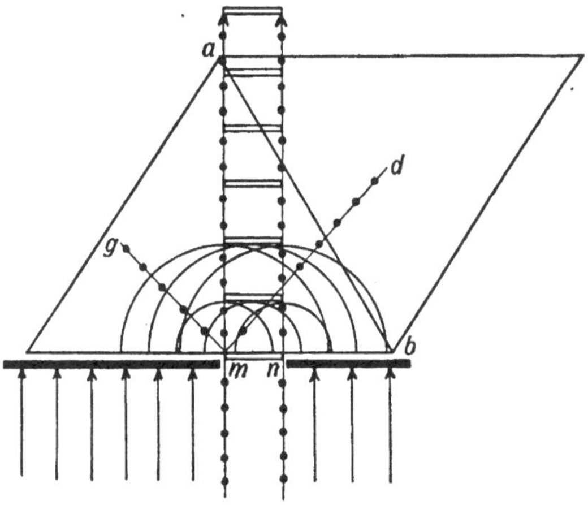

Fig. 232

the optic axis, while the portion of the wave front which travels in the direction *mg* will have its vibrations perpendicular to this axis. If, then, vibrations parallel to *ab* travel faster than do such as are perpendicular to *ab*, the wave which originates at any point on *mn* will travel faster in the direction *md* than in the direction *mg*, and will consequently have an elliptical rather than a spherical form, the longer axis of the ellipse being in the direction at right angles to the optic axis *ab*. The envelope of all the ellipses which originate in the points on *mn* will be the line *m'n'*: *The beam will therefore travel through the crystal in a direction other than that of the normal to its wave front;* that is, in the direction *mm'*. For the reason given in section 154 (p. 242) there will be destructive interference at all points outside of the parallels *mm'*, *nn'*.

On the other hand, the waves which start out from each point on *mn* because of the propagation into the crystal of the vibrations which were perpendicular to the plane of the paper (see Fig. 232) will be everywhere perpendicular to the optic axis, and hence will travel with equal speeds in all directions. The beam will therefore follow the usual law of refraction and will travel in a direction at right angles to its wave front, the waves from each point being now spheres instead of ellipses.

200. Construction of the Nicol prism. In order that the light which is transmitted by a crystal of Iceland spar may consist of vibrations in one plane only, it is necessary to dispose in some way either of the ordinary or extraordinary beam so as to prevent it from passing through the crystal. This was first accomplished in 1828 by the German physicist Nicol in the following way. If the beam *bc* (Fig. 233) is made to enter the face of the crystal at a certain oblique angle, the ordinary ray, being refracted more than the extraordinary (see Exp. 7), will travel in the crystal in the direction *co*, for example, while the extraordinary ray will take the direction *ce*. Now Nicol cut the crystal into two parts along the plane *aa*, and then cemented the parts together again with Canada balsam. This balsam has an index of refraction which is smaller than that of the ordinary ray, but larger

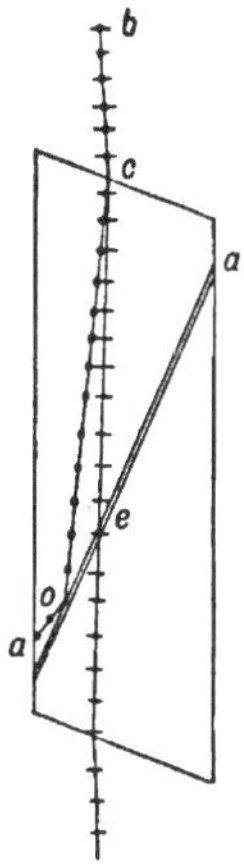

FIG. 233

than that of the extraordinary ray; hence it was possible, by using a long crystal like that shown in the figure, to choose the plane *aa* so that the ordinary ray would be totally reflected and absorbed in the blackened walls of the crystal, while the extraordinary ray would pass through. Several modifications of this device are now in use, but they are all essentially the same in principle.

201. Effects produced by the passage of polarized light through thin crystals.

Experiment 11. Arrange the polarizing apparatus precisely as in Figure 223, save that the Nicol prism replaces the mirror *m′*. Rotate the Nicol until the flame is completely extinguished. Then obtain from the instructor a so-called half-wave plate (for sodium light) of mica or selenite and place it on the slide holder *h*. You will find that, in general, the insertion of the mica causes the light to reappear. Rotate the mica about a vertical axis and note that in one revolution there are four positions, just 90° apart, at which there is extinction. These are the positions in which the plane of vibration of the light which is incident upon the mica is either parallel to or perpendicular to the plane containing the optic axis of the mica. Rotate the mica in a horizontal plane until it is just 45° from one of these positions of extinction. Then rotate the Nicol. The image of the flame will be found to disappear when the Nicol has been rotated through 90°.

We may explain this phenomenon as follows. In Figure 234 let *a* represent the plane of vibration of the light which is incident upon the sheet of mica. Let *b* represent the two vibrations into which the incident vibration is decomposed as soon as it enters the crystal, the one, *e*, in the plane of the optic axis, and the other, *o*, perpendicular to this plane. Since one of these two waves travels faster through the mica than does the other, a sheet may be chosen of just such thickness that the two beams will emerge from the crystal with *o* just one-half wave length behind *e*. Since, moreover, the beams are broad and the mica very thin, *o* and *e* are not separated from one another as they were in the case of the two narrow beams used in Experiment 5. The same portion of the ether must therefore transmit simultaneously the two beams after emergence from the mica. The fact that one of them has lost one-half wave length with respect to the other has been indicated

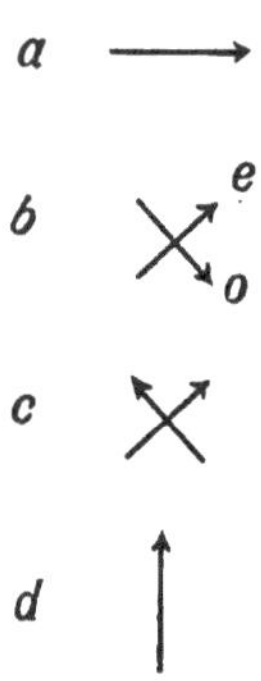

FIG. 234

at *c* (Fig. 234), by changing the position of the arrowhead on the line which represents this vibration from one end to the other. These two vibrations will recombine, upon emergence into air, into one single vibration, which is represented by Figure 234, *d*. The light which emerges from the crystal will therefore be plane polarized, but the direction of its vibration will be at right angles to the direction of vibration of the beam when it entered the crystal. Since the Nicol was originally set so as to cut out vibrations in the direction shown in Figure 234, *a*, it will, of course, transmit with maximum intensity vibrations in the direction shown in Figure 234, *d*. In order to extinguish this vibration it should be necessary to rotate the Nicol through 90°, as we found in our experiment was the case.

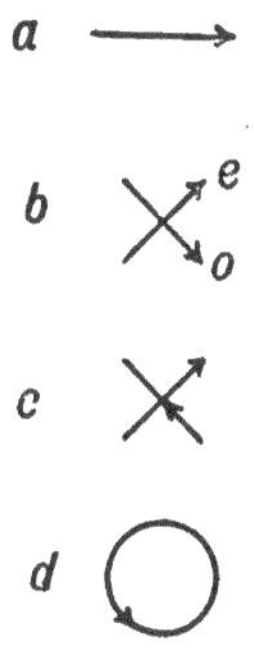

FIG. 235

If the mica plate had been just one half as thick as it was, the two components would have emerged from the crystal one-fourth wave length instead of one-half wave length apart. They would have then recombined into a circular vibration (see Fig. 235), since two equal simple harmonic forces 90° apart, acting simultaneously upon the same particle, must cause it to describe a circular path. The analyzing Nicol should obviously transmit the same fraction of this circular vibration, no matter into what plane it is turned.

Experiment 12. Replace the half-wave plate by one half as thick, that is, by a quarter-wave plate. Set it at first so that it is 45° from the point of extinction, the Nicol being set in the position corresponding to extinction when no plate is interposed; then rotate the Nicol and note that no change takes place in the intensity of the transmitted light because of this rotation.

Light of this sort produced by passing plane polarized light through a quarter-wave plate is said to be *circularly polarized.*

If the plate is of such thickness as to produce a retardation (of one component with respect to the other) of one wave length, two wave lengths, three wave lengths, and so on, the light will obviously emerge from the crystal as plane polarized light vibrating in the same plane as that in which it entered the crystal. Hence it will be completely cut out by the Nicol when the latter is set in the position of extinction for the case in which no crystal is interposed.

If the plate is of such thickness as to cause retardation intermediate between a quarter wave and a half wave, or a half wave and a whole wave, or a whole wave and a wave and a quarter, and so on, the light which emerges from the crystal is *elliptically polarized*, that is, the vibrations of the ether particles take place in the form of an ellipse (see Fig. 236). The major and minor axes of this ellipse may be easily found by observing in what direction the analyzing Nicol must be turned in order to obtain a maximum or a minimum of transmitted light.

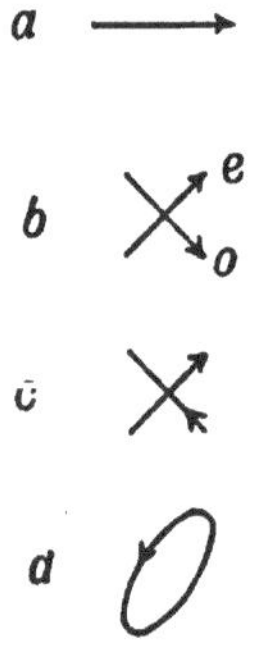

Fig. 236

202. Colors produced by thin crystals in polarized light.

Experiment 13. Set the polariscope in a window in the position shown in Figure 237, the black paper being removed from the mirror *n*. If *m* has precisely the same inclination which was given it in Experiment 1 (Fig. 223), then, when the polariscope is so turned that the prolongation of the line *ma* meets the clear sky, the white light from the sky will strike the lower side of *m* at the polarizing angle, be reflected to the mercury mirror *n*, and return with little loss as a plane polarized beam to the Nicol *N*. Set *N* so that this beam is extinguished. Place a sheet of mica about twice as thick as a half-wave plate for sodium light upon *h* and turn it until it is just 45° from a position of extinction. When viewed through the Nicol it will be found to be brilliantly colored. Rotate the Nicol slowly and notice that a rotation of 45° causes the color to disappear, but that a rotation of 90° causes a color which is the complement of the first color to appear. Further rotation through 90° will cause the first color to return, and so on. When sheets of mica of different thicknesses are used different colors will be produced, but a rotation of the Nicol through 90° will always cause the color to change to that of the complement of the original color.

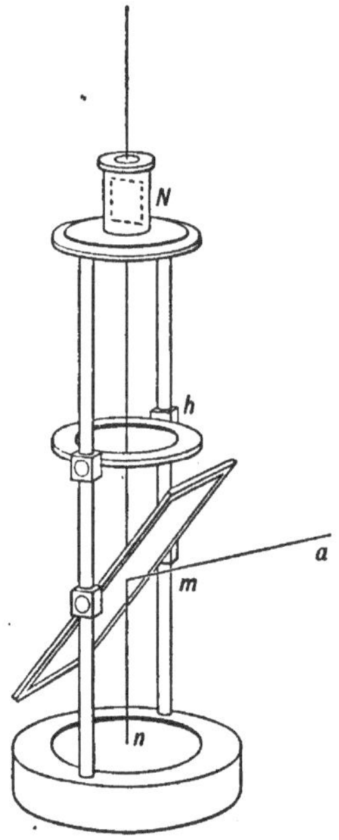

Fig. 237

In order to understand the cause of this phenomenon suppose, for simplicity, that the mica is just thick enough to produce a retardation of one-half wave length of the longest red wave.

Since the shortest violet waves have about one half the wave length of the longest red, this same plate will produce a retardation in the violet of one whole wave length. The violet wave will therefore emerge from the crystal with both of its components in the same phase, and these components will recombine into a plane vibration precisely like that which entered the crystal, namely the vibration represented in Figure 234, *a*. The red ray, however, will emerge from the crystal with one of its components one-half wave length behind the other, and these two components will recombine into a vibration at right angles to that of the entering ray, namely one of the form shown in Figure 234, *d*. If, then, the Nicol is in the position for extinction when no crystal is interposed, it will cut out all of the violet in the incident white light and transmit all of the red, so that if these red and violet waves fell alone upon the crystal, a rotation of the Nicol would cause red and violet to appear alternately. As a matter of fact, however, if the incident light is white, all of the colors between the red and the violet will be present, and the vibrations of the transmitted light which correspond to them will be ellipses of some form. However, the wave lengths which are close to the red, namely orange and yellow, will be largely transmitted by the Nicol along with the red, and will have but small components to be transmitted with the violet, while the wave lengths near the violet, namely the blues and the shorter greens, will be mainly transmitted with the violet. Hence the light which passes through the Nicol in its first position will be some shade of red, because it will have most of the shorter wave lengths subtracted from it; while, when the Nicol is turned through 90°, all of the wave lengths which were before cut out will be transmitted. The color will therefore be exactly the complement of the first color, that is, it will be some shade of blue.

A crystal which is too thin to produce one-half wave length retardation of the shortest visible rays, namely the violet, cannot show any marked color effects in polarized light, since no wave length can be entirely cut out for any position of the Nicol. On the other hand, a crystal which is so thick as to produce retardation of very many wave lengths of any one color will produce also a retardation of an exact number of wave lengths for each of many other colors scattered throughout the spectrum. These colors will all be cut out by the Nicol, and the transmitted light will likewise consist of wave lengths which are taken from all parts of the spectrum, and will therefore reproduce the effect of white light. Hence *these color phenomena in polarized light can be observed only with crystals which produce a small number of wave lengths of retardation.* By scraping crystals down to proper thicknesses in different parts, color patterns of much beauty are often produced when the crystals so treated are viewed in the polarized light. All the colors, of course, change to the complements upon rotation of the Nicol through 90°.

Experiment 14. Place a number of these designs in selenite or mica upon the slide holder *h*, and observe the appearance of the complementary colors in different portions of the design as the Nicol is rotated.

Experiment 15. Observe in convergent polarized light a crystal of Iceland spar, say 1 mm. thick, the upper and lower faces of which are planes perpendicular to the optic axes. The beam of convergent light is most easily obtained by placing the crystal very close to the Nicol in the arrangement of Figure 237, so that the observer looks down through the crystal upon a field of considerable width, from all parts of which polarized light is approaching the eye. If the Nicol was originally set for extinction, you will observe a dark center surrounded by a series of brilliantly colored rings upon which is superposed a black cross (see Fig. 238). Rotating the Nicol will cause the black cross to change to a white one, and all of the colors to change to their complements (see Fig. 239).

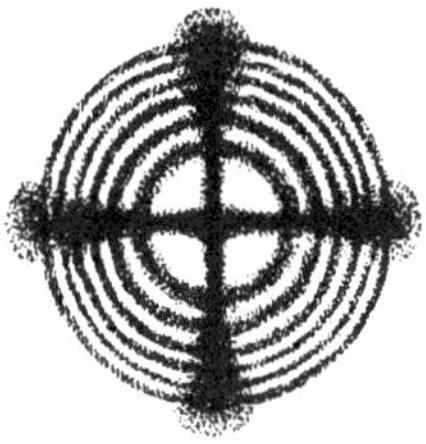

Fig. 238

Fig. 239

These effects may be explained as follows. The central rays pass through the crystal in a direction parallel to the optic axis. They therefore suffer no resolution into ordinary and extraordinary components, and hence no change in the character of their vibration. They are cut out by the analyzing Nicol, hence the black center. The rays, however, which converge upon the eye after passing through the outer edges of the crystal have traveled in directions slightly oblique to the axis, and have therefore suffered decomposition into ordinary and extraordinary rays, which have undergone different retardations. A given retardation of one ray with respect to the other corresponds to a given color precisely as explained above. A given color must, of course, be symmetrically distributed about the axis of the converging beam, since the thickness of the crystal is so distributed; hence the concentric rings of color. The black cross is superposed upon these rings because in two particular planes, namely those for which the incident vibration is respectively in and perpendicular to the plane containing the axis and the ray, even these oblique rays are not split up into components, but pass through vibrating in their original direction, and are therefore cut out by the Nicol. Upon rotating the Nicol through 90° all of these extinguished rays are, of course, transmitted; hence the white cross.

203. Rotary polarization.

Experiment 16. Arrange the polarizing apparatus as in Figure 223, save that m' is replaced by a Nicol, and place upon h a crystal of quartz, say 5 mm. thick, the upper and lower faces of which are made by planes which are at right angles to the optic axis of the quartz. When the Nicol is set for extinction the introduction of the quartz into the path of the beam will be found, in general, to cause the extinguished image of the flame to reappear. Rotate the Nicol, and measure the amount of rotation required to cause the yellow flame to disappear again. According to accepted results this rotation for sodium light should be 21.7° per millimeter of thickness of the quartz. Replace the sodium flame by the ordinary violet flame of the Bunsen burner and repeat, rotating this time until all trace of the violet color in the flame has disappeared. The rotation will be found to be nearly double that found for sodium light. The rotation in the case of light filtered through red glass will be found to be less than that in the case of yellow light.

These experiments show, first, that plane polarized light which passes through quartz in the direction of its optic axis remains plane polarized after transmission, and, second, that the plane of polarization of the light is rotated by the quartz, the amount of the rotation being greater for the short wave lengths than for the long. The discovery that quartz is able to produce these effects was made by Arago in 1811.

From the difference in the amount of rotation of different colors it follows that if plane polarized white light is incident upon the lower face of the crystal of quartz, the light which is transmitted by the analyzing Nicol will be colored, since this Nicol will extinguish completely only those vibrations which are perpendicular to its transmitting plane. It follows, further, that if the Nicol is rotated through 90°, all of the components of the white light which were before extinguished will be now transmitted and vice versa, and hence that the rotation through 90° will cause the color to change to the complement of the first color.

Experiment 17. Verify the above predictions by setting the polariscope so that light from the sky falls upon it in the manner indicated in Figure 237. To show that the lights transmitted by the analyzer in planes 90° apart are complementary, it is best to replace the Nicol by a thick crystal of Iceland spar, or some other form of double-image prism, so that both of the lights to be compared may be transmitted at once. As the analyzer is rotated, the overlapping portions of the images will be found to maintain the color of white light, while the opposite non-overlapping portions will be of complementary colors.

It has been found that there are two kinds of quartz crystals, one of which produces rotation to the right, the other to the left. This difference in optical behavior corresponds also to a difference in crystalline structure which makes it easy to distinguish the so-called *right-handed* from the *left-handed* quartz crystals without actually making the optical test.

Furthermore, it was discovered by Biot, in 1815, that there are certain liquids which possess the same property shown in the above experiments by quartz. Of these, solutions of cane sugar have received most attention, for the reason that the amount of rotation produced by a column of sugar solution of fixed length is taken as the commercial test of the strength of the solution. As in the case of quartz, there are found to be two kinds of cane sugar of precisely the same chemical constitution, but of slightly different crystalline form, which rotate the plane of polarization in opposite directions. The form which rotates to the right is called *dextrose,* the other *levulose.* It is possible to convert dextrose to levulose sugar by acting upon it with hydrochloric acid. This conversion is actually made in sugar testing, the rotation due to the conversion being the quantity directly measured.

CHAPTER XXVIII

RADIO-ACTIVITY

204. Types of radiation. Up to about 1895 the only types of radiation which were generally recognized were those comprised under the two heads,—sound waves and ether waves. Ether waves included light waves, of wave length .00038 mm. to .00076 mm.; ultraviolet waves, studied by means of their photographic effects and having wave lengths extending from .00038 down to an unknown limit, but definitely explored down to .0001; ultrared rays, studied by means of their heating effects, and having wave lengths extending from .00078 mm. up to an unknown limit, but explored up to .06 mm.; and electrical waves studied by means of electrical resonators, and having observed wave lengths extending from 3 mm. up to infinity. All of these forms of ether waves have been definitely proved to travel with the same velocity in a vacuum and to be essentially alike except in the matter of wave length.

Since 1895 three new types of radiation have gained recognition. They are cathode rays, X rays, and the radiations produced by radio-active substances. Cathode rays were observed as early as 1859 and named in 1876, but their character was not understood until 1897.

205. Cathode rays. When an electrical discharge is forced through a glass tube from which the air is being exhausted, the appearance of the discharge is found to vary continuously as the exhaustion progresses. When the pressure within the tube has been reduced to say .01 mm. of mercury, a greenish fluorescence is found to have made its appearance on the walls of the tube about the cathode. As the exhaustion is carried still farther all appearance of a discharge through the gas within the tube vanishes and the walls become suffused with a greenish fluorescence.

This fluorescence seems to be due to some sort of radiation which is emitted by the cathode in a direction normal to its surface; for whenever an object is placed within the tube a sharp shadow of it is found to be cast upon the walls at a point directly opposite the cathode, no matter where the anode may be placed. This is well illustrated by an experiment devised by Sir William Crookes in 1879, and represented in Figure 240, a shadow of the Maltese cross *M* being evident upon the wall of the tube opposite the cathode *C*.

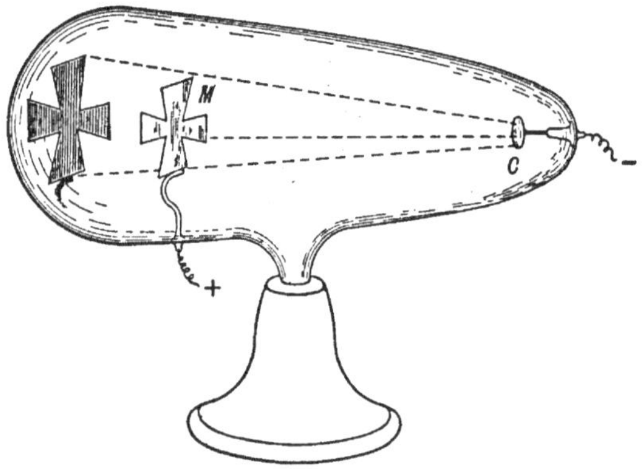

FIG. 240

206. Theory of cathode rays. The nature of these so-called cathode rays was a subject of much dispute between the years 1880 and 1895. Some thought them to be some type of ether waves emitted by the cathode, while others, following the lead of Crookes, were convinced that they consisted of material particles projected with large velocities from the surface of the cathode. The most convincing evidence in favor of the latter view was found in the fact, brought out clearly by Crookes in 1879, that the cathode rays are deflected by a magnet in precisely the way in which negatively charged particles projected normally from the surface of the cathode ought to be deflected. The experiment illustrating this deflection is shown in Figure 241. The cathode beam from

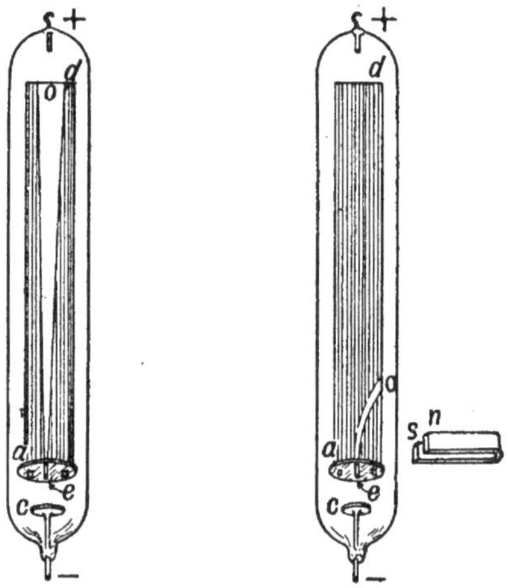

FIG. 241

c, the direction of which, after it passes through the slit *e*, is indicated by the fluorescence which it produces in the zinc sulphide screen *ad*, along which it grazes as it progresses through the tube, is deflected by the magnet *ns*, in the position shown, precisely as the motor rule indicates that it should be deflected if the cathode beam consisted of a negative current flowing from the bottom toward the top of the tube, that is, of a positive current flowing from top to bottom.

The strongest objection to the projected-particle theory was made on the ground of experiments conducted by Lenard, in Bonn, Germany, in 1893. These experiments show that the cathode rays can pass out of the vacuum tube through a thin aluminum window and travel several centimeters in air at atmospheric pressure before being absorbed. Figure 242 shows the tube commonly used for repeating this experiment. In this figure *c* represents the cathode, *A* the anode, in this case a cylinder behind the cathode, *w* a very minute and very thin piece of aluminum foil, and *E* the cathode rays diffusing in air at atmospheric pressure in the manner indicated. The presence of these rays at any point is detected by the fluorescence which they produce in an exploring zinc sulphide screen.

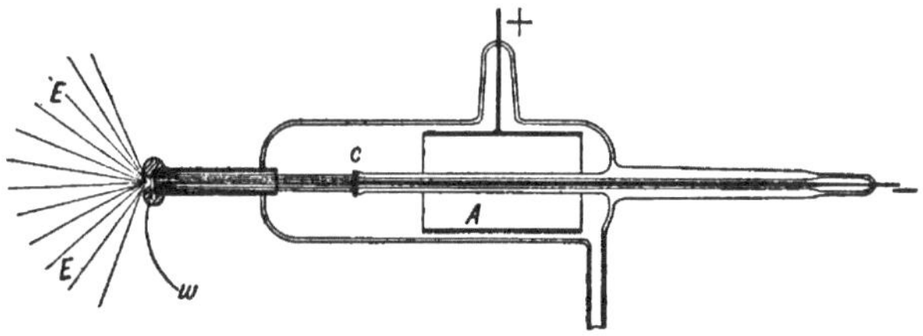

FIG. 242

This experiment shows that whatever cathode rays may be, they are able to pass through a sheet of aluminum which will not permit the passage of the molecules of gas, since otherwise the vacuum in the tube could not be maintained; and that they are also able to pass through the million or so of molecules which any straight line which is drawn through several centimeters of air at atmospheric pressure must encounter. It is evident that this experiment is not easily reconciled with the projected-particle theory if the projected particles are assumed to be atoms or molecules of ordinary matter; but all difficulty vanishes if we assume that the projected particles are exceedingly small in comparison with ordinary atoms, and if we assume further that the atoms of ordinary matter form some sort of infinitesimal stellar systems, the distances between whose members are large in comparison with the size of the members themselves.

In 1897 it was definitely shown, both by J. J. Thomson and Lenard, that the cathode rays are deflected, not only by a magnetic field, but also by an electrostatic field, in precisely the way in which they should be deflected if they consist of negatively charged particles. Furthermore, by comparing, with the aid of mathematical analysis, the amounts of deflection produced by magnetic and electric fields of known strengths it was found that the mass of the projected particles of the cathode rays, if this theory were adopted, came out but 1/1800 of the mass of the smallest known atom, namely that of hydrogen. This epoch-making discovery removed all the serious difficulties of the projected-particle theory. The ether theory, on the other hand, has now been entirely abandoned, since it does not seem possible to reconcile it in any way with the magnetic and electrostatic deflectibility of the rays. Furthermore, the cathode rays were definitely proved in 1896, by Perrin, of France, to impart negative charges to any objects upon which they fall. This accords with the projected-particle hypothesis, but it is wholly unlike any property possessed by known forms of ether waves.

207. Hypothesis as to the nature of the atom. The acceptance of the projected-particle theory as to the nature of cathode rays makes it necessary to assume that these minute negatively charged corpuscles, or *electrons* as they have been named, are constituents of the atoms of all kinds of matter, since cathode rays of the same magnetic and electric deflectibility can be produced with all sorts of electrodes and all sorts of residual gases in the discharge tube. This assumption met with very striking support in a discovery announced by Zeeman, of Amsterdam, in 1897. This observer found that the spectral lines produced by incandescent vapors become doubled, and otherwise modified, when the incandescent vapor which is being examined is placed between the poles of a powerful electro-magnet. From the character and magnitude of this effect it was found that the radiations emitted by these incandescent vapors must be produced by the vibrations, or rotations about a center, of negatively rather than positively charged particles, and that, within the limits of experimental error, the mass of these rotating negative particles is precisely that of the cathode-ray particles.

In view, then, of cathode-ray phenomena, the Zeeman effect, and radio-activity (sect. 210), the atoms of all substances are now regarded as containing negatively charged particles, or *electrons*, of about 1/1800 the mass of a hydrogen atom. These electrons, escaping from the material of the cathode under the influence of the powerful electric fields produced in cathode-ray tubes, stream away in straight lines from the surface of the cathode with velocities which have been definitely measured and found to lie between one tenth and one third the velocity of light (3×10^{10} cm. per second), the velocity being determined solely by the strength of the field which is employed to produce the discharge.

In order to account for the fact that the absorbing power of different kinds of matter for the cathode rays is nearly directly proportional to density, it is sometimes assumed that the atoms of all substances are merely aggregates of electrons, so that the total weight of a cubic centimeter of any substance is determined solely by the number of electrons contained in it, and the total absorbing power of this cubic centimeter for cathode rays is determined by the number of electrons which a given cathode-ray particle would encounter in passing through this cubic centimeter. This view would be in accordance with Lenard's experiment on the absorbing powers of different substances for cathode rays, but there is not enough evidence to substantiate it fully, and it must be held at present, therefore, merely as a speculation.

Since, however, the atoms certainly contain electrons, and since these electrons are negatively charged, in order to account for neutral atoms we must assume a total positive charge in each atom equal to the sum of the negative charges carried by its electrons. Since no definite evidence has as yet been brought forward that positive bodies of the minute mass of the negative electrons are ever set free from atoms, it has been suggested by J. J. Thomson that the atom may consist of a nucleus of positive electricity about which, or within which, the negative electrons are rotating.

208. The ionization of gases. One of the most notable properties possessed by cathode rays is the property of rendering any gas through which they pass a conductor of electricity. Thus if a charged electroscope is placed anywhere within a foot or so of the window *w* (Fig. 242), it is found to lose its charge rapidly. The way in which this discharge takes place has been very convincingly shown to be as follows. The cathode rays in passing through the gas produce in it, in some way, both positively and negatively charged particles. If, then, the electroscope is positively charged, for example, the negative particles produced in the gas by the cathode rays are drawn to the positively charged electroscope, give up their charges to it as soon as they touch it, and thus reduce its positive charge. If the electroscope is negatively charged, it is the positive particles of the gas which discharge it.

The origin of these positive and negative particles is supposed to be as follows. The neutral molecules of the gas, each of which contains negative electrons and an amount of positive electricity equal to the sum of the negative charges on its electrons, have one or more of these electrons knocked off by the bombardment of the rapidly moving cathode-ray particles. The loss of an electron by a neutral molecule of the gas leaves it positively charged. This positively charged molecule, with a few neutral molecules which it probably attaches to itself, constitutes one of the positive ions of the gas. On the other hand, the electrons, which have been knocked off from the neutral molecules, probably soon attach themselves to other neutral molecules and thus impart negative charges to them. These negatively charged molecules, with neutral molecules which become attached to them, constitute the negative ions of the gas. When two of these oppositely charged ions approach near enough to each other, they doubtless are drawn together, their opposite charges neutralized, and the molecules restored to their original condition. It is in this way at least that we attempt to account for the fact that the gas rapidly loses its conducting power when the cathode rays are removed. If, however, the ions are formed in a sufficiently strong electrical field, they may be drawn to the electrodes before they have any opportunity to unite with any other oppositely charged ions. In this case the rate at which either of the electrodes loses its charge is a measure of the total number of ions formed per second.

The ionization of a gas then differs from the ionization of liquids, studied in Chapter XVI, in the following respects. While the ionization of liquids occurs spontaneously, the ionization of gases occurs only through the action of an outside ionizing agent. The ionization of liquids occurs only in the case of a mixture of different substances, that is, in the case of solutions, but the ionization of a gas occurs as well with simple, homogeneous gases as with mixed ones.

209. X rays. It was in December, 1895, that Professor Röntgen announced the discovery of a new type of radiation which he named *X rays.* These rays have their origin in cathode-ray tubes, and are like cathode rays in the fluorescent, photographic, and ionizing effects which they produce, but differ from cathode rays, first, in that they possess a very much higher penetrating power; second, in that they are not deflected in the slightest degree either by a magnetic or an electrostatic field; third, in that they do not impart negative charges to bodies upon which they fall; and fourth, in that different bodies do not absorb them in amounts proportional to the absorption coefficients of these same bodies for cathode rays.

That the X rays have a much greater penetrating power than the cathode rays is shown by the fact that while the latter can only be obtained outside a cathode-ray tube by passing them through a very thin aluminum window, as Lenard did, the former pass with considerable ease through the glass walls of the ordinary X-ray tube. It is because of their relatively great penetrating power, and because of the fact that, in general, they are absorbed in different amounts by substances which have different densities, that it is possible to take with them shadow pictures of the bones of the body, or of any objects of different density inclosed in coverings opaque to ordinary light. It was doubtless on account of this property that the rays attracted so immediate and so widespread attention.

Subsequent study of X rays has proved conclusively that they originate in the surface of any body upon which the cathode rays fall. In the ordinary X-ray tube a platinum plate *P* (Fig. 243) is used as a target for the cathode rays which are emitted at right angles to the surface of the concave cathode *C* and come to a focus at a point on this plate. This is the point at which the X rays are generated and from which they radiate in all directions. This statement can be readily verified by turning about an X-ray tube in action, placed at a distance of a few feet from a charged electroscope, and noticing that the electroscope loses its charge rapidly when it is in any position such that rays radiating from the middle of the front face of *P* could fall upon it, but that the loss of charge ceases as soon as the tube is turned into such a position that rays coming from this point could not fall upon the electroscope.

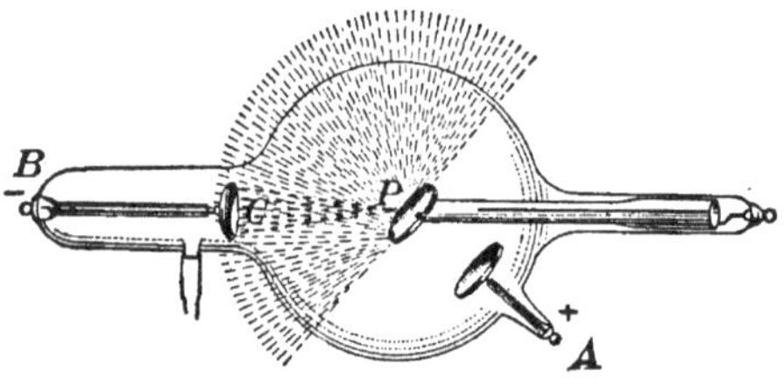

FIG. 243

The nature of the X rays is not yet definitely settled, but it is very generally believed that they are very violent ether pulses set up by the sudden stoppage of the cathode rays as they strike upon the target *P*. If this theory is correct, we must regard the ionization which the X rays produce as brought about by the shaking loose of electrons from the atoms of air because of the suddenness of the ether pulses which strike them. This theory is in accord with the fact that the rays show no regular reflection, refraction, or polarization.

210. Discovery of radio-activity. At the time of the discovery of X rays, and before their origin was known, it was surmised that they might be in some way connected with the fluorescence of the walls of the cathode-ray tube, and since physicists were looking for other sources of the rays, it occurred to Henri Becquerel, of Paris, that the mineral uranium, which under the influence of sunlight fluoresces in a manner not unlike the fluorescence of the glass tube under the influence of the cathode rays, might also give rise to X rays when exposed to sunlight. He therefore wrapped up a photographic plate in black paper, placed a uranium compound above it, and exposed the uranium to sunlight. He found that after an exposure of some days the photographic plate was darkened, thus showing that the uranium emitted rays which were capable of passing through opaque paper. He soon found, however, that the exposure of the uranium to sunlight was quite unnecessary. The effects could be produced just as well in a dark closet as in sunlight. Becquerel therefore drew the conclusion that uranium compounds spontaneously emit radiations which are akin to X rays in their photographic and penetrating effects. He soon found also that these rays were like X rays in their ionizing properties, but unlike them, and like the cathode rays, in that they were deflected both by magnetic and electrostatic fields, and in that they imparted negative charges to bodies upon which they fell. In a word, the rays which produce the photographic effects observed by Becquerel were found to be in every respect identical with cathode rays except that the electrons which are thus shot off *spontaneously* from uranium atoms have speeds of about one half the velocity of light. This is considerably larger than the velocities of ordinary cathode rays.

211. Complexity of the radiations from uranium. In 1899 Rutherford proved clearly that the rays from uranium are of two kinds, — the so-called *alpha* and *beta* rays. The *beta* rays are simply projected electrons, that is, cathode rays, but the *alpha* rays have a much less penetrating power and produce much stronger ionizing effects. They have since been shown to consist of positively charged bodies of atomic size, their mass being either that of the hydrogen molecule or the helium atom. These *alpha* rays are completely absorbed by a layer of air three or four centimeters thick. The name *radio-activity* has been given to the property, found to be possessed by quite a number of substances, of spontaneously projecting bodies of atomic or subatomic size with such velocity as to ionize a gas or produce an impression upon a photographic plate.

The method which has been most commonly used for studying the radiations of uranium, and of other substances which have been found to possess similar properties, is the following. The uranium, or other substance to be tested, is placed in a shallow metal vessel between two plates, *A* and *B* (Fig. 244), to one of which is attached a gold leaf or other indicator of an electrical charge. The rays from the radio-active substance ionize the gas between the plates, and if the plate to which the gold leaf is attached is originally charged, it loses this charge at a rate which is proportional to the number of ions formed per second in the gas between the plates because of the presence of the radio-active substance. The relative radio-activities of two substances are then compared by measuring the relative rates of discharge of the electroscope when equal weights of the two substances are spread over equal areas. When tested by this means the activity of uranium is found to be exceedingly constant; hence uranium is commonly taken as a standard, and the activities of all substances are rated in terms of the activity of uranium.

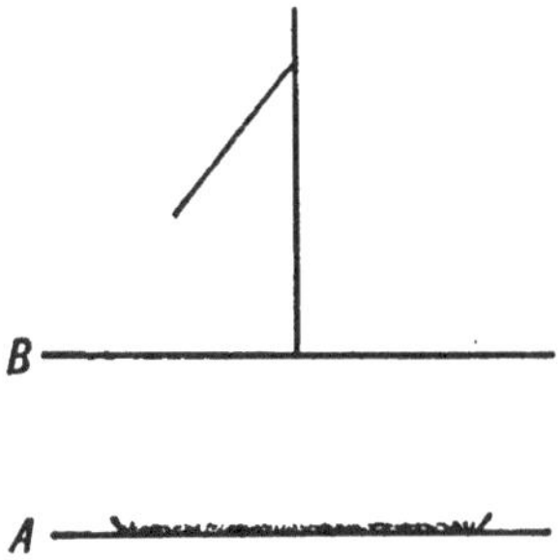

FIG. 244

The proof first given by Rutherford of the complexity of the radiations from uranium was as follows. He laid layers of aluminum foil successively over the uranium and observed the diminution in the rate of discharge of the electroscope produced thereby. The following table represents one set of his observations.

THICKNESS OF ALUMINUM FOIL .005 MM.

Number of layers	Discharge per minute in scale divisions	Ratio of successive rates of discharge
0	182	
1	77	42
2	33	43
3	14.6	44
4	9.4	65
12	7	

It will be seen that at first each succeeding layer cuts down the rate of discharge of the electrometer by approximately the same fraction of the value which it had before the introduction of that layer; in other words, the ratio of each two successive values of the leak is constant. This is, of course, as it should be if there is but one type of radiation which is being absorbed by successive layers. But it will be seen that after the third layer this relation breaks down completely, and the insertion of eight new layers after the fourth reduces the leak only from 9.4 to 7. This is interpreted as meaning that the *alpha* rays have been practically all absorbed by the first four layers, while the *beta* rays are but very slightly absorbed by the introduction of eight more layers. In order to cut the remaining leak down to one-half value, i.e. to 3.5, it was found necessary to introduce a thickness of aluminum equal to 100 more sheets. Since one layer cuts down the *alpha* radiation to less than one-half value, the *beta* radiation must be more than one hundred times as penetrating as the *alpha* radiation.

The above example illustrates one method of separating the two types of rays. Another method is to pass the radiations through a strong magnetic field. The *beta* rays are found to be easily deflected, while the *alpha* rays are deflected exceedingly little and in the opposite direction. It is this experiment which shows that the two types of rays are oppositely charged.

212. Other radio-active substances. Immediately after Becquerel's discovery of uranium rays M. and Mme. Curie, of Paris, made a careful study of all the then known elements by the electrical method, with a view of determining whether or not the property of radio-activity was possessed by other substances than uranium. They found that thorium and all the thorium compounds had about the same degree of activity as uranium and its compounds, but that no other of the then known elements behaved in this way. However, in investigating the ores of uranium they found that pitchblende, which is composed largely of uranium oxide, had about four times the activity of chemically prepared uranium oxide. Since, in general, the chemically prepared compounds of uranium were found to have activities which were proportional to the amounts of uranium present, it appeared clear that the activity of uranium was a property of the uranium atom and was not affected by the sort of combination in which the atom was found. From these facts the Curies concluded that the large activity of pitchblende must be due to the presence of some hitherto unknown element of very great activity. They at once applied themselves to the task of separating this element. Their methods were those commonly used in chemical analysis, except that, after each separation by precipitation, both the filtrate and the precipitate were evaporated to dryness and tested for their activities by the electrical method. It was thus possible to find whether the active substance had been precipitated out of the solution, or whether it had remained in the filtrate. The result of this search was the definite separation from a ton of pitchblende of a few centigrams of a new element which was named *radium* because it had the tremendous radio-activity of about two million as compared with uranium.

In chemical properties radium is closely allied to barium, from which it was found very difficult to separate it. Another radioactive substance was later extracted from pitchblende and named *actinium*. This substance is chemically allied to thorium.

The radiations from radium and actinium, like those from uranium, have been found to consist of *alpha* and *beta* rays. A third type of radiation, however, called the *gamma* rays, was also found to be emitted by radium and actinium. This is characterized by a penetrating power a hundred times greater than that of the *beta* rays and by its entire freedom from deflectability by electrostatic or magnetic fields. These *gamma* rays are commonly supposed to be X rays, which in any case, according to the theory of section 209, would of necessity accompany the *beta* rays as they do in a cathode-ray tube. In the case of uranium also these *gamma* rays are doubtless present, but are so weak as to escape detection under the ordinary conditions of experiment.

213. The spinthariscope. The most striking, though perhaps not the most conclusive, evidence that radio-active substances are continually projecting particles from their atoms with enormous velocity is furnished by an experiment devised by Crookes in 1903. When a bit of radium is placed near a screen of zinc sulphide it produces fluorescence in the screen precisely as do cathode rays or X rays. In order to study more carefully this fluorescence Crookes placed a tiny speck of radium R (Fig. 245) about a millimeter above a zinc sulphide screen S and then viewed the latter through a lens L which produced from 10 to 20 diameters magnification. The continuous soft glow of the screen, which is all that one sees with the naked eye, is, under these circumstances, resolved by the microscope into a thousand tiny flashes of light. The appearance is as though the screen were being vigorously bombarded by a continuous rain of projectiles, each impact being marked by a flash of light. The flashes are probably due indirectly to the impacts of the *alpha* particles. Since these *alpha* particles have been shown by measurements upon their electrostatic and magnetic deflectability to have about the mass of a helium atom, and to have velocities of about one tenth that of light, it will be obvious that they would strike an object with tremendous energy.

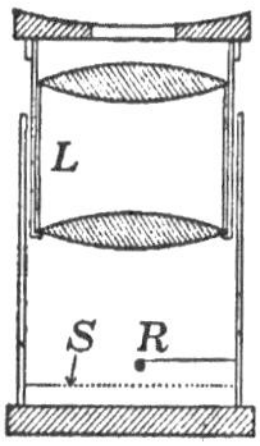

FIG. 245

214. Radio-activity an atomic property. Whatever the cause of this ceaseless emission of particles by radio-active substances, it is now fairly well established that the activity of all radio-active substances is proportional to the amount of the radio-active element present, and has nothing whatever to do with the nature of the chemical compound in which the active element is found. Thus uranium, for example, may be changed from a nitrate to a chloride, or from a chloride to a sulphide or an oxide, without producing any change whatever in its activity. Furthermore, radio-activity has been found to be equally independent of all changes in physical as well as chemical condition. The activity of uranium, for example, is not altered at all by reducing its temperature to that of liquid air. This is strong evidence in favor of the view that radio-active change, that is, the change, whatever it be, which is responsible for the emission of the *alpha* and *beta* particles, involves a change in the nature of the atom itself.

215. Radio-active transformations. The first direct evidence that the atoms of radio-active substances are continually transmuting themselves into atoms of new physical and chemical properties was contained in some experiments made in 1900 by Sir William Crookes, but first interpreted by Rutherford. Crookes found that when he precipitated an aqueous solution of uranium nitrate with ammonium carbonate and then redissolved the precipitated uranium by an excess of the reagent, there remained behind an undissolved precipitate which contained a large part of the original activity possessed by the uranium nitrate. He called this undissolved precipitate (or better, the portion of it which was responsible for the activity, for when chemically tested it showed nothing but iron, aluminum, and other impurities) *uranium X*. He soon afterwards discovered that the uranium nitrate, which had lost a large part of its activity through the separation from it of uranium X, had, in the course of a few months, completely regained its original activity, while the uranium X had lost its power of radiating. This result pointed very strongly to the conclusion that a radio-active substance is being continually produced by uranium. Not long after this Rutherford found that a product analogous to uranium X could be separated from thorium by a similar method. He called this *thorium X*. The essential difference between uranium X and thorium X lies in the fact that the activity of the first falls to half value in twenty-two days, while that of the second falls to half value in about four days.

The examination of radium revealed a behavior exactly similar to that of uranium, for it, too, was found to be continually producing a radio-active substance which, when separated from the radium, slowly lost its activity, while the radium from which it was separated regained, at a like rate, its radiating power. In the case of radium this new substance, unlike uranium X and thorium X, could be distinguished by other physical properties besides its activity. Thus Rutherford found it to be of the nature of a gas which could be separated from a radium salt by heating the latter or by dissolving it in water. The radium which had been so treated lost, for the time being, all but one fourth of its original radiating power, the other three fourths being found in the gas, or *emanation*, as Rutherford called it. This gas could be set away in bottles and the change in its activity watched from day to day. It could be condensed by passage through tubes immersed in liquid air, its presence being, in general, detected by the ionization which it imparted to the air with which it was mixed. It has recently been obtained in sufficient quantities to show a characteristic spectrum and other qualities common to gases. This gas, however, like uranium X and thorium X, has but a transitory existence, for the fact that it gradually loses its activity shows that it passes on into something else.

Nor did physicists have long to look in order to discover this substance into which the emanation from radium is transformed. They found that when the gas comes into contact with a solid object, this object, especially if it is negatively charged, becomes coated with a film of radio-active matter which can be dissolved with hydrochloric or sulphuric acid, and which is left in the dish when the acid is evaporated, or which may be rubbed off with leather and found, by means of the property of activity which it possesses, in the ash of the leather after the leather has been burned. This active deposit, generally obtained by placing a negatively charged wire in a vessel containing radium, is so infinitesimal in amount that it has never been detected in any other way than through its activity. At first it might look as though it were nothing but the active gas itself condensed on the surface of the solid object, but since the rate at which it loses its activity is altogether different from the rate at which the emanation decays; since, furthermore, the emanation atom is not positively charged, and therefore does not tend to collect on a negative wire; and since the active deposit is found to emit both *alpha* and *beta* rays, while the emanation emits only *alpha* rays, it seems necessary to conclude that this film of active matter is the product of the emanation rather than the emanation itself. In fact, it appears to bear in all respects the same relation to the emanation which the emanation bears to radium; that is, *it is the result of the disintegration of the atom of the emanation, just as the emanation is the result of the disintegration of the atom of radium.*

Thorium is found to be precisely like radium in that it gives rise to a gaseous emanation, and in that this emanation disintegrates into something else which collects upon a negatively charged wire. It has been definitely shown, however, that the thorium emanation is not the direct product of thorium, but rather of thorium X.

In order to collect the active deposit from the thorium emanation it is only necessary to thrust a wire through a cork which closes a small tin can in which some thorium oxide is placed, and make the wire the negative terminal of a battery of 100 volts or more, the can being made the positive terminal (see Fig. 246).* In the course of an hour or more the wire will have become appreciably active.

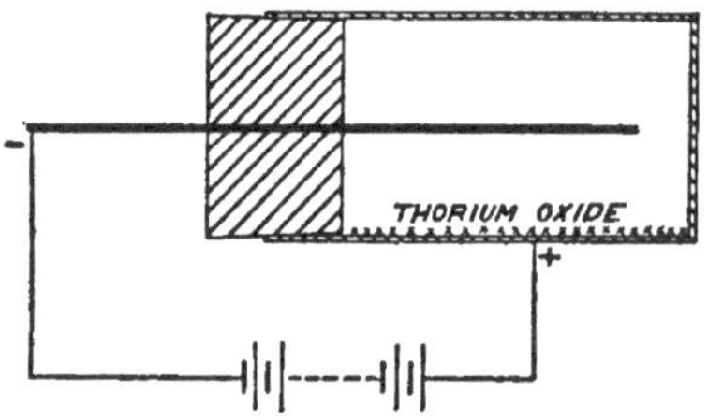

Fig. 246

The fact that this active material is deposited on a negatively charged wire throws some light on the probable nature of the disintegration of the emanation atom, for this means, of course, that the atoms which are thus deposited are positively charged. In order to account for this we must assume, that although the emanation atom produces only *alpha* rays, yet it must lose one or more electrons in passing over into the atom of this active deposit. The reason that these electrons do not produce *beta*-ray effects may be that they are projected with insufficient velocities to ionize the air.

If a wire is exposed to the thorium emanation for several days in the manner indicated above, upon removal its activity is found to decay fairly regularly, falling to half value in about 11 hours, but if it is exposed to a large amount of the emanation for only a few minutes, its activity is found to increase regularly for about $3\frac{1}{2}$ hours and then to decrease in the manner indicated in Figure 247. The initial increase in activity shows that an active substance must be forming on the wire for some time after its removal from the vessel containing thorium, and the fact that the initial activity of the wire is practically zero shows that the material from which this active substance is

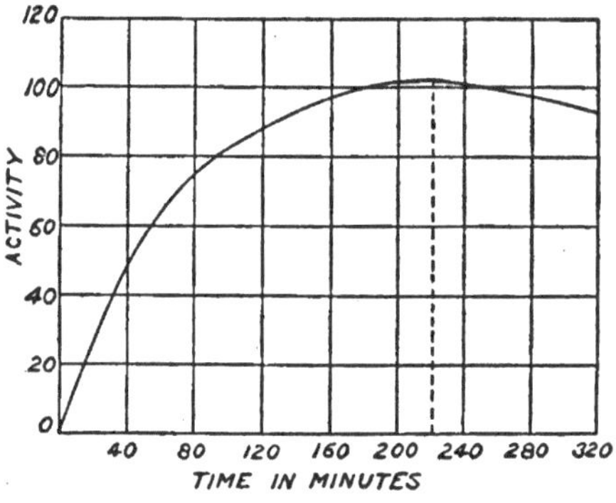

Fig. 247

* Figures 246 and 247 are taken from Rutherford's "Radio-active Transformations."

formed is not itself active. The curve is completely explained if we assume that the atom of the thorium emanation disintegrates into the atom of some nonactive substance called by Rutherford *thorium A*, which deposits upon the wire, while this nonactive substance then disintegrates into an active substance, *thorium B*.

Enough has been said to show how, one after another, the successive products of radioactive change have been discovered and studied. The following chart (Fig. 248), taken from Rutherford's "Radio-active Transformations," shows the various transformations which the radio-active elements are supposed to undergo, the time given beneath each product representing the time required for this substance to lose half its activity. It is now pretty well established,

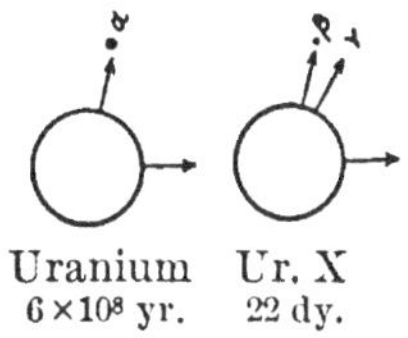

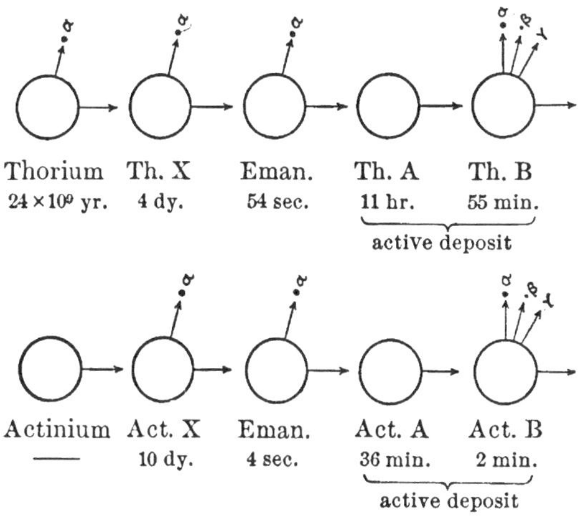

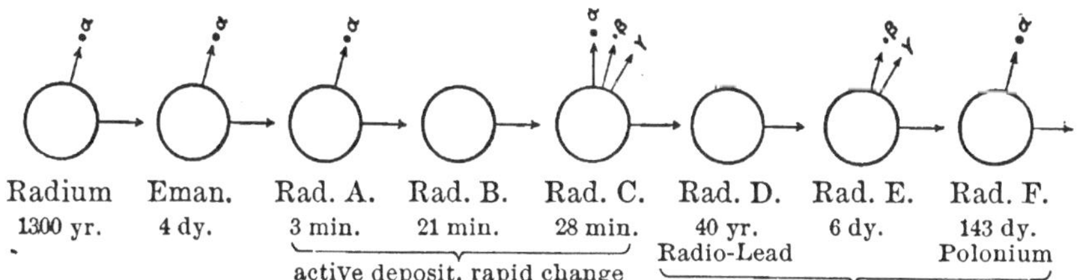

FIG. 248

however, that radium is itself only a transformation product of uranium, being produced not by the immediate disintegration of uranium X, but by some rayless change in an intermediate product called *ionium* into which the uranium X immediately disintegrates. It seems probable, too, that actinium is also a disintegration product of uranium. If this should turn out to be the case, there would be but two strongly radio-active families, namely the uranium family and the thorium family. It is worthy of remark that the atoms of uranium and thorium are the heaviest atoms known, their atomic weights being 239 and 232 respectively. These atoms, then, seem to be continually disintegrating into atoms of lower atomic weight, the process of disintegration apparently consisting in the expulsion of an *alpha* or a *beta* particle, or of both together. The residue of the atom after each such expulsion appears to have new physical and chemical properties, that is, to be a new chemical substance. This new substance is itself in general unstable, and after a time passes over into something else, with the expulsion of other particles. What are the ultimate products of this series of radio-active changes is not yet definitely known. It is certain that helium is one of them, for the spectrum of helium has been found to grow out of the emanation of radium. Rutherford regards the *alpha* particles as themselves helium atoms. It is not impossible that lead and some other common elements are products of this disintegration. At any rate, many of these elements are regularly found in uranium ores. J. J. Thomson has recently (1907) shown that the same *alpha* particle which is projected from radio-active substances is thrown off from all sorts of substances in a highly exhausted tube under the influence of very intense electrical fields. If any of these common substances are spontaneously emitting these *alpha* particles at a very slow rate, and with velocities too small to ionize the surrounding air, we should have no means of detecting the fact. Whether, then, the continuous transmutation of one element into another is confined to a few radio-active substances, or whether it is a general phenomenon of nature, is a question which must be left for the future to decide.

EXPERIMENT 28

(A) Object. To compare the radio-activities of black uranium oxide, pitchblende, thorium nitrate, and uranium nitrate.

Directions. With a mortar and pestle reduce to a fluffy powder each of the substances to be compared. Weigh out, say, 4 g. of each, and spread uniformly over a metal surface of about 25 sq. cm. This may be done either by pressing down the powder with some flat object in a shallow vessel, of, say, 3 mm. depth, until the surface is smooth and the thickness uniform, or by making an emulsion, in a small test tube, of the powder in chloroform, alcohol, or some other liquid in which the powder will not dissolve, and then pouring the emulsion very quickly into the shallow vessel and setting away until the liquid has evaporated.

Charge the electroscope (Fig. 249) by drawing a charged rod of sealing wax over the projecting end *e* of the metal rod which supports the gold leaf, and which is insulated from the metal frame *E* of the electroscope by means of an amber plug *p*.

When a deflection of from 30° to 45° has been produced, focus the telescope, or microscope, upon the gold leaf and set the eyepiece so that the leaf in discharging moves as nearly as possible at right angles to the scale in the eyepiece.

Insert the uranium oxide into the electroscope through one of the glass sides, and earth the metal case *E* by running a small wire from it to a gas or water pipe. Place the cylindrical metal tube *b* over *e* so as to cut off all outside inductive effects (see p. 12), and make two or three observations with a stop watch of the time required for the leaf to pass over a given five divisions of the scale in the eyepiece, recharging the leaf with the sealing wax between each set of readings. Replace the uranium by the other substances to be compared, being careful to place them all in just the same position, and make similar sets of readings for each. Express the activities of all the substances in terms of that of the black uranium oxide.

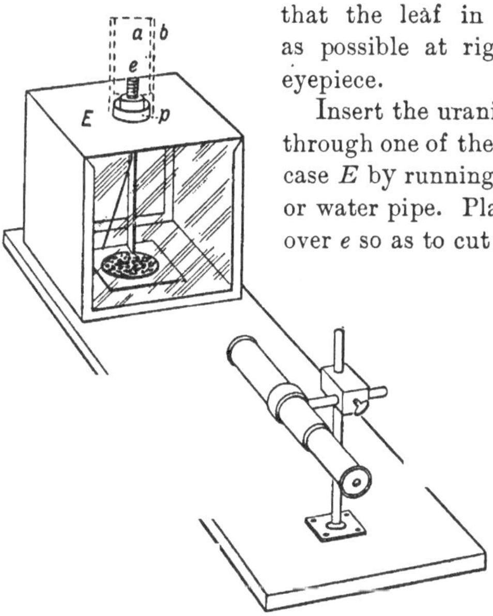

FIG. 249

(B) Object. To determine the number of ions produced per second in a gas by the radiations from a uranium oxide film of a given weight per square centimeter.

Directions. Replace the uranium oxide film in the electroscope, but this time screw upon *e* the solid brass cylinder *a*, and set over this, as coaxially as possible, the brass tube *b*. This operation places a condenser in parallel with the electroscope, so that the capacity of the system is greatly increased. Since the uranium produces the same number of ions per second as before, the same charge will be taken per second from the electroscope, but its capacity being now much increased, the rate of fall of the leaves will be much reduced. Let t_1 and t_2 represent the two times of fall of the gold leaf between the given limits before and after the condenser is added; let *PD* represent the fall in potential corresponding to the observed fall of the leaf; let c be the capacity of the electroscope alone, and C the capacity of the condenser; let Q_1 and Q_2 be the charges removed from the electroscope in the times t_1 and t_2 respectively.

Then $$\frac{Q_1}{Q_2} = \frac{t_1}{t_2}. \qquad (1)$$

But by Chapter VIII; $$Q_1 = PD \times c, \text{ and } Q_2 = PD \times (c + C). \qquad (2)$$

Hence $$\frac{Q_1}{Q_2} = \frac{c}{c + C} = \frac{t_1}{t_2}, \qquad (3)$$

or $$c = C \frac{t_1}{t_2 - t_1}. \qquad (4)$$

This gives the capacity of the electroscope in terms of t_1 and t_2, and the capacity C of the condenser. This last quantity may be computed from the mean area of the inner and outer cylindrical surfaces and their distance apart* (see eq. 17, p. 109). If now we can obtain the P.D. corresponding to the observed change in deflection of the gold leaf, we have at once the quantity of electricity passing per second to the gold-leaf system. Calling this quantity i, we have

$$i = \frac{Q_1}{t_1} = \frac{PD \times c}{t_1}. \qquad (5)$$

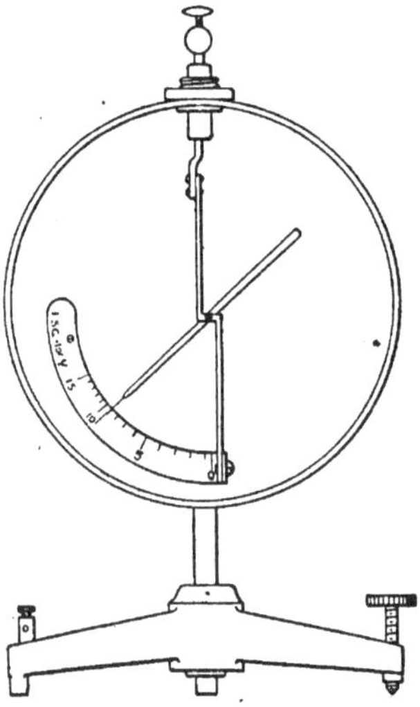

FIG. 250

To obtain *PD* in volts, attach the electroscope in parallel with a Braun electrometer (Fig. 250), or other electrostatic voltmeter, and charging both by means of the sealing wax, find by direct comparison the P.D. corresponding to the observed change in deflection of the gold leaf. Divide this by 300 $\left(\text{i.e. } \frac{3 \times 10^{10}}{10^8}\right)$, to reduce it to absolute electrostatic units. Express this current also in electro-magnetic units by dividing by 3×10^{10}. The charge on each ion of an ionized gas is known to be the same as the charge on the hydrogen ion

* The rigorous formula for a cylindrical condenser of this kind is $C = \dfrac{l}{2 \log_e \frac{b}{a}}$, in which l is the length of the inner cylinder, a the external diameter of this cylinder, and b the internal diameter of the outer tube. This is not appreciably different from the above unless a and b differ considerably.

in electrolysis, namely about 4×10^{-10} absolute electrostatic units. Hence the number of positive ions produced per second by a square centimeter of the uranium film is simply $\frac{i}{4 \times 10^{-10}}$.

EXAMPLE

(A) When 4 g. each of uranium oxide, pitchblende, uranium nitrate, and thorium nitrate were spread uniformly over circular vessels of 5.5 cm. diameter, and having rims 3 mm. in height, the observations on the times required for the leaf to fall from division 5 to division 10 were:

	1st	*2d*	*3d*	*Mean*
Uranium oxide	19.6	19.8	19.4	19.6
Pitchblende	6.4	6.4	6.6	6.5
Uranium nitrate	54.6	54.8	54.6	54.7
Thorium nitrate	16.4	16.6	16.8	16.6

In terms of the uranium oxide the activity of the pitchblende was therefore 3.17, that of the uranium nitrate .36, and that of the thorium nitrate 1.18.

(B) When a cylindrical condenser 5.89 cm. long, the diameter of whose inner and outer cylindrical surfaces were 1.594 cm. and 1.90 cm. respectively, was added, the mean time of fall of the leaf through 5 divisions when the uranium oxide was in the electroscope was 123.4 seconds. The capacity C was found to be 16.8 absolute electrostatic units. Hence $c = 3.19$ electrostatic units. The loss of PD due to fall of the gold leaf through the given 5 divisions was 68 volts. The uranium oxide was spread over a circular area having a diameter of 5.5 cm. Hence $i = \frac{Q}{t_1} = \frac{PD \times c}{t_1} = \frac{68 \times 3.19}{300 \times 19.6} = .0369$ electrostatic units, and i per sq. cm. $= \frac{.0369}{\pi (2.75)^2} = .00155$ electrostatic units $= 5.17 \times 10^{-14}$ electro-magnetic units. Therefore the number of ions formed per second by the radiation from 1 sq. cm. = 3,900,000.

PROBLEMS

CHAPTERS I AND II

1. Find the intensity of the magnetic field due to an isolated south pole of 560 units strength at a point 20 cm. from the pole. If the value of H, the horizontal component of the earth's field is 0.24 in this neighborhood, what will be the resultant field intensity and its direction if the point considered is due magnetic north of the pole?

2. A magnet has a length of 10 cm. and a pole strength of 450 units. Calculate the direction of the field with respect to the axis of the magnet, and the intensity at a point 20 cm. from one pole and 15 cm. from the other.

3. What is the field intensity due to a magnet of moment M at a point distant r cm. along a perpendicular to the magnet at its middle point?

4. If a magnet vibrates with a period of 5 sec. where the horizontal intensity is 0.18, what is its period where the intensity is 0.24?

5. A very short magnetic needle is suspended 17.32 cm. below the center of a bar magnet of pole strength 200 units. The length of the bar magnet is 20 cm. between poles. Its axis (S to N) makes an angle of 90° with the direction of the horizontal component ($H = 0.24$) of the earth's field. What angle does the axis of the magnetic needle make with H when in equilibrium?

6. If the period of oscillation of the magnetic needle in Problem 5 is 3 sec., what is it when the bar magnet has been rotated through 90°?

7. A magnetic needle is suspended above and parallel to a horizontal bar magnet lying in the magnetic meridian. When the north end of the bar points northward the period of the magnet is 7 sec., but when the bar magnet is reversed the period is 5 sec. What would be the period in the earth's field alone?

8. A bar magnet suspended to rotate about a vertical axis has a period of 4 sec. in the earth's field. Find the period of the system if a bar of aluminum is attached to the magnet so as to rotate with it about the same axis, the aluminum bar having one half the length and one half the width of the magnet, but the same vertical thickness. The density of steel is 7.7 and of aluminum 2.7. Find also the value of the moment of inertia of the bar if the magnetic moment is 1200 units and H is 0.22.

CHAPTER III

9. What current must be passed through a tangent galvanometer coil 45 cm. in diameter and consisting of 5 turns, in order that the needle may be deflected 30° in a neighborhood where $H = 0.18$? What current for 45° and for 60°?

10. What must be the radius of a single coil which, when carrying a given current, would produce at its center a field intensity equal to that caused by the same current in passing through two concentric coaxial coils of 15 cm. and 45 cm. diameters?

11. A circular wire coil of 15 turns and 50 cm. diameter must carry how much current to produce at its center a field of 0.60?

12. A current of 0.331 ampere is passed through two concentric, coaxial coils of 12 and 24 turns respectively, first when connected so that their magnetic fields are in opposite directions, and second so that they are in the same direction. The fields thus established are 0.20 and 0.30 respectively. Find the radii of the two coils.

13. Draw the curve representing the values of the tangent for all angles between 0° and 90°. Decide from a study of the curve for what value of the deflection in a tangent galvanometer the smallest error is introduced into the final result by a constant observational error in reading the deflection.

14. The reading of a tangent galvanometer was 46° uncorrected for torsion. When the torsion head was twisted through that angle, the needle turned 1.1° against the restoring force of the earth's field. Find the per cent of error introduced into the final result by using the uncorrected instead of the corrected value of the deflection.

15. A current which gives a reading of 0.27 ampere on a milliammeter deposits 0.2008 g. of silver in 10 min. 42 sec. What is the error in the ammeter reading?

16. How much water should be decomposed by a current of 0.80 ampere in one hour?

17. Given the electro-chemical equivalent of zinc as 0.03367 g. per coulomb, find how much zinc is consumed in a Daniell cell in generating a current of 0.5 ampere for 150 hr. How much copper is deposited under the conditions of the problem?

18. A current is sent through three cells, one containing acidulated water, the second copper sulphate, and the third silver nitrate. How much copper will have been deposited in the second cell while 2 g. of silver are deposited in the third cell? What will be the volume, at 76 cm. of mercury pressure and 20° C., of the mixed gases liberated in the first cell?

19. How many ounces of aluminum are deposited from a suitable solution of bauxite, a natural aluminum oxide, by the passage of 150 amperes for one hour? (Aluminum oxide = Al_2O_3 : atomic weight of Al = 27.4.)

20. Hydrogen given off at an electrode of a hydrogen voltameter displaces entirely the volume of 40 cc. of water contained by the vessel placed over the electrode. The atmospheric pressure is 747 mm. of mercury at the room temperature of 23° C. Find the weight of the liberated gas.

21. The atomic weights of three elements are m, n, and p. They all form sulphates of the form M_2SO_4. What is the ratio of their electro-chemical equivalents?

22. An element which forms oxides of the form E_2O_3 and EO has an electro-chemical equivalent of e for the compound of higher valency. What is the electro-chemical equivalent for the lower valency?

CHAPTERS IV AND V

23. When electrical energy costs 8 cents per kilowatt hour, how much does it cost to operate an incandescent lamp that takes 0.5 ampere from 110 volt mains?

24. A vessel of 30 g. water equivalent contains 1600 g. of water. Its radiation constant is such that its temperature falls at the rate of 6° C. per minute at 90° C. How much current must be passed through a wire of 10 ohms' resistance immersed in the water that the temperature shall be maintained at 90° C.?

25. From Experiment 4 it is seen that the rate at which heat is developed in a wire is $(PD)I$. Show that this is also equal to RI^2.

26. A wire immersed in water generates heat at the rate of 2 calories per second when carrying 0.5 ampere of current. Find the power in watts and in horse power expended in the wire and its resistance.

27. Show from Ohm's law that in two wires of the same circuit the ratio of the P.D.'s between the terminals of the wires is the same as the ratio of their resistances.

28. Neglecting the loss due to radiation, what will be the rise in temperature in 10 sec. of a No. 20 copper wire of mass 45.5 g. and resistance 0.3293 ohm, connected across constant potential mains of 25 volts P.D.?

29. If a wire of the same mass as that in Problem 28, but of half the diameter, is supplied with current from the same mains, what will be its rise in temperature in 10 sec.?

30. If the wire is of the same diameter as that in Problem 28, but of half the length, and hence half the mass, what will be its rise in 10 sec.?

31. If equal lengths of wires of different diameters but of the same material are placed across constant potential mains, which will burn out the sooner, the small wire or the large one? In answering this question consider radiation.

32. If the ammeter in Figure 1 reads 0.5 ampere, the voltmeter 50 volts, and if the resistance of the voltmeter is 300 ohms, what is the resistance of the coil R?

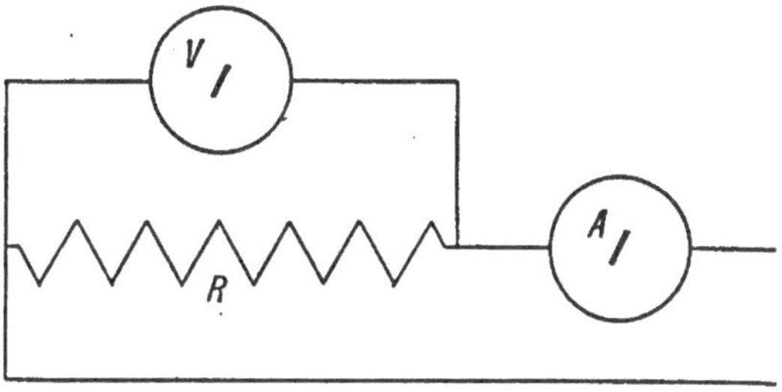

FIG. 1

33. Three coils of resistance 2, 3, and 5 ohms are connected in series across 110 volt mains. Find the current through the circuit, the P.D. across the terminals of each coil, and the power expended in each coil.

34. A 2- and a 7-ohm coil are part of a circuit in which 18 amperes are flowing. Calculate the P.D. across the terminals of this parallel circuit, the combined resistance of the two coils, and the current in each coil.

35. A room is lighted by a chandelier containing 11 incandescent lamps in parallel between wires of resistance negligible as compared to the lamp resistances. Each lamp has a resistance of 220 ohms. What is the combined resistance which they offer?

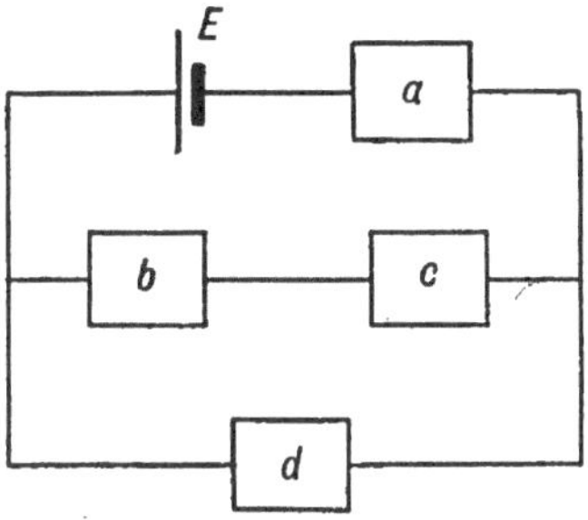

FIG. 2

36. A P.D. of 1.06 volts is maintained at the terminals of a battery E in Figure 2. If resistances of 10,000, 1100, 900, and 1 ohms are connected at a, b, c, and d respectively, what is the current through the resistance c?

37. Given the resistance of the voltmeter used in Experiment 4 as 16,000 ohms, and the resistance of the coil of platinum wire as 13.4 ohms, find the per cent of error introduced in assuming that all the current indicated by an ammeter reading of 4.08 amperes passes through the coil.

CHAPTERS VI AND VII

38. One mil is a thousandth of an inch. One circular mil is the area of a circle one mil in diameter. The area of a circle d mils in diameter is therefore d^2 circular mils. The resistance of pure soft copper such as is used in making insulated wires is given as 10.38 ohms per mil-foot at 75° F. What in C.G.S. units is the specific resistance of such copper?

39. What is the resistance at 75° F. of 1000 feet of wire 31.96 mils in diameter?

40. The resistance of aluminum wire is given as 17.03 ohms per mil-foot at 75° F. What is the ratio of the diameters of two wires, one copper and the other aluminum, such that for the same length the resistances are the same?

41. The specific resistance of copper at 0° C. is 1650 absolute C.G.S. units. Find the resistance of a trolley wire one kilometer long and one centimeter in diameter.

42. If in Figure 62, page 82, $R_1 = 10{,}000$, $R_2 = 100$, $R_3 = 100$, $G = 100$, and $V = 1.06$, what current was flowing through G?

43. The mean temperature coefficient of copper is not exactly 0.0042. The mean rate of increase of resistance for Matthiessen's pure copper between 0° and t° C. is best expressed as $\alpha = (a + bt + ct^2)$ where t is the temperature entering into the relation $R_t = R_0(1 + \alpha t)$. The constants a and b have the values 0.004019 and 0.00000214 respectively, for values of t between 0° and 100°. Find the value of α to use at 25° C. and at 50° C.

44. Compute the error introduced in finding the value of a resistance at 50° from an observation of its resistance at 25° by using the average value $\alpha = 0.0042$ instead of the value of α as given in Problem 43.

45. What is the rise in temperature of the field coil of a dynamo which at the beginning of a run had a temperature of 25° C. and a resistance of 290 ohms if its final resistance is 346 ohms? Use $\alpha = 0.0042$.

46. The P.D. between the two wires of the circuit shown in Figure 3 is 2 volts greater at a than at b. A current of 10 amperes is flowing in the circuit. What is the resistance of the line?

FIG. 3

47. If it is desired to transmit power so that 30 amperes may be delivered at the end of a line at a P.D. of 110 volts, what must be the resistance of the line in order that the "line drop" shall be 2 per cent of the voltage at the transmitting end? What power is expended in the transmission?

48. A group of incandescent lamps takes 15 amperes. The line loss is not to exceed 2 volts. What must be the size of the copper wire to be used if the lamps are 2000 feet from the transmitting end of the line? (Given No. 5 wire, 182 mils diameter, and No. 6 wire, 162 mils diameter.) How many watts are lost in the line?

49. No. 20 German silver wire is to be used in constructing a coil of resistance 5 ohms at 20° C. If the diameter of the wire is 31.96 mils and the specific resistance of the specimen is 0.00002076 per centimeter cube at 0° C., how many centimeters of the wire will be necessary? ($\alpha = 0.0004$.)

50. The average temperature coefficient of platinum is 0.00366. What is the temperature of a furnace in which the coil of a platinum thermometer has a resistance of 1020 ohms? The resistance at 0° C. is 300 ohms.

51. The resistance of the platinum thermometer coil of Problem 50 when placed in a bath of liquid air is 98 ohms. What is the temperature of the bath?

52. What would be the resistance of the coil at the limiting temperature of the absolute zero?

53. A standard resistance of 0.0001 ohm is connected in a circuit, and a millivoltmeter connected to the terminals of the standard resistance indicates a P.D. of 0.137 volt. Neglecting the resistance of the millivoltmeter, what is the current flowing in the circuit?

54. A universal galvanometer shunt has the form shown in Figure 4. The conductor k attached to the galvanometer G may be connected to the points b, c, or d. What must be the ratio of the resistances ab, bc, and cd in order that the ratio of the deflections of the galvanometer for connections at b, c, and d respectively shall be as 1:10:100? (Consider that the resistance ad is small compared to the galvanometer resistance G.)

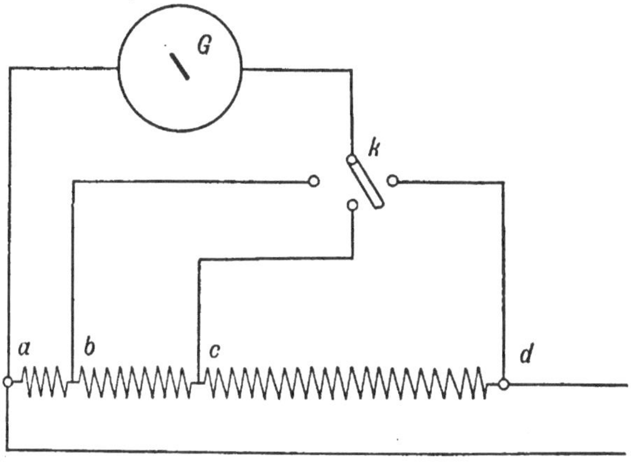

FIG. 4

55. A galvanometer having a resistance of 225 ohms is shunted by a resistance of 25 ohms. Compare the values of the current through the galvanometer with and without the shunt.

56. A milliammeter reading to 0.100 ampere has a resistance of 5 ohms. It is desired to use this instrument to measure current of from 0 to 10 amperes. What must be the resistance of a shunt such that the instrument may be read directly in amperes by applying a multiplier of 100 to its reading?

57. Two voltmeters of resistance 20,000 and 16,000 ohms respectively are connected in series across 220 volt mains. What current flows through the system, and what is the reading of each voltmeter?

58. If the voltmeters of Problem 57 are connected in parallel across 220 volt mains, what current flows through each, and what are the readings?

59. A millivoltmeter has a resistance of 20 ohms. What resistance must be connected in series with it in order that it may be used to read volts directly by multiplying its readings by 100?

CHAPTERS VIII AND IX

60. A condenser is made of two flat metal plates separated by air. If the area of the plates is 500 sq. cm., and the average distance between them is 0.01 cm., what is the capacity of the condenser in electrostatic and in electro-magnetic units?

61. If a plate condenser similar to that described in Problem 60 has a capacity of 0.002 microfarad, and another condenser with glass for a dielectric has twice the area of plates and four times the distance of separation, and has a capacity of 0.0046 microfarad, what is the dielectric constant of the glass used?

62. What quantity of electricity would charge each condenser to 110 volts?

63. What is the ratio of the resistance of a conductor as measured in the E.S. and the E.M. systems respectively?

64. Show from a consideration of the equation $C = \frac{Q}{PD}$ that when there is a given charge upon the plates of a condenser, the value of the field between the plates is inversely proportional to the dielectric constant of the medium which fills the space between the plates.

65. The charge carried by the hydrogen ion in electrolysis is about 4×10^{-10} absolute electrostatic units. Express this charge in electromagnetic units.

66. An insulated gold-leaf electroscope shows a charge which corresponds to a potential of 450 volts. When an insulated sphere, the capacity of which is 15 electrostatic units, is connected with the gold leaf, the potential falls to the point corresponding to 350 volts. Find the capacity of the gold-leaf electroscope.

67. When the terminals of a dynamo are applied directly to the terminals of an electrostatic voltmeter, the P.D. indicated is 500 volts. An oil condenser consisting of plates 25 sq. cm. in area and 1 mm. apart is placed in series with the electroscope. When the terminals of the same dynamo are applied to the extremities of this condenser combination, the voltmeter indicates 50 volts. If the dielectric constant of the oil is 4, what is the capacity of the electrostatic voltmeter in microfarads? in electrostatic units?

68. If a $\frac{1}{3}$ microfarad condenser, charged by means of a 2-volt cell, produces a throw of 10 cm. in a ballistic galvanometer, what number of microcoulombs passed through the galvanometer when it was thrown 6 cm. by the earth inductor?

CHAPTERS X AND XI

69. A battery of E.M.F. 1.07 volts and internal resistance 1.8 ohms is connected to a coil of resistance 6 ohms. What is the P.D. at the battery terminals, and what current flows in the circuit?

70. A tangent galvanometer of resistance 1.5 ohms is deflected 54° when connected to a battery. The deflection due to the same battery is only 42° when an additional resistance of 2 ohms is connected in the circuit. What is the resistance of the battery?

71. A galvanometer of 20 ohms resistance gives a deflection of 5° when connected to a battery of E.M.F. 1.08 volts and 2 ohms internal resistance. When connected to a second battery it gives a deflection of 8°. When shunted by a wire of 20 ohms resistance and connected to this second battery it gives a deflection of 7.5. Find the E.M.F. and internal resistance of the second battery.

72. Six Daniell cells, each having an E.M.F. of 1.08 volts and an internal resistance of 2 ohms, are to be connected to a circuit of 6 ohms resistance by wires of negligible resistance. What is the current through the external circuit (*a*) when all the cells are in series? (*b*) when all the cells are in parallel? (*c*) when they are arranged in three groups in parallel, each group containing two cells in series? (*d*) when arranged in two groups in parallel, each containing three cells in series? What is the P.D. at the terminals of the battery thus formed for each of the above arrangements?

73. If battery cells of E.M.F. 2.16 volts and internal resistance 1 ohm each are connected as in Problem 72, for which arrangement is the current in the external circuit a maximum, and what is its value?

74. A 15 per cent solution of copper sulphate has a conductivity of 39×10^{-12} in C.G.S. units. Calculate the current due to a P.D. of 2 volts across the terminals of an electrolytic cell of this solution if the electrodes are flat plates 20×30 cm. and 3 cm. apart.

75. The resistance of an electrolytic cell similar to that of Problem 74 is 1.94 ohms at 18° C. and 0.80 at 100° C. Find the per cent increase in conductivity per degree.

76. Three Daniell cells of E.M.F. 1.1 volts and internal resistance 1.4 ohms each are connected in series with a storage battery of unknown internal resistance by leads of 0.3 ohm. The current observed is 1.17 amperes. The storage battery terminals are then reversed and a current of 0.26 ampere is observed to flow. What is the E.M.F. of the storage battery, and what P.D. would a voltmeter across its terminals indicate for each connection?

CHAPTERS XII AND XIII

77. The field intensity in the gap space between the poles and the armature of a dynamo is 5000 units. Wires 25.5 cm. in length and 11 cm. from the axis of the armature are imbedded in the iron of the armature core. The armature makes 1400 revolutions per minute. Find the E.M.F. in volts induced in each armature conductor as it passes under a pole face.

78. A copper disk 20 cm. in diameter rotates 30 times per second in a uniform field of intensity 700 units. What E.M.F. is induced between the center and the circumference of the disk?

79. A coil of wire of 30 turns and average area 15 sq. cm., carrying a current of 2 amperes, is suspended from the beam of a sensitive balance so that the plane of the coil is horizontal. If the intensity of the earth's magnetic field is 0.57 and the angle of dip is 63°, what is the difference in weight of the coil produced by reversing the direction of the current passing through it?

80. A horizontal wire 90 cm. long lies in a northeast direction. How many lines of force does it cut if moved vertically upward for 200 cm.? The value of H is 0.26 and the value of V is 0.51.

CHAPTERS XIV AND XV

81. The core of an induction coil carries 100,000 lines of magnetic flux when a given current flows in the primary of the coil. Allowing 0.004 sec. for the flux to decrease to 10,000 lines after the primary current has been interrupted, how many turns of wire are necessary on the secondary of the coil in order that an average E.M.F. of 20,000 volts may be induced?

82. A coil of inductance 0.025 henry and 1600 turns of wire has how many lines of force per average turn for a current of 35 amperes?

83. An iron ring of 2.5 sq. cm. cross section and 20 cm. circumference has a primary of 80 turns and a secondary of 4600 turns. If the permeability of the iron is 1600 for a current of 1 ampere, find the E.M.F. induced in the secondary by reversing the primary current in 0.08 sec.

84. An iron ring of 0.75 cm. radius has a test coil wound on it of 50 turns of wire, which is connected to a ballistic galvanometer. The ballistic galvanometer circuit has a resistance of 300 ohms, and the galvanometer throws 9 scale divisions when 0.000072 coulomb passes through it. What throw is caused when the induction in the iron ring changes suddenly from 2400 lines in one direction to 3600 in the opposite direction?

CHAPTERS XVII AND XVIII

85. Amagat found that a volume of a cubic centimeter of alcohol at 14° C. was decreased by .000101 cc. for each atmosphere increase in the pressure to which it was subjected. If the density of alcohol is 0.79, what is the velocity of sound in this medium?

86. Assuming that the aural impression of a sound persists for 0.1 sec., calculate the distance of a person speaking 4 syllables per second from a reflecting surface in order that the echo may be distinct. Temperature 20° C.

87. Find the change in pitch observed by a person standing on a railway platform from which a locomotive with whistle blowing is receding at the uniform rate of 30 miles per hour.

88. Notes of 225 and 336 vibrations per second are sounded simultaneously. If the even overtones only are present in the first note, and both odd and even overtones in the second note, how many beats per second will occur, and to what overtones will they be due?

89. Two musicians stationed some distance apart are playing slightly out of tune, so that 4 beats per second are noticeable. How fast must a person travel from one toward the other in order that no beats are noticeable? Calculate for notes of 256 and 384 vibrations per second at a temperature of 0° C.

90. If the velocity of a compressional wave in a gas is 320 m. per second at 20° C., what will be the velocity at 50° and twice the pressure?

91. How much must an organ pipe be heated from 0° C. in order that the note may be changed by a semitone?

92. A whistle blown normally with air is blown with hydrogen of density 0.0692 as compared to air at the same temperature and pressure. What change is thus produced, and what is the interval between the notes?

CHAPTERS XIX AND XX

93. A brass rod 2 m. long stroked longitudinally is in tune with a 25-cm. length of a given sonometer wire. A steel rod 3 m. long stroked longitudinally is in tune with a 26-cm. length of the same sonometer wire. Find the relative velocities of sound in steel and brass.

94. The *e* string of a violin has a mass of 0.125 g. and a length of 33 cm. What is the tension in kilograms weight which it exerts when tuned to 640 vibrations? If the *g*, *d*, and *a* strings were to be made out of the same material and stretched by the same force, what would be the ratio of their diameters to that of the *e* string?

CHAPTERS XXI AND XXII

95. The theoretical limit of resolution of a lens with a circular aperture is 1.22 times the wave length of light divided by the diameter of the aperture. Mizar, the larger of the two stars at the bend of the handle of the Great Dipper, is a double star. Its two components are separated by 14.5″ of arc. What is the smallest aperture of telescope that can resolve this doublet?

96. If the distance between the images of the third order produced by a transmission grating placed 3 m. from a sodium flame is 40 cm., what is the number of lines per centimeter in the grating?

97. The two *D* lines (sodium) are separated by an angle of 53″ of arc in the first-order spectrum of a plane diffraction grating. What is the grating space?

CHAPTERS XXIII AND XXIV

98. If the absolute index of refraction of glass is 1.55 and of water 1.33, what is the minimum deviation produced in a beam of light by a prism of glass immersed in water? The angle of the prism is 60°.

99. The deviation produced by a prism with a small angle is commonly written $D = (\mu - 1)A$. Justify this equation.

100. When a layer of liquid 5 cm. deep is placed over a dot on a glass plate, the position of the dot as found by changing the focus of a microscope is 1.45 cm. above the plate. What is the index of refraction of the liquid?

101. From the law "angle of incidence equals angle of reflection" deduce the fact that a rotating mirror turns through one half the angle through which the reflected ray is rotated.

102. What is the focal length of a lens which has conjugate focal lengths of 25 cm. and 35 cm.?

103. What is the length of the real image of a line 4 cm. long and distant 40 cm. from a lens of focal length 25 cm.?

104. If a concave mirror has a curvature of 0.05, what is its focal length? How far from the mirror will be formed the real image of an object distant 35 cm. from the mirror? What will be its size relative to that of the object?

105. The lens L forms at p' a real image of p (Fig. 5). When L is moved to L' it again forms a real image of p at p'. If LL' is 60 cm. and pp' is 110 cm., what is the focal length of lens used?

FIG. 5

106. Parallel rays from the sun passing through the openings a and b (Fig. 6) in a screen placed before a concave lens L illuminate a second screen at c and d. The distance of this second screen from the lens is 15 cm. The distances ab and cd are 3 cm. and 7.5 cm. respectively. Find the focal length of the lens.

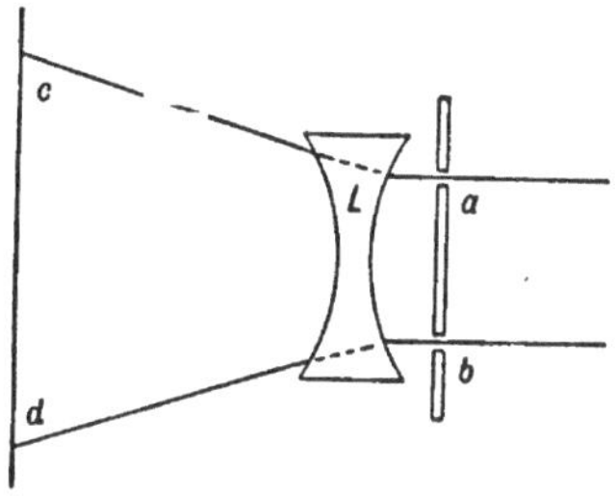

FIG. 6

107. A combination of a double concave lens and a more powerful convex lens of focal length 15 cm. has a focal length of 80 cm. Find the focal length of the concave lens.

108. An object A is placed 50 cm. from a concave lens L of unknown focal length (Fig. 7). A concave mirror M placed upon the opposite side of the lens forms a real image of the object at B, at a distance of 45 cm. from the mirror. The focal length of the mirror is 30 cm., and the distance from the mirror to the lens is 65 cm. Find the focal length of the lens.

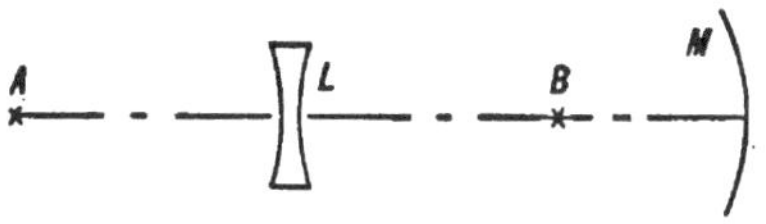

FIG. 7

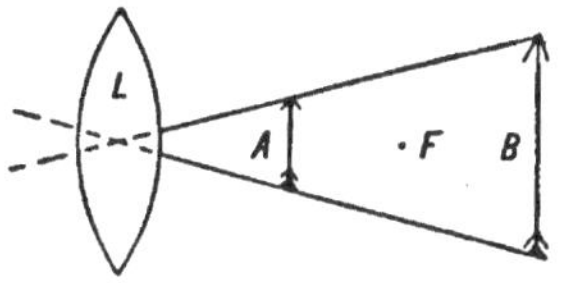

FIG. 8

109. Figure 8 shows the relative position of an object A, nearer to the lens L than its principal focus F, and its virtual image B. If A is one third the focal length from F, what are the relative sizes of A and B?

CHAPTERS XXV AND XXVI

110. At what distance from a photometer must a Hefner lamp be placed in order that it may produce the same intensity of illumination as a standard candle distant 85 cm. from the photometer?

111. Neglecting the motion of the earth, find an expression for the apparent wave length λ' of light of wave length λ from a distant star moving toward the earth with a velocity of s. Hence write an expression for the velocity of the star in the line of sight. From this equation find the velocity in the line of sight of a mass of hydrogen in the neighborhood of a sun spot that gave in the spectrometer an apparent wave length of 6566×10^{-8} instead of 6563×10^{-8}, which corresponded for a stationary source to the dark band observed in the solar spectrum. (See Young, "Manual of Astronomy," pp. 231–233.)

112. If to an observer on the earth, who is by the earth's motion moving through space at the rate of 18.5 miles per second, light from a distant star seems to come in a direction making an angle of 20.5″ with the true direction of the star, find the velocity of light in miles per hour. $\sin 20.5'' = 0.005842$, $\cos 20.5'' = 0.999987$. (See Young, "Manual of Astronomy," p. 150.)

CHAPTERS XXVII AND XXVIII

113. At what angle must a beam of light be incident upon a smooth water surface in order that the reflected beam shall be plane polarized?

114. How thick should be a piece of quartz with its parallel faces cut perpendicular in order that plane polarized light incident upon it shall, after transmission, have the same plane of vibration as before?

115. The index of refraction of the ordinary ray in Iceland spar is 1.6543. The index of Canada balsam is 1.536. At what angle must the ordinary ray in a Nicol prism meet the interface in order that it may just suffer total reflection?

116. If an α particle projected from uranium produces on the average 100,000 positive ions, how many α particles are required to produce the ionization current given in the example at the end of Experiment 28?

TABLES

Table 1

SATURATED WATER VAPOR

Showing pressure P (in mm. of mercury) and density D of aqueous vapor saturated at temperature t; or showing boiling point t of water and density D of steam corresponding to an outside pressure P.

t	P	D	t	P	D	t	P	D
—10	2.2	2.3×10^{-6}	30	31.5	30.1×10^{-6}	88.5	496.2	
— 9	2.3	2.5 "	35	41.8	39.3 "	89	505.8	
— 8	2.5	2.7 "	40	54.9	50.9 "	89.5	515.5	
— 7	2.7	2.9 "	45	71.4	65.3 "	90	525.4	428.4×10^{-6}
— 6	2.9	3.2 "	50	92.0	83.0 "	90.5	535.5	
— 5	3.2	3.4 "	55	117.5	104.6 "	91	545.7	
— 4	3.4	3.7 "	60	148.8	130.7 "	91.5	556.1	
— 3	3.7	4.0 "	65	187.0	162.1 "	92	566.7	
— 2	3.9	4.2 "	70	233.1	199.5 "	92.5	577.4	
— 1	4.2	4.5 "	71	243.6		93	588.3	
0	4.6	4.9 "	72	254.3		93.5	599.6	
1	4.9	5.2 "	73	265.4		94	610.6	
2	5.3	5.6 "	74	276.9		94.5	622.0	
3	5.7	6.0 "	75	288.8	243.7 "	95	633.6	511.1 "
4	6.1	6.4 "	75.5	294.9		95.5	645.4	
5	6.5	6.8 "	76	301.1		96	657.4	
6	7.0	7.3 "	76.5	307.4		96.5	669.5	
7	7.5	7.7 "	77	313.8		97	681.8	
8	8.0	8.2 "	77.5	320.4		97.5	694.2	
9	8.5	8.7 "	78	327.1		98	707.1	
10	9.1	9.3 "	78.5	333.8		98.2	712.3	
11	9.8	10.0 "	79	340.7		98.4	717.4	
12	10.4	10.6 "	79.5	347.7		98.6	722.6	
13	11.1	11.2 "	80	354.9	295.9 "	98.8	727.9	
14	11.9	12.0 "	80.5	362.1		99	733.2	
15	12.7	12.8 "	81	369.5		99.2	738.5	
16	13.5	13.5 "	81.5	377.0		99.4	743.8	
17	14.4	14.4 "	82	384.6		99.6	749.2	
18	15.3	15.2 "	82.5	392.4		99.8	754.7	
19	16.3	16.2 "	83	400.3		100	760.0	606.2 "
20	17.4	17.2 "	83.5	408.3		100.2	765.5	
21	18.5	18.2 "	84	416.5		100.4	771.0	
22	19.6	19.3 "	84.5	424.7		100.6	776.5	
23	20.9	20.4 "	85	433.2	357.1 "	100.8	782.1	
24	22.2	21.6 "	85.5	441.7		101	787.7	
25	23.5	22.9 "	86	450.5		102	816.0	
26	25.0	24.2 "	86.5	459.3		103	845.3	
27	26.5	25.6 "	87	468.3		105	906.4	715.4 "
28	28.1	27.0 "	87.5	477.4		107	971.1	
29	29.7	28.5 "	88	486.8		110	1075.4	840.1 "

Table 2

DENSITY OF DRY AIR AT TEMPERATURE t AND PRESSURE H mm. OF MERCURY

t	$H = 720$	730	740	750	760	770	DIFFERENCE PER mm.
10°	.001181	.001198	.001214	.001231	.001247	.001263	**16**
11	1177	1194	1210	1226	1243	1259	1 2
12	1173	1189	1206	1222	1238	1255	2 3
13	1169	1185	1202	1218	1234	1250	3 5
14	1165	1181	1197	1214	1230	1246	4 6
15°	.001161	.001177	.001193	.001209	.001225	.001242	5 8
16	1157	1173	1189	1205	1221	1237	6 10
17	1153	1169	1185	1201	1217	1233	7 11
18	1149	1165	1181	1197	1213	1229	8 13
19	1145	1161	1177	1193	1209	1224	9 14
20°	.001141	.001157	.001173	.001189	.001204	.001220	**15**
21	1137	1153	1169	1185	1200	1216	1 2
22	1133	1149	1165	1181	1196	1212	2 3
23	1130	1145	1161	1177	1192	1208	3 4
24	1126	1141	1157	1173	1188	1204	4 6
25°	.001122	.001138	.001153	.001169	.001184	.001200	5 7
26	1118	1134	1149	1165	1180	1196	6 9
27	1114	1130	1145	1161	1176	1192	7 10
28	1110	1126	1142	1157	1172	1188	8 12
29	1107	1122	1138	1153	1169	1184	9 13
30°	.001103	.001119	.001134	.001149	.001165	.001180	

Correction for Moisture in Above Table

Dew-point	Subtract	Dew-point	Subtract	Dew-point	Subtract	Dew-point	Subtract
—10°	.000001	0°	.000003	+10°	.000006	+20°	.000010
— 8	.000002	+2	.000003	+12	.000006	+22	.000012
— 6	.000002	+4	.000004	+14	.000007	+24	.000013
— 4	.000002	+6	.000004	+16	.000008	+26	.000015
— 2	.000003	+8	.000005	+18	.000009	+28	.000016

Table 3

DENSITY OF WATER

TEMP. C°	DENSITY	TEMP. C°	DENSITY
0°	0.999884	13°	0.999443
1	0.999941	14	0.999312
2	0.999982	15	0.999173
3	1.000004	16	0.999015
3.95	1.000000	17	0.998854
4	1.000013	18	0.998667
5	1.000003	19	0.998473
6	0.999983	20	0.998272
7	0.999946	22	0.997839
8	0.999899	24	0.997380
9	0.999837	26	0.996879
10	0.999760	28	0.996344
11	0.999668	30	0.995778
12	0.999562	100	0.958860

Table 4

DENSITY OF MERCURY

TEMP. C°	DENSITY
0°	13.596
10	13.572
12	13.567
14	13.562
16	13.557
18	13.552
20	13.547
22	13.542
24	13.537
26	13.532
28	13.528
30	13.523
32	13.518
34	13.513

Table 5

DENSITIES AND ELECTRICAL CONDUCTIVITIES AT 18° C. OF NORMAL SOLUTIONS (FROM KOHLRAUSCH)

A = equivalent weight (O = 16.00) or concentration in gram equivalents per liter.
p = number of grams of substance to 100 grams of the solution.
d = density in grams per cubic centimeter at 18° C.
σ = electrical conductivity, i.e. reciprocal of specific resistance in ohms per cubic centimeter
$\Delta\sigma$ = relative increase in σ per degree in neighborhood of 18° C.

	A	p	d	$10^3\sigma$	$\frac{\Delta\sigma}{\sigma}$
KOH	56.16	5.359	1.0479	184.0	.0186
KCl	74.60	7.139	1.0449	98.3	.0193
KNO_3	101.19	9.544	1.0602	80.5	.0200
NH_4Cl	53.52	5.271	1.0153	97 0	.0194
NaOH	40.06	3.844	1.0420	160.0	.0197
NaCl	58.50	5.629	1.0392	74.3	.0212
$NaC_2H_3O_2$	82.05	7.897	1.0400	41.2	.0250
$\frac{1}{2}Na_2SO_4$	71.08	6.703	1.0604	50.8	.0236
LiCl	42.48	4.157	1.0226	63.4	.0220
$\frac{1}{2}CaCl_2$	55.45	5.313	1.0436	67.8	.0207
$\frac{1}{2}ZnCl_2$	68.15	6.442	1.0578	55.0	.0220
$\frac{1}{2}ZnSO_4$	80.73	7.483	1.0789	26.6	.0220
$\frac{1}{2}CuSO_4$	79.83	7.408	1.0776	25.8	.0220
$AgNO_3$	169.97	14.91	1.1400	67.8	.0210
HCl	36.46	3.587	1.0165	300.0	.0159
HNO_3	63.05	6.107	1.0325	299.0	.0150
$\frac{1}{2}H_2SO_4$	49.04	4.758	1.0307	197.0	.0120
$C_2H_4O_2$	60.03	5.959	1.0074	—	—
Sugar	342.2	30.30	1.1294	—	—

Table 6

VISCOSITIES OF AQUEOUS SOLUTIONS OF VARYING CONCENTRATIONS

If z is the viscosity of a dilute solution in terms of the viscosity of water, and x the concentration in gram molecules per liter, then, according to Arrhenius,

$$z = A^x.$$

A, therefore, represents the viscosity of a normal solution divided by the viscosity of pure water.

The values in the following table refer to viscosities at 25° C.

SUBSTANCE	A	SUBSTANCE	A
Alcohol $\{C_2H_5OH\}$	1.030	Potassium Sulphate $\{K_2SO_4\}$	1.0982
Cane Sugar $\{C_{12}H_{22}O_{11}\}$	1.046	Sodium Acetate $\{Na_2C_2H_3O_2\}$	1.3998
Copper Nitrate $\{Cu(NO_3)_2\}$	1.1729	Sodium Chloride $\{NaCl\}$	1.0986
Copper Sulphate $\{CuSO_4\}$	1.3533	Sodium Nitrate $\{NaNO_3\}$	1.0522
Ether $\{(C_2H_5)_2O\}$	1.026	Sodium Sulphate $\{Na_2SO_4\}$	1.2253
Hydrochloric Acid $\{HCl\}$	1.0699	Zinc Nitrate $\{Zn(NO_3)_2\}$	1.1666
Nitric Acid $\{HNO_3\}$	1.0233	Zinc Sulphate $\{ZnSO_4\}$	1.3613
Potassium Nitrate $\{KNO_3\}$	0.9664		

Table 7

ATOMIC WEIGHTS (O = 16.00)

(International 1904)

Aluminum	Al	27.1	Neodymium	Nd	143.6
Antimony	Sb	120.2	Neon	Ne	20.
Argon	A	39.9	Nickel	Ni	58.7
Arsenic	As	75.0	Nitrogen	N	14.04
Barium	Ba	137.4	Osmium	Os	191.
Bismuth	Bi	208.5	Oxygen	O	16.00
Boron	B	11.	Palladium	Pd	106.5
Bromine	Br	79.96	Phosphorus	P	31.0
Cadmium	Cd	112.4	Platinum	Pt	194.8
Cæsium	Cs	132.9	Potassium	K	39.15
Calcium	Ca	40.1	Praseodymium	Pr	140.5
Carbon	C	12.00	Radium	Rd	225.
Cerium	Ce	140.25	Rhodium	Rh	103.0
Chlorine	Cl	35.45	Rubidium	Rb	85.4
Chromium	Cr	52.1	Ruthenium	Ru	101.7
Cobalt	Co	59.0	Samarium	Sm	150.
Columbium	Cb	94.	Scandium	Sc	44.1
Copper	Cu	63.6	Selenium	Se	79.2
Erbium	Er	166.	Silicon	Si	28.4
Fluorine	F	19.	Silver	Ag	107.93
Gadolinium	Gd	156.	Sodium	Na	23.05
Gallium	Ga	70.	Strontium	Sr	87.6
Germanium	Ge	72.5	Sulphur	S	32.06
Glucinum	Gl	9.1	Tantalum	Ta	183.
Gold	Au	197.2	Tellurium	Te	127.6
Helium	He	4.	Terbium	Tb	160.
Hydrogen	H	1.008	Thallium	Tl	204.1
Indium	In	114.	Thorium	Th	232.5
Iodine	I	126.85	Thulium	Tm	171.
Iridium	Ir	193.0	Tin	Sn	119.0
Iron	Fe	55.9	Titanium	Ti	48.1
Krypton	Kr	81.8	Tungsten	W	184.0
Lanthanum	La	138.9	Uranium	U	238.5
Lead	Pb	206.9	Vanadium	V	51.2
Lithium	Li	7.03	Xenon	Xe	128.
Magnesium	Mg	24.36	Ytterbium	Yb	173.0
Manganese	Mn	55.0	Yttrium	Yt	89.0
Mercury	Hg	200.0	Zinc	Zn	65.4
Molybdenum	Mo	96.0	Zirconium	Zr	90.6

Table 8

ELECTRO-CHEMICAL EQUIVALENTS

SUBSTANCE	VALENCY	GRAMS PER COULOMB	SUBSTANCE	VALENCY	GRAMS PER COULOMB
Aluminum .	3	.0009357	Magnesium .	2	.0001261
Bromine . .	1	.0008276	Manganese .	2	.0002849*
Cadmium .	2	.000582*	Mercury . .	1	.002075
Chlorine . .	1	.0003672	Mercury . .	2	.001037
Cobalt . .	2	.0003056*	Nickel . . .	2	.0003042
Copper . .	1	.0006588	Nitrogen . .	3	.0000485*
Copper . .	2	.0003294	Oxygen . .	2	.0000829
Fluorine .	1	.0001968*	Platinum . .	4	.000500*
Gold . . .	3	.000681	Potassium .	1	.0004054
Hydrogen .	1	.0000104	Silver . . .	1	.001118
Iodine . .	1	.001314	Sodium . .	1	.0002387
Iron . . .	2	.0002895	Tin	2	.0006163
Iron . . .	3	.0001930	Tin	4	.0003081
Lead . . .	2	.001072	Zinc	2	.0003387

* Starred values are calculated.

Table 9

SPECIFIC RESISTANCES AND TEMPERATURE COEFFICIENTS

Metals

SUBSTANCE	RESISTANCE AT 0° C. IN OHMS PER CENTIMETER CUBE	MEAN TEMPERATURE COEFFICIENT 0° – 100° C.
Aluminum	2.906×10^{-6}	.00435
Antimony	35.42×10^{-6}	
Bismuth	130.9×10^{-6}	
Copper (annealed)	1.584×10^{-6}	.0042
Copper (hard drawn)	1.619×10^{-6}	
Iron (annealed)	9.693×10^{-6}	.00625
Iron (hard drawn)	$15. \times 10^{-6}$	
Gold	2.088×10^{-6}	.00377
Mercury	94.34×10^{-6}	.0009
Nickel	12.35×10^{-6}	.00622
Platinum	9.035×10^{-6}	.00367
Silver	1.561×10^{-6}	.00400
Tin	10.5×10^{-6}	.00440
Zinc	5.75×10^{-6}	.00406
German Silver	20.89×10^{-6}	.00027
Platinum Silver (Pt 33%, Ag 66%) . .	31.6×10^{-6}	
Platinum Iridium (Pt 80%, Ir 20%) .	30.9×10^{-6}	.00082
Platinum Rhodium (Pt 90%, Rh 10%) .	21.1×10^{-6}	.00143
Manganin (Cu 84%, Mn 12%, Ni 4%) .	46.7×10^{-6}	.0000

Table 9 (*continued*)

Electrolytes

R = resistance in ohms per centimeter cube; α = per cent decrease in resistance per degree Centigrade; solutions given in per cent by weight of salts free from water of crystallization

	5%		10%		15%	
	R	α	R	α	R	α
KCl	14.5	2.	7.34	1.9	4.95	1.9
NaCl	14.9	2.2	8.27	2.1	6.10	2.1
NH_4Cl	10.9	2.	5.62	1.9	3.86	1.7
$CuSO_4$	52.6	2.2	31.2	2.2	23.8	2.3
$ZnSO_4$	52.6	2.2	31.3	2.2	23.8	2.2
$MgSO_4$	38.4	2.3	24.4	2.4	20.8	2.5
$AgNO_3$	38.5	2.2	20.8	2.2	14.7	2.2
KOH	5.81	1.9	3.17	1.9	2.35	1.9
HCl	2.53	1.58	1.59	1.56	1.34	1.55
HNO_3	3.88	1.5	2.17	1.45	1.63	1.4
H_2SO_4	4.78	1.21	2.55	1.28	1.84	1.36

Table 10

DIELECTRIC CONSTANTS

Air	1.00	Quartz	4.5
Alcohol	26.5	Shellac	2.7 – 3.7
Ebonite	2 – 3	Sulphur	2. – 4.
Common glass	3 – 5	Benzol	2.2 – 2.3
India rubber	2.2 – 2.7	Turpentine	2.2
Mica	2. – 8.	Water	81.
Paraffin	1.9 – 2.	Gases	.9998 – 1.015
Petroleum oil	2.1 – 2.2	Vacuum	0.9985

Table 11

YOUNG'S MODULUS

Approximate Values in Dynes per Square Centimeter

Cast iron	12×10^{11}
Brass wire	10×10^{11}
Copper wire	12×10^{11}
Steel wire	$19 - 20 \times 10^{11}$
Aluminum	6.5×10^{11}
Nickel	20×10^{11}
Silver	7.3×10^{11}
Glass	6.5×10^{11}
Oak (cut longitudinally)	0.9×10^{11}
Pine (cut longitudinally)	$0.5 - 0.6 \times 10^{11}$

Table 12

DENSITIES

Solids

Aluminum	2.58	Iron (wrought)	7.86
Bismuth	9.80	Lead	11.3
Brass (about)	8.5	Nickel	8.9
Brick	2.1	Oak	0.8
Copper	8.92	Pine	0.5
Cork	0.24	Platinum	21.50
Diamond	3.52	Quartz	2.65
Glass (common crown)	2.6	Silver	10.53
" (flint)	3.0–6.3	Sugar	1.6
Gold	19.3	Sulphur	2.07
Ice at 0° C.	0.9167	Tin	7.29
Iron (cast)	7.4	Zinc	7.15

Liquids

Alcohol at 20° C.	0.789	Gasoline	.79
Carbon bisulphide	1.29	Mercury	13.596
Ethyl ether at 0° C.	0.735	Sulphuric acid	1.85
Glycerin	1.26	Hydrochloric acid	1.27
Turpentine	0.87	Nitric acid	1.56
Benzol	.88	Olive oil	0.91

Gases at 0° C. 76 cm. of Mercury Pressure

Acetylene (C_2H_2)	0.001185	Hydrogen (H_2)	0.0000895
Ammonia (NH_3)	0.000770	Marsh gas (CH_4)	0.000715
Carbon monoxide (CO)	0.001252	Nitrogen (N_2)	0.001257
Carbon dioxide (CO_2)	0.001974	Oxygen (O_2)	0.001430

Table 13

AVERAGE SPECIFIC HEATS

Alcohol	at 40°	0.648	Lead	0° – 100°	0.0315
Aluminum	0° – 100°	0.2185	Magnesium		0.251
Bismuth		0.0303	Mercury	25° – 50°	0.0333
Brass		0.094	Nickel	10° – 100°	0.1128
Copper	0° – 100°	0.095	Paraffin		0.683
Carbon bisulphide	40°	0.2429	Petroleum	21° – 58°	0.511
Ebonite		0.33	Platinum	0° – 100°	0.0323
German silver		0.0946	Quartz	20° – 100°	0.19
Glass		0.20	Silver	0° – 100°	0.0568
Gold		0.0316	Steel		0.118
Graphite	11°	0.160	Tin	0° – 100°	0.0559
Ice		0.504	Turpentine		0.467
Iron	0° – 100°	0.1130	Zinc		0.0935

Table 14

WAVE LENGTH IN MICRONS (μ) OR THOUSANDTHS MILLIMETERS

FRAUNHOFER LINE	SOURCE	WAVE LENGTH IN μ	FRAUNHOFER LINE	SOURCE	WAVE LENGTH IN μ
	Potassium	.7699	b_1	Magnesium	.5184
	Potassium	.7665	b_2	Magnesium	.5173
A	Sun	.7604	b_3	Iron	.5169
a	Sun	.7185	b_4	Magnesium, iron	.5168
B	Oxygen	.6870		Cadmium	.5086
	Lithium	.6708		Barium	.4934
C	Hydrogen	.6563		Calcium	.4878
	Strontium	.6550	F	Hydrogen	.4861
	Calcium	.6499		Strontium	.4607
	Cadmium	.6438		Barium	.4554
	Strontium	.6408		Mercury	.4358
α	Oxygen	.6278	f	Hydrogen	.4340
	Mercury	.6152	G'	Iron	.4326
	Lithium	.6104	G	Iron, calcium	.4308
D_1	Sodium	.5896		Calcium	.4227
D_2	Sodium	.5890	h	Hydrogen	.4102
D_3	Helium	.5876		Mercury	.4078
	Mercury	.5790		Mercury	.4047
	Mercury	.5769		Potassium, iron	.4046
	Mercury	.5461	H	Hydrogen, calcium	.3968
	Cadmium	.5379			
	Cadmium	.5338			
E	Iron, calcium	.5270			
	Iron	.5270			

Table 15

REFRACTIVE INDICES FOR DIFFERENT COLORS

		RED (C)	YELLOW (D)	BLUE (F)
Water		1.3317	1.3335	1.3377
Alcohol		1.3606	1.3624	1.3667
Carbon bisulphide		1.6198	1.6293	1.6541
Crown glass	light	1.5127	1.5153	1.5214
	heavy	1.6126	1.6152	1.6213
Flint glass	light	1.6038	1.6085	1.6200
	heavy	1.7434	1.7515	1.7723
Iceland spar	ordinary ray	1.6545	1.6585	1.6679
	extraordinary ray	1.4846	1.4864	1.4908
Quartz	ordinary ray	1.5418	1.5442	1.5496
	extraordinary ray	1.5509	1.5533	1.5589

Table 16

REDUCTION OF BAROMETRIC HEIGHT TO 0° C.

(The table corrections represent the number of millimeters to be subtracted from the observed height h. They are obtained from the formula $(.000181 - .000019)ht$, the first number being the cubical expansion coefficient of mercury, the second the linear coefficient of brass.)

t	OBSERVED HEIGHT IN mm.									
	680	690	700	710	720	730	740	750	760	770
	mm.	mm.	mm.	mm.	mm.	mm.	mm.	mm.	mm.	mm.
10°	1.10	1.12	1.13	1.15	1.17	1.18	1.20	1.22	1.23	1.25
11	1.21	1.23	1.25	1.27	1.28	1.30	1.32	1.34	1.35	1.37
12	1.32	1.34	1.36	1.38	1.40	1.42	1.44	1.46	1.48	1.50
13	1.43	1.45	1.47	1.50	1.52	1.54	1.56	1.58	1.60	1.62
14	1.54	1.56	1.59	1.61	1.63	1.66	1.68	1.70	1.72	1.75
15	1.65	1.68	1.70	1.73	1.75	1.77	1.80	1.82	1.85	1.87
16	1.76	1.79	1.81	1.84	1.87	1.89	1.92	1.94	1.97	2.00
17	1.87	1.90	1.93	1.96	1.98	2.01	2.04	2.07	2.09	2.12
18	1.98	2.01	2.04	2.07	2.10	2.13	2.16	2.19	2.22	2.25
19	2.09	2.12	2.15	2.19	2.22	2.25	2.28	2.31	2.34	2.37
20	2.20	2.24	2.27	2.30	2.33	2.37	2.40	2.43	2.46	2.49
21	2.31	2.35	2.38	2.42	2.45	2.48	2.52	2.55	2.59	2.62
22	2.42	2.46	2.49	2.53	2.57	2.60	2.64	2.67	2.71	2.74
23	2.53	2.57	2.61	2.65	2.68	2.72	2.76	2.79	2.83	2.87
24	2.64	2.68	2.72	2.76	2.80	2.84	2.88	2.92	2.95	2.99
25	2.75	2.79	2.84	2.88	2.92	2.96	3.00	3.04	3.08	3.12

Table 17

USEFUL CONSTANTS AND RELATIONS

$\pi = 3.1416$. $\pi^2 = 9.8696$, $\frac{1}{\pi} = 0.31831$. logarithm $\pi = .49715$.

Naperian base $e = 2.7183$. $\log_{10} e = .43429$. $\log_e N = \frac{\log_{10} N}{\log_{10} e} = 2.3026 \log_{10} N$.

1 inch = 25.4 millimeters. 1 meter = 39.37 inches. 1 mile = 1.609 kilometers. 1 kilogram = 2.2 pounds. 1 pound = 453.59 grams. 1 ounce = 28.35 grams. 1 grain = 64.8 milligrams. Mechanical equivalent of 1 calorie (15°) = 4.19×10^7 ergs.

1 horse power = 746 watts = 33000 foot pounds per minute.

1 watt = 1 joule per second. 1 joule = 10^7 ergs = 0.7373 foot pounds.

1 radian = 57.3 degrees. 180° = π radians.

Half diameter of the earth, equatorial, 6378.2 kilometers; polar, 6356.5 kilometers; mean, 6367.4 kilometers.

Mean density of the earth, 5.5270.

Table 18

Resistances of Copper and of German Silver Wire

Brown and Sharp Gauge

Number	Diameter in Mils ($\frac{1}{1000}$ in.)	Pure Copper Ohms per 1000 ft.	18% German Silver Ohms per 1000 ft.
000	409.64	.064	——
00	364.80	.081	——
0	324.95	.102	——
1	289.30	.129	——
2	257.63	.163	——
3	229.42	.205	——
4	204.31	.259	——
5	181.94	.326	——
6	162.02	.411	——
7	144.28	.519	——
8	128.49	.654	——
9	114.43	.824	11.832
10	101.89	1.040	18.720
11	90.74	1.311	23.598
12	80.81	1.653	29.754
13	71.96	2.084	37.512
14	64.08	2.628	47.304
15	57.07	3.314	59.652
16	50.82	4.179	75.222
17	45.26	5.269	94.842
18	40.30	6.645	119.610
19	35.89	8.617	155.106
20	31.96	10.566	190.188
21	28.46	13.323	239.814
22	25.35	16.799	302.382
23	22.57	21.185	381.330
24	20.10	26.713	480.834
25	17.90	33.684	606.312
26	15.94	42.477	764.586
27	14.20	53.563	964.134
28	12.64	67.542	1215.756
29	11.26	85.170	1533.060
30	10.03	107.391	1933.038
31	8.93	135.402	2437.236
32	7.95	170.765	3073.770
33	7.08	215.312	3875.616
34	6.30	271.583	4888.494
35	5.61	342.443	6163.974
36	5.00	431.712	7770.816
37	4.45	544.287	9797.166
38	3.97	686.511	12357.198
39	3.53	865.046	15570.828
40	3.14	1091.865	19653.570

Table 19

RÉSUMÉ OF DEFINING EQUATIONS AND UNITS

MAGNITUDE	DEFINING EQUATION	SYMBOL FOR IN C.G.S. UNITS (E. M. SYSTEM)	RATIO E. M. UNIT TO E. S. UNIT	RATIO PRACTICAL UNIT TO C.G.S. UNIT	NAME
Magnet pole	$F = \frac{mM}{r^2}$ *	$\frac{g.^{\frac{1}{2}}\, cm.^{\frac{3}{2}}}{sec.}$			
Magnetic field strength	$\mathcal{H} = \frac{F}{m}$ *	$\frac{g^{\frac{1}{2}}}{cm.^{\frac{1}{2}}\, sec.}$			gauss
Magnetic flux . . .	$\Phi = \mathcal{H} a$ †	$\frac{g.^{\frac{1}{2}}\, cm.^{\frac{3}{2}}}{sec.}$			maxwell
Magnetic moment . .	$M = ml$	$\frac{g.^{\frac{1}{2}}\, cm.^{\frac{5}{2}}}{sec.}$			
Magnetization . . .	$\mathcal{J} = \frac{M}{V} = \frac{m}{a}$	$\frac{g.^{\frac{1}{2}}}{cm.^{\frac{1}{2}}\, sec.}$			
Magnetic induction .	$\mathcal{B} = \frac{\Phi}{a} = \mathcal{H} + 4\pi \mathcal{J}$ ‡	$\frac{g.^{\frac{1}{2}}}{cm.^{\frac{1}{2}}\, sec.}$			
Current	$F = IlH$	$\frac{g.^{\frac{1}{2}}\, cm.^{\frac{1}{2}}}{sec.}$	v §	10^{-1}	ampere
Quantity	$Q = It$	$g.^{\frac{1}{2}}\, cm.^{\frac{1}{2}}$	v	10^{-1}	coulomb
Potential	$PD = \frac{W}{Q}$	$\frac{g.^{\frac{1}{2}}\, cm.^{\frac{3}{2}}}{sec.^2}$	v^{-1}	10^8	volt
Resistance	$I = \frac{PD}{R}$	$\frac{cm.}{sec.}$	v^{-2}	10^9	ohm
Capacity	$C = \frac{Q}{PD}$	$\frac{sec.^2}{cm.}$	v^2	10^{-9}	farad
Self-induction . . .	$\Phi = LI$	cm.		10^9	henry

* For air.
† For air, otherwise $\Phi = \mathcal{B} a$.
‡ $\mathcal{B} = \mathcal{H}$ for air, since $\mathcal{J} = 0$.
§ $v = 3 \times 10^{10}$.

NATURAL SINES

Angle	.0	.1	.2	.3	.4	.5	.6	.7	.8	.9	Complement		Difference
0°	0.0000	0017	0035	0052	0070	0087	0105	0122	0140	0157	0175	**89°**	
1	0175	0192	0209	0227	0244	0262	0279	0297	0314	0332	0349	**88**	
2	0349	0366	0384	0401	0419	0436	0454	0471	0488	0506	0523	**87**	
3	0523	0541	0558	0576	0593	0610	0628	0645	0663	0680	0698	**86**	
4	0698	0715	0732	0750	0767	0785	0802	0819	0837	0854	0872	**85**	
5	0.0872	0889	0906	0924	0941	0958	0976	0993	1011	1028	1045	**84**	
6	1045	1063	1080	1097	1115	1132	1149	1167	1184	1201	1219	**83**	
7	1219	1236	1253	1271	1288	1305	1323	1340	1357	1374	1392	**82**	
8	1392	1409	1426	1444	1461	1478	1495	1513	1530	1547	1564	**81**	
9	1564	1582	1599	1616	1633	1650	1668	1685	1702	1719	1736	**80**	
10	0.1736	1754	1771	1788	1805	1822	1840	1857	1874	1891	1908	**79**	
11	1908	1925	1942	1959	1977	1994	2011	2028	2045	2062	2079	**78**	
12	2079	2096	2113	2130	2147	2164	2181	2198	2215	2233	2250	**77**	17
13	2250	2267	2284	2300	2317	2334	2351	2368	2385	2402	2419	**76**	
14	2419	2436	2453	2470	2487	2504	2521	2538	2554	2571	2588	**75**	
15	0.2588	2605	2622	2639	2656	2672	2689	2706	2723	2740	2756	**74**	
16	2756	2773	2790	2807	2823	2840	2857	2874	2890	2907	2924	**73**	
17	2924	2940	2957	2974	2990	3007	3024	3040	3057	3074	3090	**72**	
18	3090	3107	3123	3140	3156	3173	3190	3206	3223	3239	3256	**71**	
19	3256	3272	3289	3305	3322	3338	3355	3371	3387	3404	3420	**70**	
20	0.3420	3437	3453	3469	3486	3502	3518	3535	3551	3567	3584	**69**	
21	3584	3600	3616	3633	3649	3665	3681	3697	3714	3730	3746	**68**	
22	3746	3762	3778	3795	3811	3827	3843	3859	3875	3891	3907	**67**	
23	3907	3923	3939	3955	3971	3987	4003	4019	4035	4051	4067	**66**	16
24	4067	4083	4099	4115	4131	4147	4163	4179	4195	4210	4226	**65**	
25	0.4226	4242	4258	4274	4289	4305	4321	4337	4352	4368	4384	**64**	
26	4384	4399	4415	4431	4446	4462	4478	4493	4509	4524	4540	**63**	
27	4540	4555	4571	4586	4602	4617	4633	4648	4664	4679	4695	**62**	
28	4695	4710	4726	4741	4756	4772	4787	4802	4818	4833	4848	**61**	
29	4848	4863	4879	4894	4909	4924	4939	4955	4970	4985	5000	**60**	
30	0.5000	5015	5030	5045	5060	5075	5090	5105	5120	5135	5150	**59**	15
31	5150	5165	5180	5195	5210	5225	5240	5255	5270	5284	5299	**58**	
32	5299	5314	5329	5344	5358	5373	5388	5402	5417	5432	5446	**57**	
33	5446	5461	5476	5490	5505	5519	5534	5548	5563	5577	5592	**56**	
34	5592	5606	5621	5635	5650	5664	5678	5693	5707	5721	5736	**55**	
35	0.5736	5750	5764	5779	5793	5807	5821	5835	5850	5864	5878	**54**	
36	5878	5892	5906	5920	5934	5948	5962	5976	5990	6004	6018	**53**	14
37	6018	6032	6046	6060	6074	6088	6101	6115	6129	6143	6157	**52**	
38	6157	6170	6184	6198	6211	6225	6239	6252	6266	6280	6293	**51**	
39	6293	6307	6320	6334	6347	6361	6374	6388	6401	6414	6428	**50**	
40	0.6428	6441	6455	6468	6481	6494	6508	6521	6534	6547	6561	**49**	
41	6561	6574	6587	6600	6613	6626	6639	6652	6665	6678	6691	**48**	13
42	6691	6704	6717	6730	6743	6756	6769	6782	6794	6807	6820	**47**	
43	6820	6833	6845	6858	6871	6884	6896	6909	6921	6934	6947	**46**	
44°	6947	6959	6972	6984	6997	7009	7022	7034	7046	7059	7071	**45°**	
Complement		.9	.8	.7	.6	.5	.4	.3	.2	.1	.0	Angle	

NATURAL COSINES

NATURAL SINES

Angle	.0	.1	.2	.3	.4	.5	.6	.7	.8	.9	Complement		Difference
45°	0.7071	7083	7096	7108	7120	7133	7145	7157	7169	7181	7193	44°	
46	7193	7206	7218	7230	7242	7254	7266	7278	7290	7302	7314	43	12
47	7314	7325	7337	7349	7361	7373	7385	7396	7408	7420	7431	42	
48	7431	7443	7455	7466	7478	7490	7501	7513	7524	7536	7547	41	
49	7547	7559	7570	7581	7593	7604	7615	7627	7638	7649	7660	40	
50	0.7660	7672	7683	7694	7705	7716	7727	7738	7749	7760	7771	39	
51	7771	7782	7793	7804	7815	7826	7837	7848	7859	7869	7880	38	11
52	7880	7891	7902	7912	7923	7934	7944	7955	7965	7976	7986	37	
53	7986	7997	8007	8018	8028	8039	8049	8059	8070	8080	8090	36	
54	8090	8100	8111	8121	8131	8141	8151	8161	8171	8181	8192	35	
55	0.8192	8202	8211	8221	8231	8241	8251	8261	8271	8281	8290	34	10
56	8290	8300	8310	8320	8329	8339	8348	8358	8368	8377	8387	33	
57	8387	8396	8406	8415	8425	8434	8443	8453	8462	8471	8480	32	
58	8480	8490	8499	8508	8517	8526	8536	8545	8554	8563	8572	31	
59	8572	8581	8590	8599	8607	8616	8625	8634	8643	8652	8660	30	9
60	0.8660	8669	8678	8686	8695	8704	8712	8721	8729	8738	8746	29	
61	8746	8755	8763	8771	8780	8788	8796	8805	8813	8821	8829	28	
62	8829	8838	8846	8854	8862	8870	8878	8886	8894	8902	8910	27	8
63	8910	8918	8926	8934	8942	8949	8957	8965	8973	8980	8988	26	
64	8988	8996	9003	9011	9018	9026	9033	9041	9048	9056	9063	25	
65	0.9063	9070	9078	9085	9092	9100	9107	9114	9121	9128	9135	24	
66	9135	9143	9150	9157	9164	9171	9178	9184	9191	9198	9205	23	7
67	9205	9212	9219	9225	9232	9239	9245	9252	9259	9265	9272	22	
68	9272	9278	9285	9291	9298	9304	9311	9317	9323	9330	9336	21	
69	9336	9342	9348	9354	9361	9367	9373	9379	9385	9391	9397	20	6
70	0.9397	9403	9409	9415	9421	9426	9432	9438	9444	9449	9455	19	
71	9455	9461	9466	9472	9478	9483	9489	9494	9500	9505	9511	18	
72	9511	9516	9521	9527	9532	9537	9542	9548	9553	9558	9563	17	
73	9563	9568	9573	9578	9583	9588	9593	9598	9603	9608	9613	16	5
74	9613	9617	9622	9627	9632	9636	9641	9646	9650	9655	9659	15	
75	0.9659	9664	9668	9673	9677	9681	9686	9690	9694	9699	9703	14	
76	9703	9707	9711	9715	9720	9724	9728	9732	9736	9740	9744	13	4
77	9744	9748	9751	9755	9759	9763	9767	9770	9774	9778	9781	12	
78	9781	9785	9789	9792	9796	9799	9803	9806	9810	9813	9816	11	
79	9816	9820	9823	9826	9829	9833	9836	9839	9842	9845	9848	10	
80	0.9848	9851	9854	9857	9860	9863	9866	9869	9871	9874	9877	9	3
81	9877	9880	9882	9885	9888	9890	9893	9895	9898	9900	9903	8	
82	9903	9905	9907	9910	9912	9914	9917	9919	9921	9923	9925	7	
83	9925	9928	9930	9932	9934	9936	9938	9940	9942	9943	9945	6	2
84	9945	9947	9949	9951	9952	9954	9956	9957	9959	9960	9962	5	
85	0.9962	9963	9965	9966	9968	9969	9971	9972	9973	9974	9976	4	
86	9976	9977	9978	9979	9980	9981	9982	9983	9984	9985	9986	3	1
87	9986	9987	9988	9989	9990	9990	9991	9992	9993	9993	9994	2	
88	9994	9995	9995	9996	9996	9997	9997	9997	9998	9998	9998	1	
89°	9998	9999	9999	9999	9999	1.0000	1.0000	1.0000	1.0000	1.0000	1.0000	0°	0
Complement		.9	.8	.7	.6	.5	.4	.3	.2	.1	.0	Angle	

NATURAL COSINES

NATURAL TANGENTS

Angle	.0	.1	.2	.3	.4	.5	.6	.7	.8	.9	Complement Difference		
0°	0.0000	0017	0035	0052	0070	0087	0105	0122	0140	0157	0175	**89°**	
1	0175	0192	0209	0227	0244	0262	0279	0297	0314	0332	0349	**88**	
2	0349	0367	0384	0402	0419	0437	0454	0472	0489	0507	0524	**87**	
3	0524	0542	0559	0577	0594	0612	0629	0647	0664	0682	0699	**86**	
4	0699	0717	0734	0752	0769	0787	0805	0822	0840	0857	0875	**85**	
5	0.0875	0892	0910	0928	0945	0963	0981	0998	1016	1033	1051	**84**	
6	1051	1069	1086	1104	1122	1139	1157	1175	1192	1210	1228	**83**	
7	1228	1246	1263	1281	1299	1317	1334	1352	1370	1388	1405	**82**	
8	1405	1423	1441	1459	1477	1495	1512	1530	1548	1566	1584	**81**	
9	1584	1602	1620	1638	1655	1673	1691	1709	1727	1745	1763	**80**	
10	0.1763	1781	1799	1817	1835	1853	1871	1890	1908	1926	1944	**79**	18
11	1944	1962	1980	1998	2016	2035	2053	2071	2089	2107	2126	**78**	
12	2126	2144	2162	2180	2199	2217	2235	2254	2272	2290	2309	**77**	
13	2309	2327	2345	2364	2382	2401	2419	2438	2456	2475	2493	**76**	
14	2493	2512	2530	2549	2568	2586	2605	2623	2642	2661	2679	**75**	
15	0.2679	2698	2717	2736	2754	2774	2792	2811	2830	2849	2867	**74**	
16	2867	2886	2905	2924	2943	2962	2981	3000	3019	3038	3057	**73**	19
17	3057	3076	3096	3115	3134	3153	3172	3191	3211	3230	3249	**72**	
18	3249	3269	3288	3307	3327	3346	3365	3385	3404	3424	3443	**71**	
19	3443	3463	3482	3502	3522	3541	3561	3581	3600	3620	3640	**70**	
20	0.3640	3659	3679	3699	3719	3739	3759	3779	3799	3819	3839	**69**	
21	3839	3859	3879	3899	3919	3939	3959	3979	4000	4020	4040	**68**	20
22	4040	4061	4081	4101	4122	4142	4163	4183	4204	4224	4245	**67**	
23	4245	4265	4286	4307	4327	4348	4369	4390	4411	4431	4452	**66**	
24	4452	4473	4494	4515	4536	4557	4578	4599	4621	4642	4663	**65**	21
25	0.4663	4684	4706	4727	4748	4770	4791	4813	4834	4856	4877	**64**	
26	4877	4899	4921	4942	4964	4986	5008	5029	5051	5073	5095	**63**	
27	5095	5117	5139	5161	5184	5206	5228	5250	5272	5295	5317	**62**	22
28	5317	5340	5362	5384	5407	5430	5452	5475	5498	5520	5543	**61**	
29	5543	5566	5589	5612	5635	5658	5681	5704	5727	5750	5774	**60**	23
30	0.5774	5797	5820	5844	5867	5890	5914	5938	5961	5985	6099	**59**	
31	6009	6032	6056	6080	6104	6128	6152	6176	6200	6224	6249	**58**	24
32	6249	6273	6297	6322	6346	6371	6395	6420	6445	6469	6494	**57**	
33	6494	6519	6544	6569	6594	6619	6644	6669	6694	6720	6745	**56**	25
34	6745	6771	6796	6822	6847	6873	6899	6924	6950	6976	7002	**55**	
35	0.7002	7028	7054	7080	7107	7133	7159	7186	7212	7239	7265	**54**	26
36	7265	7292	7319	7346	7373	7400	7427	7454	7481	7508	7536	**53**	27
37	7536	7563	7590	7618	7646	7673	7701	7729	7757	7785	7813	**52**	28
38	7813	7841	7869	7898	7926	7954	7983	8012	8040	8069	8098	**51**	28
39	8098	8127	8156	8185	8214	8243	8273	8302	8332	8361	8391	**50**	29
40	0.8391	8421	8451	8481	8511	8541	8571	8601	8632	8662	8693	**49**	30
41	8693	8724	8754	8785	8816	8847	8878	8910	8941	8972	9004	**48**	31
42	9004	9036	9067	9099	9131	9163	9195	9228	9260	9293	9325	**47**	32
43	9325	9358	9391	9424	9557	9490	9523	9556	9590	9623	9657	**46**	33
44°	9657	9691	9725	9759	9793	9827	9861	9896	9930	9965	1.0000	**45°**	34
Complement		.9	.8	.7	.6	.5	.4	.3	.2	.1	.0	Angle	

NATURAL COTANGENTS

NATURAL TANGENTS

Angle	.0	.1	.2	.3	.4	.5	.6	.7	.8	.9	Dif.
45°	1.0000	1.0035	1.0070	1.0105	1.0141	1.0176	1.0212	1.0247	1.0283	1.0319	36
46	1.0355	1.0392	1.0428	1.0464	1.0501	1.0538	1.0575	1.0612	1.0649	1.0686	37
47	1.0724	1.0761	1.0799	1.0837	1.0875	1.0913	1.0951	1.0990	1.1028	1.1067	38
48	1.1106	1.1145	1.1184	1.1224	1.1263	1.1303	1.1343	1.1383	1.1423	1.1463	40
49	1.1504	1.1544	1.1585	1.1626	1.1667	1.1708	1.1750	1.1792	1.1833	1.1875	41
50	1.1918	1.1960	1.2002	1.2045	1.2088	1.2131	1.2174	1.2218	1.2261	1.2305	43
51	1.2349	1.2393	1.2437	1.2482	1.2527	1.2572	1.2617	1.2662	1.2708	1.2753	45
52	1.2799	1.2846	1.2892	1.2938	1.2985	1.3032	1.3079	1.3127	1.3175	1.3222	47
53	1.3270	1.3319	1.3367	1.3416	1.3465	1.3514	1.3564	1.3613	1.3663	1.3713	49
54	1.3764	1.3814	1.3865	1.3916	1.3968	1.4019	1.4071	1.4124	1.4176	1.4229	52
55	1.4281	1.4335	1.4388	1.4442	1.4496	1.4550	1.4605	1.4659	1.4715	1.4770	54
56	1.4826	1.4882	1.4938	1.4994	1.5051	1.5108	1.5166	1.5224	1.5282	1.5340	57
57	1.5399	1.5458	1.5517	1.5577	1.5637	1.5697	1.5757	1.5818	1.5880	1.5941	60
58	1.6003	1.6066	1.6128	1.6191	1.6255	1.6319	1.6383	1.6447	1.6512	1.6577	64
59	1.6643	1.6709	1.6775	1.6842	1.6909	1.6977	1.7045	1.7113	1.7182	1.7251	68
60	1.7321	1.7391	1.7461	1.7532	1.7603	1.7675	1.7747	1.7820	1.7893	1.7966	72
61	1.8040	1.8115	1.8190	1.8265	1.8341	1.8418	1.8495	1.8572	1.8650	1.8728	77
62	1.8807	1.8887	1.8967	1.9047	1.9128	1.9210	1.9292	1.9375	1.9458	1.9542	82
63	1.9626	1.9711	1.9797	1.9883	1.9970	2.0057	2.0145	2.0233	2.0323	2.0413	88
64	2.0503	2.0594	2.0686	2.0778	2.0872	2.0965	2.1060	2.1155	2.1251	2.1348	94
65	2.145	2.154	2.164	2.174	2.184	2.194	2.204	2.215	2.225	2.236	10
66	2.246	2.257	2.267	2.278	2.289	2.300	2.311	2.322	2.333	2.344	11
67	2.356	2.367	2.379	2.391	2.402	2.414	2.426	2.438	2.450	2.463	12
68	2.475	2.488	2.500	2.513	2.526	2.539	2.552	2.565	2.578	2.592	13
69	2.605	2.619	2.633	2.646	2.660	2.675	2.689	2.703	2.718	2.733	14
70	2.747	2.762	2.778	2.793	2.808	2.824	2.840	2.856	2.872	2.888	16
71	2.904	2.921	2.937	2.954	2.971	2.989	3.006	3.024	3.042	3.060	17
72	3.078	3.096	3.115	3.133	3.152	3.172	3.191	3.211	3.230	3.250	19
73	3.271	3.291	3.312	3.333	3.354	3.376	3.398	3.420	3.442	3.465	22
74	3.487	3.511	3.534	3.558	3.582	3.606	3.630	3.655	3.681	3.700	25
75	3.732	3.758	3.785	3.812	3.839	3.867	3.895	3.923	3.952	3.981	28
76	4.011	4.041	4.071	4.102	4.134	4.165	4.198	4.230	4.264	4.297	32
77	4.331	4.366	4.402	4.437	4.474	4.511	4.548	4.586	4.625	4.665	37
78	4.705	4.745	4.787	4.829	4.872	4.915	4.959	5.005	5.050	5.097	44
79	5.145	5.193	5.242	5.292	5.343	5.396	5.449	5.503	5.558	5.614	52
80	5.67	5.73	5.79	5.85	5.91	5.98	6.04	6.11	6.17	6.24	7
81	6.31	6.39	6.46	6.54	6.61	6.69	6.77	6.85	6.94	7.03	8
82	7.12	7.21	7.30	7.40	7.49	7.60	7.70	7.81	7.92	8.03	10
83	8.14	8.26	8.39	8.51	8.64	8.78	8.92	9.06	9.21	9.36	14
84	9.51	9.68	9.84	10.0	10.2	10.4	10.6	10.8	11.0	11.2	
85	11.4	11.7	11.9	12.2	12.4	12.7	13.0	13.3	13.6	14.0	3
86	14.3	14.7	15.1	15.5	15.9	16.3	16.8	17.3	17.9	18.5	6
87	19.1	19.7	20.4	21.2	22.0	22.9	23.9	24.9	26.0	27.3	
88	28.6	30.1	31.8	33.7	35.8	38.2	40.9	44.1	47.7	52.1	
89°	57.	64.	72.	82.	95.	115.	143.	191.	286.	573.	
Angle	.0	.1	.2	.3	.4	.5	.6	.7	.8	.9	

NATURAL TANGENTS

	0	1	2	3	4	5	6	7	8	9	1	2	3	4	5	6	7	8	9
10	0000	0043	0086	0128	0170	0212	0253	0294	0334	0374	4	8	12	17	21	25	29	33	37
11	0414	0453	0492	0531	0569	0607	0645	0682	0719	0755	4	8	11	15	19	23	26	30	34
12	0792	0828	0864	0899	0934	0969	1004	1038	1072	1106	3	7	10	14	17	21	24	28	31
13	1139	1173	1206	1239	1271	1303	1335	1367	1399	1430	3	6	10	13	16	19	23	26	29
14	1461	1492	1523	1553	1584	1614	1644	1673	1703	1732	3	6	9	12	15	18	21	24	27
15	1761	1790	1818	1847	1875	1903	1931	1959	1987	2014	3	6	8	11	14	17	20	22	25
16	2041	2068	2095	2122	2148	2175	2201	2227	2253	2279	3	5	8	11	13	16	18	21	24
17	2304	2330	2355	2380	2405	2430	2455	2480	2504	2529	2	5	7	10	12	15	17	20	22
18	2553	2577	2601	2625	2648	2672	2695	2718	2742	2765	2	5	7	9	12	14	16	19	21
19	2788	2810	2833	2856	2878	2900	2923	2945	2967	2989	2	4	7	9	11	13	16	18	20
20	3010	3032	3054	3075	3096	3118	3139	3160	3181	3201	2	4	6	8	11	13	15	17	19
21	3222	3243	3263	3284	3304	3324	3345	3365	3385	3404	2	4	6	8	10	12	14	16	18
22	3424	3444	3464	3483	3502	3522	3541	3560	3579	3598	2	4		8	10	12	14	15	17
23	3617	3636	3655	3674	3692	3711	3729	3747	3766	3784	2	4	6	7	9	11	13	15	17
24	3802	3820	3838	3856	3874	3892	3909	3927	3945	3962	2	4		7	9	11	12	14	16
25	3979	3997	4014	4031	4048	4065	4082	4099	4116	4133	2	3		7	9	10	12	14	15
26	4150	4166	4183	4200	4216	4232	4249	4265	4281	4298	2	3	5	7	8	10	11	13	15
27	4314	4330	4346	4362	4378	4393	4409	4425	4440	4456	2	3			8	9	11	13	14
28	4472	4487	4502	4518	4533	4548	4564	4579	4594	4609	2	3			8	9	11	12	14
29	4624	4639	4654	4669	4683	4698	4713	4728	4742	4757	1	3	4	6	7	9	10	12	13
30	4771	4786	4800	4814	4829	4843	4857	4871	4886	4900	1	3	4	6	7	9	10	11	13
31	4914	4928	4942	4955	4969	4983	4997	5011	5024	5038	1	3	4		7	8	10	11	12
32	5051	5065	5079	5092	5105	5119	5132	5145	5159	5172	1		4		7		9	11	12
33	5185	5198	5211	5224	5237	5250	5263	5276	5289	5302	1	3	4	5	6	8	9	10	12
34	5315	5328	5340	5353	5366	5378	5391	5403	5416	5428	1		4		6	8	9	10	11
35	5441	5453	5465	5478	5490	5502	5514	5527	5539	5551	1		4		6	7	9	10	11
36	5563	5575	5587	5599	5611	5623	5635	5647	5658	5670	1	2	4	5	6	7	8	10	11
37	5682	5694	5705	5717	5729	5740	5752	5763	5775	5786	1	2			6	7	8	9	10
38	5798	5809	5821	5832	5843	5855	5866	5877	5888	5899	1	2		5	6	7	8	9	10
39	5911	5922	5933	5944	5955	5966	5977	5988	5999	6010	1	2	3	4	5	7	8	9	10
40	6021	6031	6042	6053	6064	6075	6085	6096	6107	6117	1	2	3	4	5	6	8	9	10
41	6128	6138	6149	6160	6170	6180	6191	6201	6212	6222	1		3	4	5		7		9
42	6232	6243	6253	6263	6274	6284	6294	6304	6314	6325	1		3	4	5		7		9
43	6335	6345	6355	6365	6375	6385	6395	6405	6415	6425	1	2	3	4	5	6	7	8	9
44	6435	6444	6454	6464	6474	6484	6493	6503	6513	6522	1			4			7	8	9
45	6532	6542	6551	6561	6571	6580	6590	6599	6609	6618	1			4			7	8	9
46	6628	6637	6646	6656	6665	6675	6684	6693	6702	6712	1	2	3	4	5	6	7	7	8
47	6721	6730	6739	6749	6758	6767	6776	6785	6794	6803	1		3	4	5			7	
48	6812	6821	6830	6839	6848	6857	6866	6875	6884	6893	1		3	4	4			7	
49	6902	6911	6920	6928	6937	6946	6955	6964	6972	6981	1	2	3	4	4	5	6	7	8
50	6990	6998	7007	7016	7024	7033	7042	7050	7059	7067	1	2	3	3	4	5	6	7	8
51	7076	7084	7093	7101	7110	7118	7126	7135	7143	7152	1	2		3	4			7	8
52	7160	7168	7177	7185	7193	7202	7210	7218	7226	7235	1	2		3	4			7	7
53	7243	7251	7259	7267	7275	7284	7292	7300	7308	7316	1	2	2	3	4	5	6	6	7
54	7324	7332	7340	7348	7356	7364	7372	7380	7388	7396	1	2	2	3	4	5	6	6	7

LOGARITHMS

	0	1	2	3	4	5	6	7	8	9	1	2	3	4	5	6	7	8	9
55	7404	7412	7419	7427	7435	7443	7451	7459	7466	7474	1	2	2	3	4	5	5	6	7
56	7482	7490	7497	7505	7513	7520	7528	7536	7543	7551	1	2	2	3	4	5	5	6	7
57	7559	7566	7574	7582	7589	7597	7604	7612	7619	7627	1	2	2	3	4	5	5	6	7
58	7634	7642	7649	7657	7664	7672	7679	7686	7694	7701	1	1	2	3	4	4	5	6	7
59	7709	7716	7723	7731	7738	7745	7752	7760	7767	7774	1	1	2	3	4	4	5	6	7
60	7782	7789	7796	7803	7810	7818	7825	7832	7839	7846	1	1	2	3	4	4	5	6	6
61	7853	7860	7868	7875	7882	7889	7896	7903	7910	7917	1	1	2	3	4	4	5	6	6
62	7924	7931	7938	7945	7952	7959	7966	7973	7980	7987	1	1	2	3	3	4	5	6	6
63	7993	8000	8007	8014	8021	8028	8035	8041	8048	8055	1	1	2	3	3	4	5	5	6
64	8062	8069	8075	8082	8089	8096	8102	8109	8116	8122	1	1	2	3	3	4	5	5	6
65	8129	8136	8142	8149	8156	8162	8169	8176	8182	8189	1	1	2	3	3	4	5	5	6
66	8195	8202	8209	8215	8222	8228	8235	8241	8248	8254	1	1	2	3	3	4	5	5	6
67	8261	8267	8274	8280	8287	8293	8299	8306	8312	8319	1	1	2	3	3	4	5	5	6
68	8325	8331	8338	8344	8351	8357	8363	8370	8376	8382	1	1	2	3	3	4	4	5	6
69	8388	8395	8401	8407	8414	8420	8426	8432	8439	8445	1	1	2	2	3	4	4	5	6
70	8451	8457	8463	8470	8476	8482	8488	8494	8500	8506	1	1	2	2	3	4	4	5	6
71	8513	8519	8525	8531	8537	8543	8549	8555	8561	8567	1	1	2	2	3	4	4	5	5
72	8573	8579	8585	8591	8597	8603	8609	8615	8621	8627	1	1	2	2	3	4	4	5	5
73	8633	8639	8645	8651	8657	8663	8669	8675	8681	8686	1	1	2	2	3	4	4	5	5
74	8692	8698	8704	8710	8716	8722	8727	8733	8739	8745	1	1	2	2	3	4	4	5	5
75	8751	8756	8762	8768	8774	8779	8785	8791	8797	8802	1	1	2	2	3	3	4	5	5
76	8808	8814	8820	8825	8831	8837	8842	8848	8854	8859	1	1	2	2	3	3	4	5	5
77	8865	8871	8876	8882	8887	8893	8899	8904	8910	8915	1	1	2	2	3	3	4	4	5
78	8921	8927	8932	8938	8943	8949	8954	8960	8965	8971	1	1	2	2	3	3	4	4	5
79	8976	8982	8987	8993	8998	9004	9009	9015	9020	9025	1	1	2	2	3	3	4	4	5
80	9031	9036	9042	9047	9053	9058	9063	9069	9074	9079	1	1	2	2	3	3	4	4	5
81	9085	9090	9096	9101	9106	9112	9117	9122	9128	9133	1	1	2	2	3	3	4	4	5
82	9138	9143	9149	9154	9159	9165	9170	9175	9180	9186	1	1	2	2	3	3	4	4	5
83	9191	9196	9201	9206	9212	9217	9222	9227	9232	9238	1	1	2	2	3	3	4	4	5
84	9243	9248	9253	9258	9263	9269	9274	9279	9284	9289	1	1	2	2	3	3	4	4	5
85	9294	9299	9304	9309	9315	9320	9325	9330	9335	9340	1	1	2	2	3	3	4	4	5
86	9345	9350	9355	9360	9365	9370	9375	9380	9385	9390	1	1	2	2	3	3	4	4	5
87	9395	9400	9405	9410	9415	9420	9425	9430	9435	9440	0	1	1	2	2	3	3	4	4
88	9445	9450	9455	9460	9465	9469	9474	9479	9484	9489	0	1	1	2	2	3	3	4	4
89	9494	9499	9504	9509	9513	9518	9523	9528	9533	9538	0	1	1	2	2	3	3	4	4
90	9542	9547	9552	9557	9562	9566	9571	9576	9581	9586	0	1	1	2	2	3	3	4	4
91	9590	9595	9600	9605	9609	9614	9619	9624	9628	9633	0	1	1	2	2	3	3	4	4
92	9638	9643	9647	9652	9657	9661	9666	9671	9675	9680	0	1	1	2	2	3	3	4	4
93	9685	9689	9694	9699	9703	9708	9713	9717	9722	9727	0	1	1	2	2	3	3	4	4
94	9731	9736	9741	9745	9750	9754	9759	9763	9768	9773	0	1	1	2	2	3	3	4	4
95	9777	9782	9786	9791	9795	9800	9805	9809	9814	9818	0	1	1	2	2	3	3	4	4
96	9823	9827	9832	9836	9841	9845	9850	9854	9859	9863	0	1	1	2	2	3	3	4	4
97	9868	9872	9877	9881	9886	9890	9894	9899	9903	9908	0	1	1	2	2	3	3	4	4
98	9912	9917	9921	9926	9930	9934	9939	9943	9948	9952	0	1	1	2	2	3	3	4	4
99	9956	9961	9965	9969	9974	9978	9983	9987	9991	9996	0	1	1	2	2	3	3	3	4

INDEX

(Numbers refer to pages)

Made in United States
Troutdale, OR
11/03/2026

56847403R00226